166
Structure and Bonding

Series Editor:

D.M.P. Mingos, Oxford, United Kingdom

Editorial Board:

F.A. Armstrong, Oxford, United Kingdom
X. Duan, Beijing, China
L.H. Gade, Heidelberg, Germany
K.R. Poeppelmeier, Evanston, IL, USA
G. Parkin, NewYork, USA
M. Takano, Kyoto, Japan

Aims and Scope

The series *Structure and Bonding* publishes critical reviews on topics of research concerned with chemical structure and bonding. The scope of the series spans the entire Periodic Table and addresses structure and bonding issues associated with all of the elements. It also focuses attention on new and developing areas of modern structural and theoretical chemistry such as nanostructures, molecular electronics, designed molecular solids, surfaces, metal clusters and supramolecular structures. Physical and spectroscopic techniques used to determine, examine and model structures fallwithin the purview of *Structure and Bonding* to the extent that the focus is on the scientific results obtained and not on specialist information concerning the techniques themselves. Issues associated with the development of bonding models and generalizations that illuminate the reactivity pathways and rates of chemical processes are also relevant.

The individual volumes in the series are thematic. The goal of each volume is to give the reader, whether at a university or in industry, a comprehensive overview of an area where new insights are emerging that are of interest to a larger scientific audience. Thus each review within the volume critically surveys one aspect of that topic and places it within the context of the volume as a whole. The most significant developments of the last 5 to 10 years should be presented using selected examples to illustrate the principles discussed. A description of the physical basis of the experimental techniques that have been used to provide the primary data may also be appropriate, if it has not been covered in detail elsewhere. The coverage need not be exhaustive in data, but should rather be conceptual, concentrating on the new principles being developed that will allow the reader, who is not a specialist in the area covered, to understand the data presented. Discussion of possible future research directions in the area is welcomed.

Review articles for the individual volumes are invited by the volume editors.

In references *Structure and Bonding* is abbreviated *Struct Bond* and is cited as a journal.

More information about this series at
http://www.springer.com/series/430

Dongpeng Yan • Min Wei
Editors

Photofunctional Layered Materials

With contributions by

R. Tian · M. Wei · D. Yan · J. Liang · R. Ma · T. Sasaki · M. Shao · D.G. Evans · X. Duan · D.-H. Park · G. Choi · J.-H. Choy · T. Okada · M. Sohmiya · M. Ogawa · L. Zhang · Z. Liu

Springer

Editors
Dongpeng Yan
Beijing University of Chemical
  Technology
Beijing
China

Min Wei
Beijing University of Chemical
  Technology
Beijing
China

ISSN 0081-5993   ISSN 1616-8550 (electronic)
Structure and Bonding
ISBN 978-3-319-16990-3   ISBN 978-3-319-16991-0 (eBook)
DOI 10.1007/978-3-319-16991-0

Library of Congress Control Number: 2015940114

Springer Cham Heidelberg New York Dordrecht London
© Springer International Publishing Switzerland 2015
This work is subject to copyright. All rights are reserved by the Publisher, whether the whole or part of the material is concerned, specifically the rights of translation, reprinting, reuse of illustrations, recitation, broadcasting, reproduction on microfilms or in any other physical way, and transmission or information storage and retrieval, electronic adaptation, computer software, or by similar or dissimilar methodology now known or hereafter developed.
The use of general descriptive names, registered names, trademarks, service marks, etc. in this publication does not imply, even in the absence of a specific statement, that such names are exempt from the relevant protective laws and regulations and therefore free for general use.
The publisher, the authors and the editors are safe to assume that the advice and information in this book are believed to be true and accurate at the date of publication. Neither the publisher nor the authors or the editors give a warranty, express or implied, with respect to the material contained herein or for any errors or omissions that may have been made.

Printed on acid-free paper

Springer International Publishing AG Switzerland is part of Springer Science+Business Media (www.springer.com)

Preface

Photofunctional materials are a large family of photoactive compounds that can transfer, absorb, store, switch, or utilize light energy. Layered compounds with two-dimensional (2D) structures have recently emerged as a new platform for developing novel photofunctional materials and have already been explored and applied in the fields of light-harvesting, photoluminescence, photocatalysis, biological imaging, band structure engineering, photochromic sensors, and optoelectronic devices.

Given the rapid development of photofunctional layered materials, we believe the collection of reviews in this volume is timely. In the first chapter, Yan and colleagues review the photofunctionality of layered double hydroxide (LDH) materials, with a particular focus on the incorporation of organic chromophores, polymers, metal complexes, and quantum dots in LDHs; the fluorescence, optical sensing, and IR/UV absorption properties of the resulting composite materials are covered.

Although similar to LDH materials, layered rare earth hydroxides (LREHs)—which contain rare earth hydroxide cationic host layers—have been recognized as a new class of multifunctional materials with an intercalation ability that leads to extensive host–guest chemistry. In the second chapter, Sasaki and colleagues provide an overview of the development of LREH materials and summarize the synthetic methods, structure characterization, as well as their photoluminescence properties.

Photocatalytic materials have attracted great attention in environmental science, as well as in energy storage and transfer. In the third chapter, Wei et al. summarize the synthesis and photocatalytic applications of LDH-based materials. In particular, several important photocatalytic systems—such as pollutant degradation, water splitting for solar fuel production, and reduction of CO_2 to useful carbon materials—are reviewed and discussed.

In view of their potential benefits to human health, the fabrication of new bio-functional nanomaterials has become a fast-growing area. The use of LDHs as fluorescence imaging materials and related biomedical applications has been widely developed. In the fourth chapter, Choy et al. review the structural chemistry of bio-LDH nanohybrid systems and summarize their applications in gene and drug delivery (in both therapeutics and diagnostics).

Photoresponsive materials (such as photochromic compounds) are regarded as one of the most interesting photo-related topics. In the fifth chapter, Ogawa et al. provide an overview of the development of photochromic intercalation compounds and describe how the formation of different layered host materials allows control over the orientation and aggregation of photochromic guests as well as the resulting photoresponsive properties.

As 2D functional units, graphene and its derivatives have attracted considerable interest due to their novel optical, electronic, mechanical, and surface properties. Recently, photogenerated free radicals have been shown to provide a new way for the covalent functionalization and band structure engineering of graphene. In the final chapter of this volume, Zhang and Liu discuss the photochemistry of graphene, including its photohalogenation, photoarylation, photoalkylation, and photocatalytic oxidation.

There has been an explosive growth in publications dealing with photofunctional layered materials in recent years. The reviews in this volume have attempted to cover representative examples of such layered compounds in the literature published up to 2014. Many interesting additional examples involving layered structures and novel photofunctionalities may have been excluded, either inadvertently or due to space limitations, but we hope that the volume has captured the more important developments.

We express our thanks to all the authors for helping us capture the essence of recent research in this field.

Beijing, China
Dongpeng Yan
Min Wei

Contents

Layered Double Hydroxide Materials: Assembly and Photofunctionality 1
Rui Tian, Dongpeng Yan, and Min Wei

Layered Rare Earth Hydroxides: Structural Aspects and Photoluminescence Properties 69
Jianbo Liang, Renzhi Ma, and Takayoshi Sasaki

Layered Double Hydroxide Materials in Photocatalysis 105
Mingfei Shao, Min Wei, David G. Evans, and Xue Duan

Bio-Layered Double Hydroxides Nanohybrids for Theranostics Applications 137
Dae-Hwan Park, Goeun Choi, and Jin-Ho Choy

Photochromic Intercalation Compounds 177
Tomohiko Okada, Minoru Sohmiya, and Makoto Ogawa

Photochemistry of Graphene 213
Liming Zhang and Zhongfan Liu

Index 239

Layered Double Hydroxide Materials: Assembly and Photofunctionality

Rui Tian, Dongpeng Yan, and Min Wei

Contents

1 Introduction 3
2 Optical Properties of LDHs 4
3 Luminescent Properties of LDHs Nanoparticles 6
   3.1 Defect-Induced Luminescence 6
   3.2 Rare-Earth-Doped Luminescence 7
4 LDH-Based UV-Shielding Materials and the Applications 8
   4.1 Introduction of UV Light and UV-Shielding Materials 8
   4.2 UV-Shielding Properties of LDHs 8
   4.3 Host–Guest Interaction and UV-Shielding Performances 10
   4.4 Application of LDH-Based UV-Shielding Materials
       in Polypropylene and Asphalt 13
5 LDH-Based IR Absorption Materials and the Application 17
   5.1 Introduction of IR Absorption Materials 17
   5.2 IR Absorption Properties of LDHs 17
   5.3 Intercalation and Performances of LDH-Based IR Absorption Materials ... 19
   5.4 Application of LDH-Based IR Absorption Materials 22
6 Fabrication and Application of LDH-Based Host–Guest Luminescent Materials ... 24
   6.1 The Basic Guest Units of LDH-Based Photofunctional Materials 25
   6.2 Fabrication of LDH-Based Photofunctional Material 25
   6.3 Adjusting Photofunctionalities of LDH-Based Materials 38
7 Conclusion and Outlook 63
References 64

R. Tian • M. Wei
State Key Laboratory of Chemical Resource Engineering, Beijing University of Chemical Technology, Beijing 100029, P. R. China

D. Yan (✉)
State Key Laboratory of Chemical Resource Engineering, Beijing University of Chemical Technology, Beijing 100029, P. R. China

College of Chemistry, Beijing Normal University, Beijing 100875, P. R. China
e-mail: yandp@mail.buct.edu.cn; yandp@bnu.edu.cn

© Springer International Publishing Switzerland 2015
D. Yan, M. Wei (eds.), *Photofunctional Layered Materials*, Structure and Bonding 166, DOI 10.1007/978-3-319-16991-0_1

Abstract As a large type of inorganic layered compounds with diversity of composition in host layer and interlayer anions, layered double hydroxides (LDHs) have been intensively studied toward construction of advanced photofunctional materials. In this chapter, the optical properties of LDH materials and related potential applications—mainly involving the functionalities of LDH layers (such as tunable color, infrared radiation (IR) absorption, and ultraviolet (UV) shielding)—will be described firstly. Then, we will review the development of intercalated luminescent materials by adjusting both the host layer and guest molecules, in which the static and dynamic photofunctional modulations have been focused. Due to the host–guest and guest–guest interaction, fluorescence properties of composites can be effectively altered, and intelligent materials with stimuli-responsive performance can be further obtained. Finally, perspectives on the future development of LDH-based solid-state luminescent materials are addressed.

Keywords Assembly · Layered double hydroxides · Luminescence · Sensor · UV shielding

Abbreviations

AFM	Atomic force microscopy
CL	Chemiluminescence
ECL	Electrochemiluminescence
EDTA	Ethylenediaminetetraacetic acid
EL	Electroluminescence
FRET	Förster resonance energy transfer
HMIs	Heavy metal ions
IR	Infrared radiation
LDHs	Layered double hydroxides
LDPE	Low-density polyethylene
LED	Light-emitting diode
MMO	Mixed metal oxide
PL	Photoluminescence
RH	Relative humidity
SEM	Scanning electron microscopy
SNAS	Separate nucleation and aging steps
TEM	Transmission electron microscopy
UTF	Ultrathin film
UV-Vis	Ultraviolet-visible
VOCs	Volatile organic compounds
XRD	X-ray diffraction

1 Introduction

Layered double hydroxides (LDHs) are one type of anionic clay materials, also known as hydrotalcite compound [1, 2]. The basic structure of LDHs is derived from the brucite with edge-sharing $M(OH)_6$ octahedra, in which a fraction of the divalent cations are substituted by trivalent cations in the layer and the positive charges can be balanced by anions within the hydrated interlayer galleries [3–5]. For the common binary LDHs, the general formula can be expressed as $[M^{2+}_{1-x}M^{3+}_{x}(OH)_2] (A^{n-})_{x/n} \cdot mH_2O$, with tunable composition and relative proportions for both di- and trivalent cations as well as interlayer ions [6–8]. This versatility gives rise to a large amount of LDH-based materials with variable functionalities due to the exchange of interlayer ions and diverse properties from different elemental compositions and charge densities of host layers. The structures of LDHs have been comprehensively studied and introduced in several previous literatures [9–11], and the models for typical packing fashions (3R and 2H) of LDHs are illustrated in Scheme 1 [12].

During last few decades, LDHs have received much attention in the fields of catalysis [13], separation processes [14], photochemical cell [15], and drug carrier [16]. Moreover, based on the tunable host layer and interlayer guest molecules, the optical/luminescent properties have also been largely investigated, and a number of photofunctional materials have been developed [17]. Due to the rapid development in the design and fabrication of LDH-based photo-related materials, it is timely and

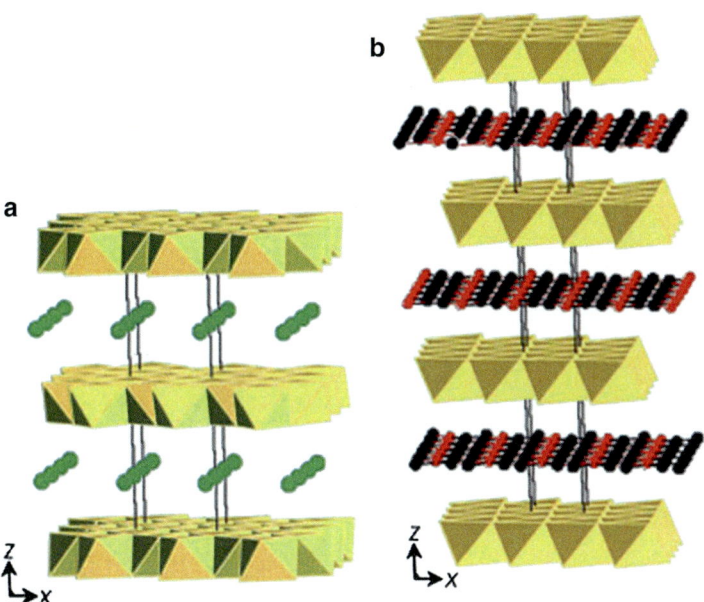

Scheme 1 Typical models of LDHs structures with two packing fashions: (**a**) 2H; (**b**) 3R [12]

necessary to summarize recent progress in this area and thus hopefully stimulate the construction of new types of layered photofunctional materials.

2 Optical Properties of LDHs

Due to different photo-absorption characteristics of the metal ions in the LDH layers, the powdered LDH materials can present different colors under daylight. Moreover, the nontransparent LDH powders can be further fabricated into continuous self-supporting transparent films through solvent evaporation or spin-coating method. The resulting films usually exhibit oriented ($00l$) reflection due to the edge–edge interaction between the adjacent LDH particles. For example, alkoxide-intercalated LDHs have been synthesized in a nonaqueous media (methanol), and the hydrolysis of these LDH derivatives resulted in a colloidal LDH suspension which can be used as a precursor for the formation of continuous transparent films via solvent evaporation [18, 19]. For another example, the fabrication of oriented LDHs films without organic solvents has been attempted, and separate nucleation and aging steps (SNAS) were adopted as a main method [20]. Due to the uniform and small crystal size (about 40 nm) of the LDH platelets, both the face-to-face and edge-to-edge interactions occur between individual LDH platelets, which facilitate the formation of densely packed c-oriented LDH film [21]. The obtained films were sufficiently thick and mechanically robust and maintained good optical transparency as well as uniform colors for ZnAl-NO_3, NiAl-NO_3, and ZnAl-Tb-EDTA (EDTA, ethylenediaminetetraacetic acid) samples (Fig. 1). Moreover, Zhang [22] has fabricated oriented LDH films by a spin-coating method, and the as-prepared

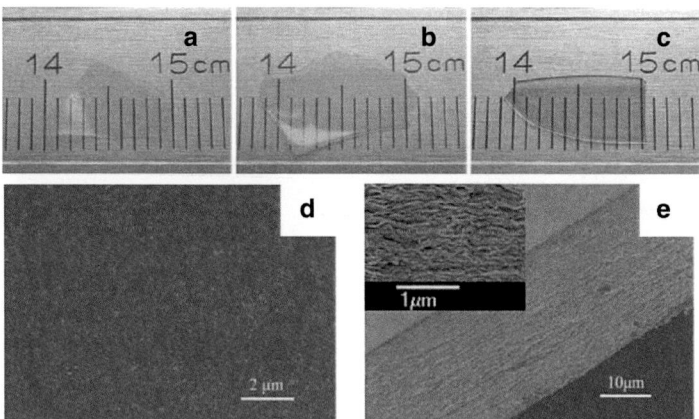

Fig. 1 Photographs of transparent self-supporting LDH films of (**a**) ZnAl-NO_3, (**b**) NiAl-NO_3, (**c**) ZnAl-Tb(EDTA) (the ruler with centimetre scale lies behind the films), and SEM images of ZnAl-NO_3: (**d**) top view and (**e**) edge view with a high-resolution image of this structure shown in the inset image [21]

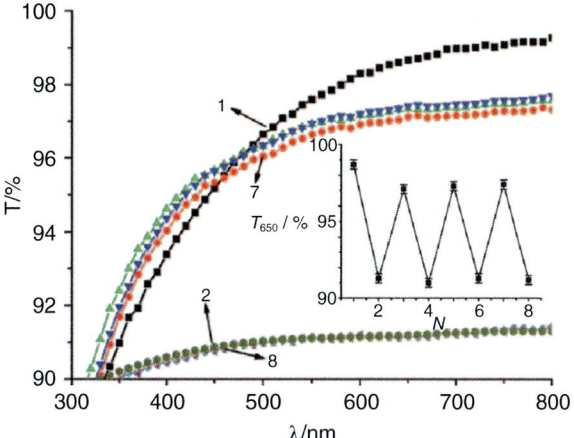

Fig. 2 UV–Vis transmittance spectra of the MMO_{10} film (1, 3, 5, 7) and the rehydrated LDH film (2, 4, 6, 8). The inset shows the transmittance at 650 nm T_{650} as a function of cycle number n; cycling occurs between MMO and LDH films [23]

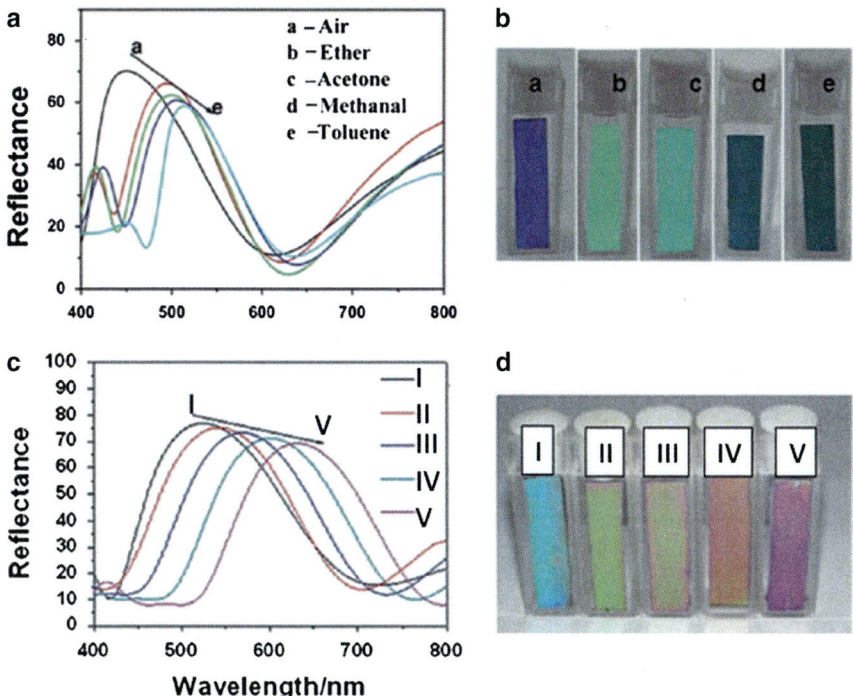

Fig. 3 Response of the $(MMO-TiO_2)_6 1DPC$ toward different VOCs and RH: (**A**) reflectance spectra and (**B**) photographs in air (a) and various VOCs: (b) ether, (c) acetone, (d) methanal, and (e) toluene; (**C**) reflectance spectra and (**D**) photographs of color variation (I–V: 5 %, 30 %, 54 %, 75 %, and 85 %, respectively) [24]

films displayed good uniformity and compact structure composed of LDH platelets stacked parallel to the substrate surface.

The calcinations of LDH materials can form mixed metal oxide (MMO) with diverse optical properties compared with the pristine LDHs. Han et al. [23] have studied the optical antireflection (AR) performance of MMO through calcinations of LDHs. Figure 2 shows the changes of transmittance by the reconstruction effect of LDH materials through recycling the calcination–rehydration procedure. These AR properties were generated by the transformation between porous (MMO film) and nonporous state (LDH film), and this phenomenon makes the MMO film an effective candidate for erasable AR coatings. Furthermore, MMO-TiO_2 one-dimensional photonic crystals (1DPC) can be fabricated through deposition of LDHs and TiO_2, followed by calcination [24]. The porous structure of MMO enabled the adsorption of volatile gas or water molecule in the mesopores of the MMO-TiO_2 structure, which results in the change of optical refractive index (Fig. 3). Thus, this system can serve as a colorimetric sensor for the detection of volatile organic compounds (VOCs) or relative humidity (RH).

3 Luminescent Properties of LDHs Nanoparticles

3.1 Defect-Induced Luminescence

Several observations have proven that the pristine LDH particles and colloids can exhibit well-defined luminescence. For example, the excitation and emission spectrum of Zn-Al-LDH has been recorded in Fig. 4 [25], and emission in the 350–550 nm can be observed. The basic mechanisms for the luminescence of the LDH colloids are attributed to the numerous surface defects of LDH nanocrystals. These surface defects can act as traps which are beneficial to the recombination of excited electrons and holes, and thus, obvious luminescence can be observed. In addition, the effects of the surface charge density (e.g., the ratio of Zn to Al) on the photoluminescence (PL) have been investigated. Figure 4 shows that higher Zn/Al ratio resulted in a stronger PL intensity, that is, different surface charge density may largely influence the PL intensity of LDH colloids. In addition, fluorescence properties of MgAl-LDHs have also been studied, and influence of MgAl-LDHs with different platelet size has been considered. The results showed that LDHs with higher specific surface areas may result in more surface defects and stronger fluorescence intensity [26].

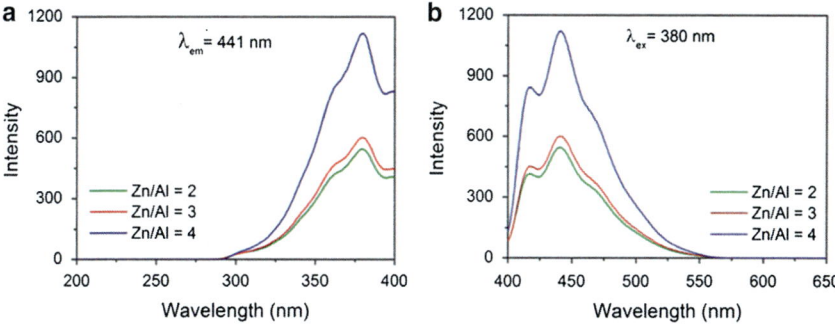

Fig. 4 Effect of Zn/Al ratio on (**a**) excitation and (**b**) emission spectra for the colloids of the pristine Zn–Al-LDHs [25]

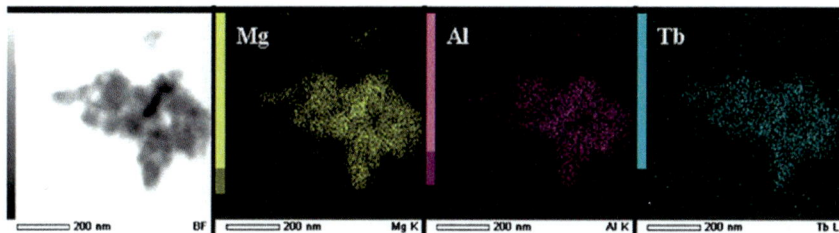

Fig. 5 TEM images and the corresponding EDS images of Mg, Al, and Tb elements in Tb^{3+} doped LDHs [29]

3.2 Rare-Earth-Doped Luminescence

One of the most striking features of LDHs is their versatility of composition, and thus, numerous metal cations can be accommodated in the host layer. It was reported that some rare-earth ions, such as lanthanide Tb^{3+}, Eu^{3+}, and Nd^{3+}, can be doped into the octahedral lattice of the brucite-like layer [27, 28]. Incorporation of Tb^{3+} has been achieved by coprecipitation method under ambient conditions [29]. Figure 5 shows the homogeneous distribution of Tb^{3+} in the lattice of LDH layers regardless of the large ionic size of Tb^{3+}, and up to 19 wt% of Tb^{3+} can be incorporated. This Tb^{3+}-doped LDH presents strong luminescence in the green region due to the intra electron transition of Tb^{3+}. Additionally, the photosensitizer could also be incorporated into the interlayer of rare-earth-doped LDHs to obtain an advanced performance. For example, Förster resonance energy transfer (FRET) process could occur between the interlayer guest and doped ions in the host layer of LDHs [30–32], which leads to enhanced fluorescence.

4 LDH-Based UV-Shielding Materials and the Applications

4.1 Introduction of UV Light and UV-Shielding Materials

UV light possesses the shortest wavelength in the solar spectrum, and it takes up about 8 % of the total solar energy. UV radiation includes UVC (220–290 nm), UVB (290–320 nm), and UVA (320–400 nm). Among them, UVC region is always absorbed by ozonosphere, and UVA and UVB in the range of 290–400 nm could reach the earth's surface. The UV light presents high energy which has severe damages to the environment and human health [33]. Some diseases, sunburn, acceleration of aging, and even cancer can be triggered due to the damage of skin fibers, proteins, and nucleic acids. And the commonly used organic materials (such as polypropylene, asphalt, rubber, and paints) are easy to degrade under the high energy UV light as a result of the covalent bonds breakage (e.g., C–H, C–C, C–Cl) [34]. Therefore, the shielding of UV irradiation is of great importance in our daily life.

UV-shielding materials could be classified into organic, inorganic, and organic/inorganic composites. The main organic UV absorbents include benzotriazole, benzophenone, salicylic acid, and esters; the shielding mechanism is the UV-induced molecular rearrangement, leading to the release of the absorbed energy [35, 36]. The inorganic UV blocking materials contain TiO_2, ZnO, $CaCO_3$, French chalk, and LDHs, and they are commonly dispersed into organic or inorganic materials to form organic/inorganic composites with enhanced UV-shielding performances [37, 38].

4.2 UV-Shielding Properties of LDHs

4.2.1 Scattering Effect of LDH Particles

According to the Lambert–Beer's law, the particle size severely affects the scattering and reflection effects of UV light. Xing et al. [38, 39] have prepared Zn-Al-LDHs through separate nucleation and aging steps (SNAS) at different temperatures, and the influences of particle sizes on the UV blocking properties have been investigated. The particle sizes calculated from XRD patterns via Scherrer expression are shown in Table 1. The results show that with increase of aging temperature, the particle size of LDHs increases in both a and c directions.

The suspension of $ZnAl-CO_3$-LDHs with different aging temperature and particle size can be obtained by dispersing the LDHs into water at 0.02 wt%, and their UV–Vis transmittance curves are shown in Fig. 6. With the increase of aging temperature, the UV-shielding performances increased obviously. In order to investigate the exact relation between particle size and UV-shielding performance,

Table 1 Particle sizes of LDHs at different aging temperatures [39]

Aging temperature (°C)	60	80	100
Particle size in a direction D_a (nm)	37.227	39.178	41.806
Particle size in c direction D_c (nm)	25.318	34.097	42.662

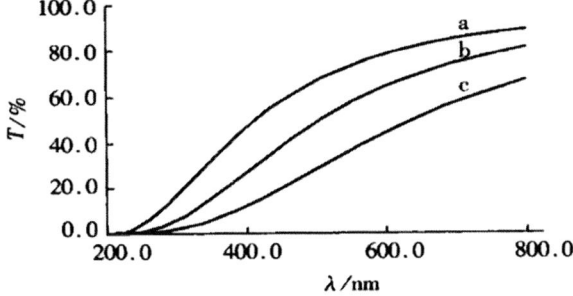

Fig. 6 UV–Vis transmittance curves of suspension of ZnAl-CO$_3$-LDHs with different aging temperature: (a) 60 °C, (b) 80 °C, and (c) 100 °C [39]

Table 2 Threshold values of particle size in a and c directions at different wavelengths [39]

Wavelength (nm)	Threshold value in a direction	Threshold value in c direction
290	32.92	9.73
320	33.77	12.79
400	34.77	16.47
600	35.56	19.35
800	35.79	20.22

the absorbance and transmittance values are analyzed at different wavelengths, and the threshold values for UV shielding at different wavelengths are shown in Table 2. It can be known that long wavelengths correspond to high threshold value in a and c directions. Therefore, different preparation conditions can be chosen to meet the requirements of transmittance in different wavelengths.

4.2.2 Reflection and Absorption of LDHs

The reflection and absorption of UV light are largely dependent on the layered structure of LDHs. The rigid LDH laminates could effectively reflect UV light, and multiple decrements of UV intensity may occur through multilayer reflection of UV light [34]. Moreover, the difference of metal elemental composition in LDHs layers may result in different performance of UV shielding. It was known that the UV-shielding properties of MgAl-LDH are highly related to the particle scattering and reflection of LDH layers. However, the replacement of Mg by Zn atom in the layer may further improve the UV-shielding properties [34, 40]. Figure 7a shows

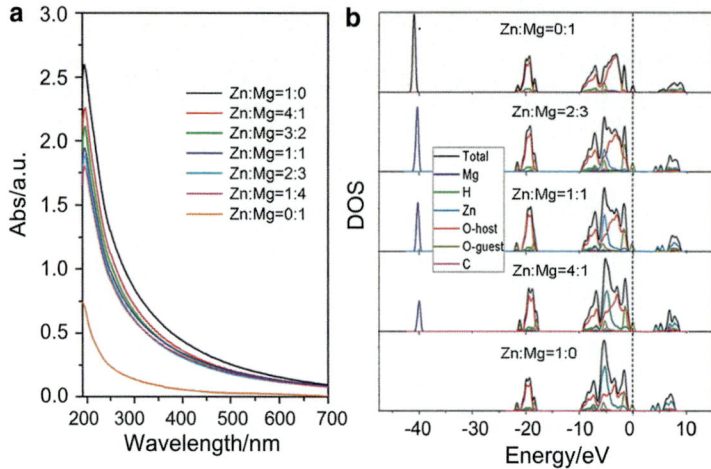

Fig. 7 (**a**) UV–Vis absorption spectra of the $((Zn_xMg_y)_2Al\text{-}CO_3\text{-}LDH/PAA)_{10}$ films ($x:y = 1:0$, 4:1, 3:2, 1:1, 2:3, 1:4 and 0:1; $x + y = 1.0$); (**b**) the density of states for the $(Zn_xMg_y)_2Al\text{-}CO_3\text{-}LDHs$ ($x:y = 0:1$, 2:3, 1:1, 4:1 and 1:0; $x + y = 1.0$) [34]

the UV–Vis absorption spectra of the $((Zn_xMg_y)_2Al\text{-}CO_3\text{-}LDH/PAA)_{10}$ films which are assembled by polyacrylate (PAA, without any UV absorption) and LDHs with the same particle size but different Zn^{2+}/Mg^{2+} ratios [34]. The results show that the UV absorption capacity enhances along with the increased Zn^{2+}/Mg^{2+} ratios. The calculated density of states (DOS) of the $(Zn_xMg_y)_2Al\text{-}CO_3\text{-}LDHs$ are shown in Fig. 7b, which indicates that the UV absorption of the pure $Mg_2Al\text{-}CO_3\text{-}LDHs$ is due to the interlayer CO_3^{2-} anions. Compared with the $Mg_2Al\text{-}CO_3\text{-}LDHs$ (4.58 eV), a decrease in the bandgap (3.33–4.16 eV) of the $(Zn_xMg_y)_2Al\text{-}CO_3\text{-}LDHs$ occurs, which facilitates the increase in photon absorption efficiency. Therefore, the employment of Zn element is beneficial to tune the transition mode, electron structure, bandgap, and UV-shielding properties of LDH materials.

4.3 Host–Guest Interaction and UV-Shielding Performances

Due to the versatility and ions exchangeable properties of LDH materials, a number of anions with strong UV absorption properties have been intercalated into the interlayer of LDHs. Therefore, the UV-shielding properties of the LDH-based materials can be effectively improved due to the synthetic effects of interlayer absorption, particle scattering, and structural reflection in the layer. Moreover, the host–guest interaction between interlayer guests with LDHs layers may further influence their shielding properties.

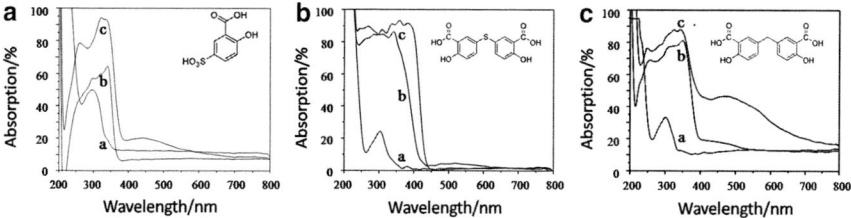

Fig. 8 UV absorption curves of (**A**) SSA-LDHs, (**B**) TDSA-LDHs, and (**C**) MDSA-LDHs composites, and the details are (a) ZnAl-NO$_3$-LDHs, (b) UV absorbents, and (c) intercalated LDHs, and the insets show the structures of UV absorbents [42–44]

4.3.1 Salicylic Acid and Esters

Salicylic acid and esters are commonly used as organic UV absorbents; Xing et al. [41] have prepared sodium salicylate (SS) intercalated LDHs by ion-exchange method. The obtained supramolecular materials exhibit advanced UV-shielding properties and promoted stability, which can be due to the electrostatic interaction and hydrogen bond between LDHs and guest ions. UV transmittance measurements indicate that the UV-shielding performance is highly improved for the ZnAl-SS-LDHs, compared with the pristine SS and ZnAl-CO$_3$-LDHs samples. In addition, some other salicylate compounds, including 5-sulfosalicylic acid (SSA) [42], 5, 5′-thiodisalicylic acid (TDSA) [43], and 5,5′-methylenedisalicylic acid (MDSA) [44], have been intercalated into LDHs, and the resulting LDHs composites exhibit promoted UV absorption performance (Fig. 8).

4.3.2 Benzophenones

He et al. [33] have prepared 2-hydroxy-4-methoxybenzophenone-5-sulfonic acid (HMBA, an organic UV absorbers) assembled Zn-Al-LDHs through ion-exchange method, and the obtained composites possess better UV absorption ability and high stability. Moreover, HMBA/LDH system can also be obtained by the SNAS method [45]. The intercalated LDHs composites had a strong UV absorption capacity, and a broad-range characteristic can be attributed to the regular arrangement of HMBA in LDHs galleries (Fig. 9A). Photostability tests were carried out by irradiating the samples under UV light for different time. The UV absorption intensity of pure HMBA has decreased for about 20 % in the 200–400 nm after UV irradiation accompanied by a significant increase in absorption in the visible region, which indicates a structural change for the HMBA (Fig. 9B). For the HMBA-LDHs composites, except for the slight change due to the degradation of absorbed HMBA on LDHs surface, there is almost no obvious change during the UV irradiation (Fig. 9C). Thus, significantly improved photostability was obtained upon the intercalation of UV absorbents.

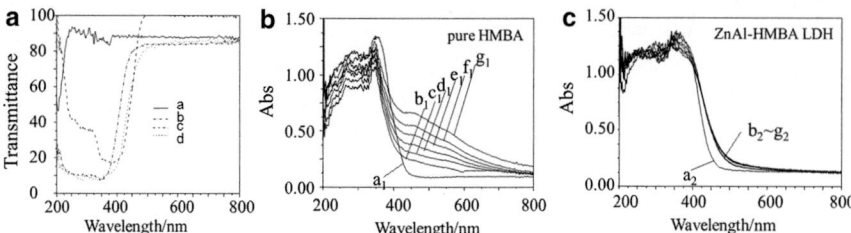

Fig. 9 (**A**) UV–Vis absorption curves of (a) ZnAl-CO$_3$-LDHs, (b) pure HMBA, (c) sodium HMBA, and (d) ZnAl-HMBA-LDHs; UV-Visible absorption curves of (**B**) pure HMBA and (**C**) ZnAl-HMBA-LDHs after different times of UV irradiation, (a)–(g): 0 min, 15 min, 30 min, 45 min, 60 min, 75 min, and 90 min [45]

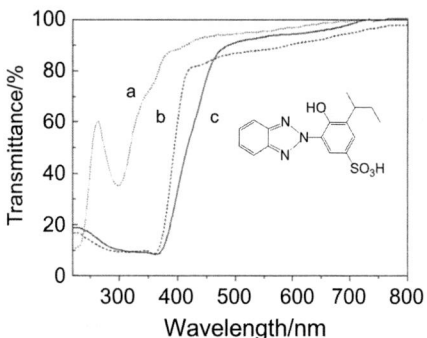

Fig. 10 UV transmittance curves of (a) ZnAl-NO$_3$-LDHs, (b) BZO, and (c) ZnAl-BZO-LDHs, and the inset shows the structure of BZO [46]

4.3.3 Benzotriazole

Tuo et al. [35, 46] have intercalated 5-benzotriazolyl-4-hydroxy-3-sec-butyl-benzenesulfonic acid (BZO) into interlayer of LDHs by ion-exchange method using ZnAl-NO$_3$-LDHs as a precursor. Figure 10 shows the UV transmittance curves of intercalated BZO-LDHs composites as well as the pure LDH and BZO precursors. UV absorption bands of ZnAl-NO$_3$-LDHs are located around 300 and 230 nm (Fig. 10a), and BZO possesses strong absorption below 400 nm (Fig. 10b). After intercalation, the BZO/LDH exhibits a broader range of UV shielding compared with the pristine BZO (Fig. 10c) and also maintains high visible light transmittance.

4.3.4 Others

Some other organic UV absorbents have also been intercalated into the interlayer galleries of LDHs, and superior UV blocking properties, visible light transmittance,

Layered Double Hydroxide Materials: Assembly and Photofunctionality

Table 3 Molecules and its structure of UV absorbents for intercalation

Interlayer guest molecules	Structure	Refs.
4-Hydroxy-3-methoxybenzoic acid		[33]
4-Hydroxy-3-methoxycinnamic acid		[33]
4,4′-Diaminostilbene-2,2′-disulfonic acid		[33]
p-Aminobenzoic acid		[33]
Urocanic acid		[33]
Cinnamic acid		[47]
p-Methoxycinnamic acid		[47]
2-Phenylbenzimidazole-5-sulfonic acid		[35, 48]
3,6-Dihydroxynaphthalene-2,7-disulfonate		[36]
2-Naphthylamine-1,5-disulfonic acid		[36, 49]
2,3-Dihydroxynaphthalene-6-sulfonic acid		[36, 50]
Aurintricarboxylic acid		[51]

and photo- and thermal stability can also be obtained. The molecules and their chemical structures are listed in Table 3.

4.4 Application of LDH-Based UV-Shielding Materials in Polypropylene and Asphalt

4.4.1 Polypropylene

Polypropylene (PP) is one of the most common plastic, but its application has been hindered due to the weak photostability. The mechanical properties of PP are usually affected by the UV photo-oxidative degradation under sunlight [43, 44]. To improve the anti-UV ability of PP, photostabilizers with

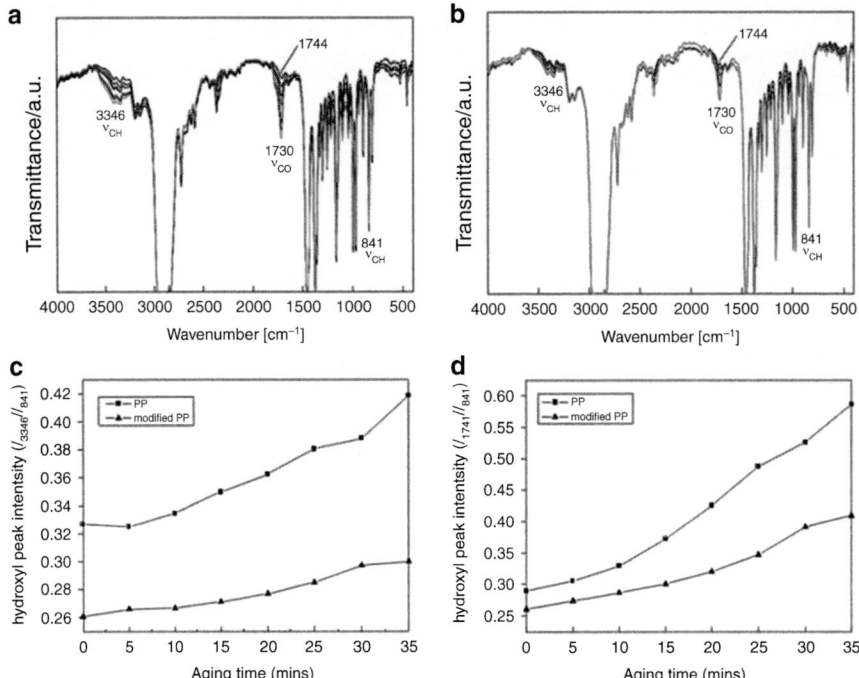

Fig. 11 FT-IR spectra of (**a**) pristine PP and (**b**) PP modified with 1 % ZnAl-BZO-LDHs under UV irradiation for 35 min, and the relative peak intensities of (**c**) hydroxyl and (**d**) carbonyl bands for pristine PP and LDH-modified PP (1 %) [35]

UV-shielding properties have been added to the PP composite [35, 42, 49–51]. Figure 11 shows the compared IR reflectance spectra of PP films with and without the addition of ZnAl-BZO-LDHs during photodegradation process [35]. Wavenumbers at 1,456, 1,376, 1,165, 974, 841, and 808 cm^{-1} are characteristic IR absorption bands of PP and ZnAl-BZO-LDHs/PP (Fig. 11a, b) samples. The absorption intensity ratios I(carbonyl)/I(C–H) and I(hydroxyl)/I(C–H) can be calculated (Fig. 11c, d), and the results show that the rate for formation of C=O and OH/OOH groups in ZnAl-BZO-LDHs/PP under UV irradiation is significantly lower than that in the pristine PP. Therefore, the addition of ZnAl-BZO-LDH has markedly blocked the UV light and prevented the degradation of PP and thus enhances the photostability. Figure 12 shows the digital photos of pristine PP and LDH-modified PP after UV exposure for 15 min. It was found that the pristine PP film becomes brittle after exposure to UV light, while the LDH-modified sample maintains high mechanical properties under the same conditions. This result further confirms that LDHs can effectively shield UV light to prevent degradation.

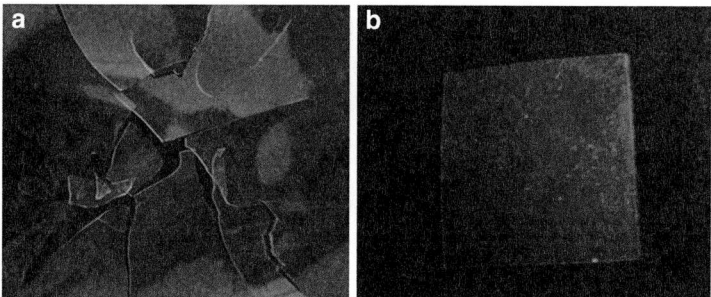

Fig. 12 Photographs of (**a**) pristine PP and (**b**) LDH-modified PP after UV exposure for 15 min [35]

Table 4 Softening point increment and VAI values for three asphalt samples [41]

Sample	Initial softening point (°C)	Softening point after 4 days (°C)	ΔS (°C)	Initial viscosity (Pa·s)	Viscosity after 4 days (Pa·s)	VAI
Pristine asphalt	49.3	55.4	6.1	0.926	1.137	0.228
Asphalt with 3 wt% $Mg_4Al_2\text{-}CO_3$-LDH	50.2	51.7	1.5	1.102	1.330	0.207
Asphalt with 3 wt% $Zn_4Al_2\text{-}CO_3$-LDH	49.2	50.4	1.2	1.064	1.254	0.179

4.4.2 Asphalt Pavement

As a highly viscous mixture of polycyclic aromatic hydrocarbon, asphalt has been widely used in road constructions. However, UV radiation always leads to the aging of asphalt, which is caused by the chemical transformation of the containing organic polymers. The aging of asphalt would subsequently affect the upper layers of the pavement, resulting in cracking, stripping, or other degradation of the pavement. Considering the serious damages of pavement, it is important to find new ways to improve the ability of asphalt to delay UV radiation.

In order to investigate the practical application, the law of solar radiation was studied, and a series of artificial tests (such as accelerated aging tests, rheological, and viscoelastic measurements) have been carried out. Important parameters in evaluating the aging resistance ability of asphalt, such as viscosity aging index (VAI) and softening point increment (ΔS) after aging, are obtained. The higher values of VAI and ΔS correspond to samples with weaker antiaging properties [52]. Wang et al. [40] have performed UV light irradiation tests of asphalt and composite of asphalt with 3 wt% LDHs in an accelerated weather tester for 4 days. Table 4 shows that the values of VAI and ΔS for the asphalt doped with LDHs are lower than those of the pristine asphalt, especially for VAI values. Digital photos in

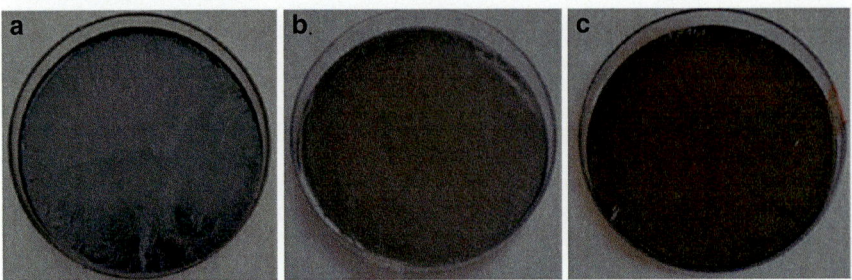

Fig. 13 Asphalt samples aged in UV light for 4 days: (**a**) pristine asphalt, (**b**) asphalt with 3 wt% MgAl-CO_3-LDH, and (**c**) asphalt with 3 wt% ZnAl-CO_3-LDH [41]

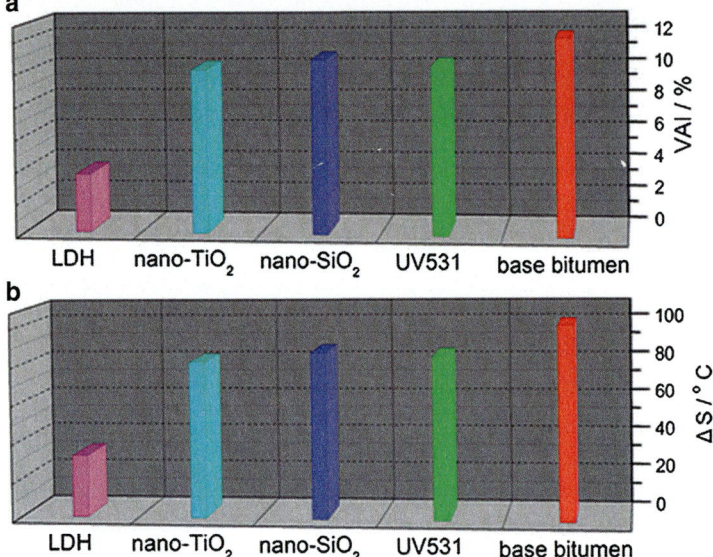

Fig. 14 (**a**) Viscosity aging index and (**b**) softening point increment of base bitumen and the four modified bitumen after irradiation with a UV lamp (500 W, 15,000 µW cm^{-2}) for 12 days [34]

Fig. 13 have further confirmed the enhanced antiaging performances of LDH-modified asphalt. Thus, it can be concluded that the employment of LDHs as anti-UV aging agents can result in enhanced photostability and prolonged service lifetime of asphalt. As well, Shi et al. [34] have investigated the UV resistance properties of LDHs. The values of VAI and ΔS show that the LDH-incorporated asphalt possesses superior performances than those modified by commercial UV absorbents (such as UV 531, SiO_2 and TiO_2), demonstrating a promising UV-resistance agent (Fig. 14).

5 LDH-Based IR Absorption Materials and the Application

5.1 Introduction of IR Absorption Materials

Solar light plays a crucial role in the photosynthesis and the growth of things on earth. However, the heat supplied by solar during the daytime would rapidly lose as heat radiation at night, which restricts the growth of plants to some extent. The heat loss from the earth surface to the surrounding is usually diffused by means of low-energy mid-infrared radiation, ranging from 7 to 25 μm (1,428–400 cm^{-1}) with the peak in the range of 9–11 μm (1,111–909 cm^{-1}). Therefore, to keep the temperature and prevent heat loss, agricultural plastic films (agri-film) [53, 54] are widely used in "green house."

Polymers, such as low-density polyethylene (LDPE), are commonly used agri-films for crop harvest and energy saving [55, 56]. However, LDPE has poor heat preservation, and the heat is rapidly lost through the film at night [57]. To effectively enhance the ability of heat storage of agri-film, some additives are needed. The heat-preservation additives should block the loss of long-wavelength IR at night and maintain transmission of visible light during the daytime [55, 58]. Some inorganic additives, such as French chalk, china clay, and metal oxides (ZnO, TiO_2, and Sb_2O_3), were incorporated as IR absorbents [54, 59, 60]. However, the contained impurities and variable particle sizes have accelerated the degradation of the film, and their transmittance of visible light and mechanical property are unsatisfactory [61]. Therefore, there is a need to develop materials with high purity and controllable particle size as alternative inorganic heat-preservation additives used in agri-films.

5.2 IR Absorption Properties of LDHs

LDHs have been used as IR absorption materials due to the uniform particle size, high crystallization degree, good photo-/thermal-stability, and tunable IR absorption abilities due to the changeable metal-oxide vibrations. By adding the LDHs particles into agri-film, the IR absorption ability, transparency, heat-preservation capacity, and the mechanical properties can be improved.

5.2.1 IR Absorption Ability

LDHs can strongly absorb IR due to the vibration of metal-oxide bonds in host layer and the interlayer anions [62]. For the typical CO_3^{2-} intercalated LDHs, the broad peak around 3,441 cm^{-1} can be ascribed to the stretching vibration of OH groups attached to the metal atoms in the LDH layers. The band at about 1,625 cm^{-1} can be

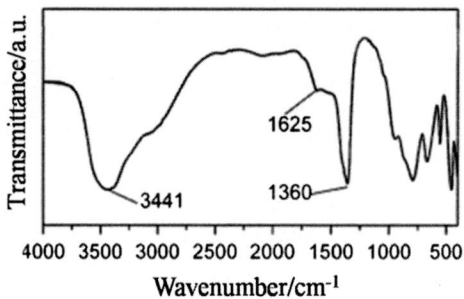

Fig. 15 FT-IR spectra of MgAl-CO$_3$-LDH [64]

Table 5 Typical heat-retention additives and their refractive index [67]

Materials	Refractive index	Materials	Refractive index
LDPE	1.51	MgAl-LDH	1.50
Silica	1.46	China clay	1.56
Light calcium carbonate	1.49	Diatomite	1.45
French chalk	1.55	Muscovite	1.58

attributed to the deformation vibration of water molecules in the interlayer domain. In addition to these absorption bands, characteristic peaks at 1,360, 840 and 671 cm^{-1} can be attributed to the ν_3 vibration of CO$_3^{2-}$ [63]. The vibrations of M-O units below 1,050 cm^{-1} (Fig. 15) are the key characteristics for the applications in IR absorption materials [64]. Moreover, in order to extend the range of IR absorption, Eu-doped LDH was investigated [58, 65]. In addition to the common IR absorption, the employment of Eu resulted in broad vibration peaks in the range of 600–900 cm^{-1} and 1,250–1,350 cm^{-1}, which has further improved the heat-preservation properties of agri-film.

5.2.2 Antireflection Property

The particle size of LDHs can be controlled due to different preparation methods [66]. Different from the traditional additives (such as the French chalk and china clay) with the particle size in the range of several μm, the LDH particles with the size distribution in the range of 50–500 nm can be obtained [64, 67]. Thus, LDHs with small size can be adopted as an alternative to promote dispersion degree in the agri-film. Moreover, the similar refractive index of LDHs and LDPE (illustrated in Table 5) has effectively reduced refraction phenomenon, leading to the low light loss and the promoted transparency in the agri-film.

5.2.3 Heat-Preservation, Photostability, and Antifogging Properties

To achieve high heat-preservation ability, visible light transparency and infrared irradiation blocking are significant. The LDH materials with high visible light transmittance enabled visible light get through the agri-film during the daytime, and the IR absorption properties of LDHs could prevent the infrared irradiation and heat loss at night.

As an additive, LDHs are highly stable under the photo-irradiation, which ensures their long service lifetime in practical use. Moreover, LDHs can absorb or isolate acidic groups and heavy metal ions to protect agri-film from degradation. Owing to the large specific surface area, porous structure, and oil absorption property of LDHs, the addition of LDHs enables controlled release of antifogging agent, which can further enhance the antifogging properties of agri-film.

However, the IR absorption of NO_3^- or CO_3^{2-} intercalated LDHs usually ranges from 7 to 8 μm, which cannot perfectly meet the strong IR loss at 9–11 μm (909–1,111 cm^{-1}). Therefore, the introduction of suitable anions with high IR absorption at 909–1,111 cm^{-1} is necessary to reduce the IR loss and further promote the capability of heat preservation of agri-film [68].

5.3 Intercalation and Performances of LDH-Based IR Absorption Materials

Several molecule-based systems possess strong IR absorption required for blocking heat irradiation. However, these molecules sometimes are acidic, and their direct use as additives may result in fast degradation or aging of the film. Therefore, the incorporation of these molecules (usually containing sulfate, carboxyl, and phosphate groups) into LDHs may be an effective way to achieve high photo- and thermal stability of the agri-film.

5.3.1 Intercalation with Inorganic Anions

Jiao et al. [63] have prepared Mg_2Al-CO_3-SO_4-LDH via the ion-exchange method, and the obtained composites have well-defined crystal structure and present higher selective IR absorption ability than the pristine Mg_2Al-CO_3-LDHs. The peaks at 1,104 and 1,194 cm^{-1} can be attributed to the ν_3 vibration of SO_4^{2-}, associated with peaks at 1,019 and 619 cm^{-1} for ν_1 and ν_4 vibrations (Fig. 16). Therefore, the Mg_2Al-CO_3-SO_4-LDHs possess better selective IR absorption properties.

Phosphate groups (such as the vibrations of P–O and P=O) present strong IR absorption in the range of 1,300–900 cm^{-1} which is exactly the region required for effective absorption of heat irradiation. Wang et al. [69, 70] have prepared phosphate groups intercalated MgAl-LDHs. As shown in Fig. 17, peaks at 1,253 cm^{-1}

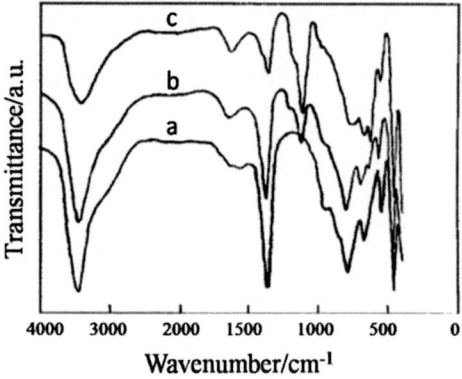

Fig. 16 IR spectra of samples exchanged by SO_4^{2-} for different time: (a) 0 h, (b) 4 h, and (c) 16 h [63]

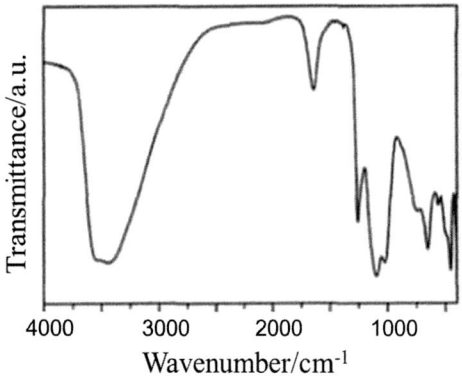

Fig. 17 IR spectra of MgAl-H_2PO_4-LDH [69]

Table 6 Average transmittances of mixed MgAl-H_2PO_4-LDHs and MgAl-CO_3-LDHs in different ranges [69]

Sample	7–14 μm	7–25 μm	9–11 μm
MgAl-CO_3-LDHs	34.5	29.7	39.5
MgAl-CO_3-LDHs:MgAl-H_2PO_4-LDHs (3:1)	28.9	26.3	37.2
MgAl-CO_3-LDHs:MgAl-H_2PO_4-LDHs (2:1)	27.1	25.2	24.6
MgAl-CO_3-LDHs:MgAl-H_2PO_4-LDHs (1:1)	23.6	21.5	19.8
MgAl-CO_3-LDHs:MgAl-H_2PO_4-LDHs (1:2)	25.0	22.6	18.5
MgAl-CO_3-LDHs:MgAl-H_2PO_4-LDHs (1:3)	25.6	23.2	18.3
MgAl-H_2PO_4-LDHs	28.2	24.2	17.8

and 1,091 cm^{-1} are attributed to the vibration of PO_2, while the peak at 1,020 cm^{-1} is due to the stretching vibration of P–OH. Further experiment was carried out by mixing MgAl-CO_3-LDH and MgAl-H_2PO_4-LDH at different ratios (Table 6), and employment of H_2PO_4 has effectively improved the selective IR properties compared with pristine MgAl-CO_3-LDH, especially in the region of 9–11 μm. In order to fabricate materials with comprehensive IR absorption properties, all of the IR regions (such as 7–25, 7–14 and 9–11 μm) should be taken into consideration. The sample with the ratio of MgAl-CO_3-LDH and MgAl-H_2PO_4-LDH at 1:1 possess optimized absorption ability in these three region. As another example, Badreddine et al. [71] have prepared a series of phosphate anions (such as PO_3^-, $H_2PO_4^-$, HPO_4^{2-}, PO_4^{3-}, $P_2O_7^{4-}$ and $P_3O_{10}^{5-}$) intercalated LDH composites, and the obtained composites displayed largely promoted IR absorption properties.

5.3.2 Intercalation with Organic Anions

Some organic anions containing phosphate groups have also been intercalated into the interlayer of LDHs [70], and the LDH-based composites with different IR absorption intensity and selectivity were obtained. For example, Wang et al. have employed *N,N*-Bis(phosphonomethyl)glycine (GLYP) [53], aminotrimethylenephosphonic acid (ATMP) [55], and *N*-phosphonomethyliminodiacetic acid (PMIDA) [59, 62] as interlayer guests (Fig. 18) to promote the IR absorption capacity in the range of 909–1,111 cm^{-1} in the composites. Thus, the obtained hybrid materials possess higher infrared absorption than the original MgAl-CO_3-LDH, and the heat loss via the radiation to the atmosphere environment can be effectively inhibited.

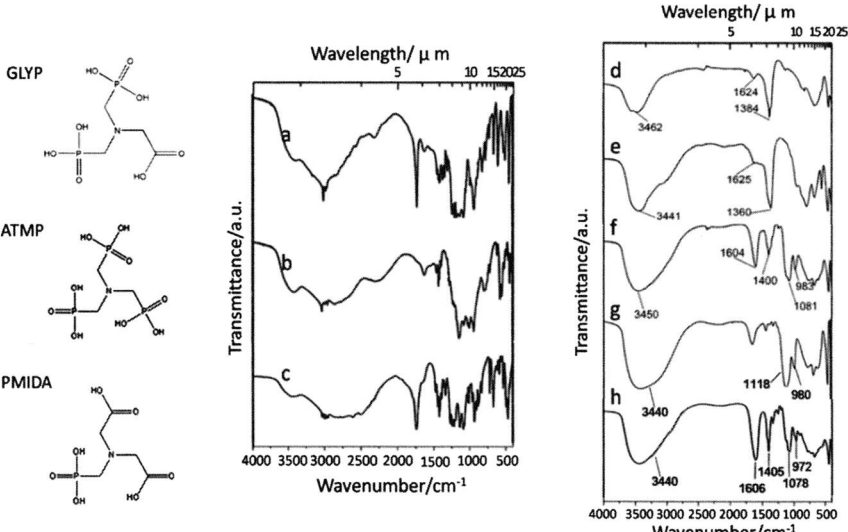

Fig. 18 Molecular structure and IR spectra of (a) GLYP, (b) ATMP, (c) PMIDA, (d) MgAl-NO_3-LDH, (e) MgAl-CO_3-LDH, (f) MgAl-GLYP-LDH, (g) MgAl-ATMP-LDH, and (h) MgAl-PMIDA-LDH [53, 55, 59]

5.4 Application of LDH-Based IR Absorption Materials

The obtained MgAl-X-LDHs, X = $H_2PO_4^-$ [69], GLYP [53], ATMP [55], PMIDA [59], and IDA [62]) were mixed with LDPE to form the LDH/LDPE composite films by the master batch technology [67, 72]. Taking the MgAl-GLYP-LDH as an example [53, 70], TEM images of LDH/LDPE composites (4 wt% loadings of GLYP-LDH and CO_3-LDH) are shown in Fig. 19. The continuous gray areas stand for the polymer matrices, while the dark dots in left pictures represent LDHs composites. It can be clearly seen that both GLYP-LDH and CO_3-LDH particles are uniformly dispersed in the polymer matrices without any aggregation of clusters.

The TG-DTA curves of LDPE, GLYP-LDH/LDPE, and CO_3-LDH/LDPE are shown in Fig. 20. The DTA curves in three samples show endothermic transition at about 130 °C (melting point of LDPE). The main weight loss is between 270 and 500 °C with an exothermic peak at 410 °C which can be attributed to the degradation of the hydrocarbon chains [73–75]. However, for GLYP-LDH/LDPE and CO_3-LDH/LDPE, the weight loss begins at 320 °C and 380 °C, respectively, and the exothermic peaks shift to 420 °C and 425 °C. Moreover, the T_{50} values (the temperature corresponding to 50 % weight loss) for LDPE, GLYP-LDH/LDPE, and CO_3-LDH/LDPE are calculated at 420 °C, 460 °C, and 450 °C, respectively. Therefore, it can be concluded that addition of LDHs composites has effectively promoted the thermal stability of LDPE film.

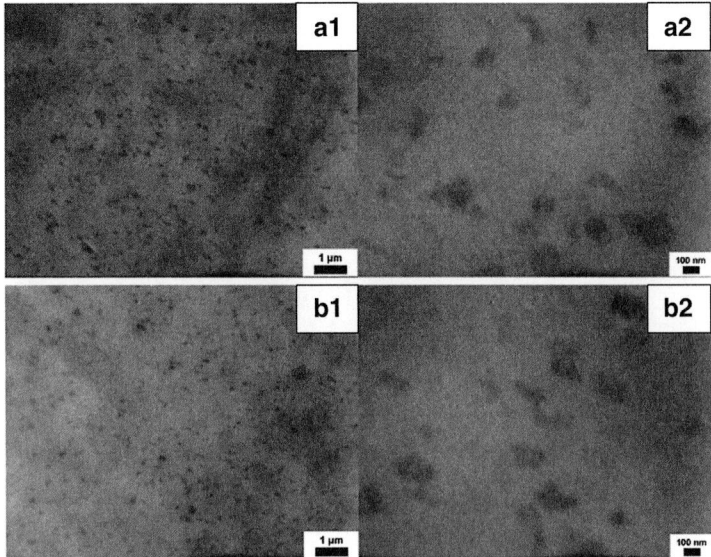

Fig. 19 TEM images of LDH/LDPE composites with 4 wt% loadings of (**a**) GLYP-LDH and (**b**) CO_3-LDH at different magnification scales (scale bar: 1 μm for *left* and 100 nm for *right pictures*) [53]

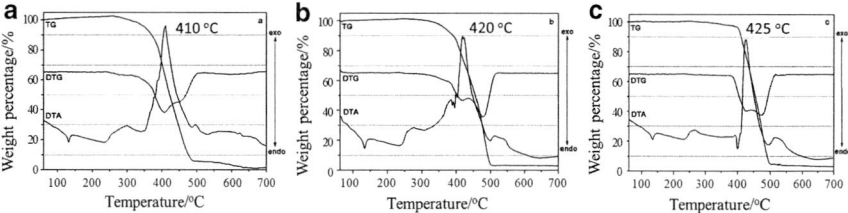

Fig. 20 TG-DTA-DTG curves of (**a**) LDPE, LDH/LDPE composites with 4 wt% loadings of (**b**) GLYP-LDH and (**c**) CO_3 LDH [73, 74]

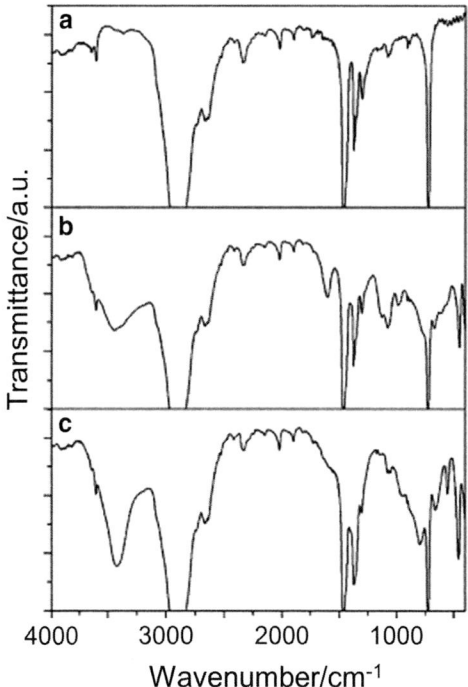

Fig. 21 FT-IR spectra of (**a**) pure LDPE film and LDH/LDPE films with 4 wt% loadings of (**b**) GLYP-LDH and (**c**) CO_3-LDH [73, 74]

Figure 21 shows the FT-IR spectra of the LDPE film, GLYP-LDH/LDPE, and CO_3-LDH/LDPE film. Compared with the pristine LDPE film, the CO_3-LDH-doped LDPE composite exhibits superior infrared radiation in the range of 1,428–400 cm^{-1}, while GLYP-LDH/LDPE displays advanced infrared absorbing ability in the range of 909–1,111 cm^{-1}. In addition, mechanical properties of incorporated LDPE films were also carried out, and the average values of breaking elongation and tensile strength are shown in Table 7. Compared with LDPE film, the breaking elongation property was promoted after the addition of LDHs, and the tensile strength was just slightly reduced. Therefore, the addition of LDHs would not influence the mechanical properties of LDPE.

Table 7 Mechanical properties of LDPE and LDH/LDPE [73, 74]

Samples	Breaking elongation (%)	Tensile strength (MPa)
LDPE	442.3	14.0
MgAl-GLYP-LDH/LDPE	447.2	13.3
MgAl-CO_3-LDH/LDPE	447.1	13.6

6 Fabrication and Application of LDH-Based Host–Guest Luminescent Materials

The photoactive guest molecules assembled into the LDHs host layers can achieve new type of host–guest composite materials with advanced photo-related performances (such as luminescence). There are several advantages for the accommodation of photofunctional species into LDHs inorganic matrices:

1. LDHs layers can provide a confined and isolated microenvironment for the intercalated guest molecules. In the confined interlayer space, the vibration and thermal motion of guests could be inhibited, and this may effectively improve the luminescent efficiency of the photoactive molecules. Furthermore, the photofunctional guests with different emission can be assembled into the host layers in a step-by-step manner, in which the rigid LDH layers can avoid the permeation between different guest molecules.
2. The regular and crystalline LDHs layers allow the ordered arrangement and high orientation of interlayer guest molecules based on host–guest interactions. By tuning the chemical composition and charge density of host layers as well as the amount, arrangement and aggregation state of the guest molecules, the photo-related performance can be finely adjusted.
3. Due to the introduction of LDHs layers, the thermal- and photostability of the intercalated composites can be largely improved. Thus, the problem involving the stability of pristine organic materials in the practical applications can be resolved effectively.
4. The combination of transparent insulating LDHs layers with photoactive molecules may form an inorganic/organic multi-quantum-well structure. The insulating LDHs can act as energy barrier layers in the quantum well, which can accelerate the formation of exciton and improve the fluorescence efficiency.
5. Based on the changes in the aggregation states and/or the conformations of the interlayer photoactive molecules induced by the external environmental stimuli, the corresponding photo-related properties can be further tuned due to the alternations of host–guest and guest–guest interactions. Therefore, the LDH-based intelligent-responsive materials can be further constructed.

According to the design strategies mentioned above, a number of host–guest photofunctional materials have been continuously reported. In the following section, we mainly focus on the host–guest photofunctional materials with tunable static and dynamic luminescence.

6.1 The Basic Guest Units of LDH-Based Photofunctional Materials

To date, both the organic and inorganic photoactive species (including polymers, complexes, small molecules, quantum dots, and polyoxometalates) can serve as good candidates as guest molecules to assemble with LDHs. And the difficulties, such as the aggregation-induced quenching, lifetime and stability are eager to be solved. Table 8 gives typical examples of the interlayer photofunctional guests and their related fabrication methods.

6.2 Fabrication of LDH-Based Photofunctional Material

6.2.1 Layer-by-Layer Assembly

Layer-by-layer (LbL) assembly is a widely used method to achieve multilayer LDH-based ultrathin films (UTFs), which involves alternative dipping a substrate in positively charged LDH nanosheets and negative-charged photofunctional molecules [129]. Based on their electrostatic interactions, the guest molecules can be accommodated into the interlayer of LDH nanosheets. The employment of LDH nanosheets has provided an opportunity to finely tune the arrangement and relative distance of the interlayer molecules, and the homogeneous distribution of photofunctional molecules can be obtained [130]. Due to the confined and ordered environment provided by the rigid LDHs nanosheets, the luminescence quenching commonly caused by the aggregation and/or accumulation of photoactive molecules can be efficiently reduced. Moreover, the existence of inorganic layers may improve the thermal and optical stability of the photofunctional molecules [76]. Therefore, LbL assembly has supplied an effective way for the fabrication of hybrid UTFs with advanced photo-related properties.

Assembly Based on π-Conjugated Polymers

Luminescent organic polymer materials have attracted extensive attention, particularly due to their applications in optoelectronic devices, such as liquid-crystal displays [131, 132]. However, the development of these polymers is severely hindered by their short lifetime and poor photo- or thermal stability [133]. Yan et al. have presented a potential solution to these problems by assembling the polymers with rigid LDH nanosheets which can serve as a new inorganic/organic quantum well structure. Several anionic polymers, such as the derivates of poly (p-phenylene) [76], poly(phenylenevinylene) [78], and polythiophene [79], have been studied, and investigation on a sulfonated poly(p-phenylene) anionic derivate (APPP) will be particularly introduced below.

Table 8 Photofunctional guests used in LDH-based composites and their fabrication methods

Materials	Categories	Molecules	Method	Refs.
Organic	Polymers	Poly(p-phenylene) anionic derivate (APPP)	LbL	[76, 77]
		Poly(5-methoxy-2-(3-sulfopropoxy)-1,4-phenylene-vinylene anionic derivate (APPV)	LbL	[77, 78]
		Sulfonated polythiophene (SPT)	LbL	[79, 80]
		Polydiacetylene (PDA)	LbL	[81]
		Poly{1-4[4-(3-carboxy-4-hydroxyphenylazo)benzenesulfonamido]-1,2-ethanediyl sodium salt} (PAZO)	LbL	[82]
	Complexes	Tris(1,10-phenanthroline-4,7-diphenyl-sulfonate)ruthenium (II) ([Ru(dpds)$_3$]$^{4-}$)	LbL	[80, 83]
		Bis(8-hydroxyquinolate)zinc, tris (8-hydroxyquinolate-5-sulfonate) aluminum	LbL/co-intercalation	[84–87]
		[Tetrakis(4-carboxyphenyl)-porphyrinato]zinc(II) (ZnTPPC)	Ion exchange	[88]
		Pd(II)-5,10,15,20-tetrakis (4-carboxyphenyl) porphyrin (PdTPPC)	Coprecipitation	[89]
		5,10,15,20-Tetrakis (4-sulfonatophenyl) porphyrin (TPPS) and its palladium complex (PdTPPS)	Coprecipitation	[89–91]
		Zinc phthalocyanines (ZnPc)	Coprecipitation	[92]
		Tris[2-(4,6-difluorophenyl) pyridinato-C^2,N]iridium(III) (Ir(F$_2$ppy)$_3$)	LbL	[93]
	Small molecules	Sulforhodamine B (SRB)	Co-intercalation	[94]
		Bis(N-methyl-acridinium) (BNMA)	LbL	[77, 95]
		3,4,9,10-Perylene tetracarboxylate (PTCB)	LbL/co-intercalation	[96, 97]
		Benzocarbazole (BCZC)	Coprecipitation	[98]
		Bis(2-sulfonatostyryl)biphenyl (BSB)	LbL	[99]
		Sulfonated derivate of cyanin (Scy)	LbL	[100]
		2,2′-(1,2-Ethenediyl)bis [5-[[4-(diethylamino)-6-[(2,5-disulfophenyl) amino]-1,3,5-triazin-2-yl]amino] benzene-sulfonicacid] hexasodium salt (BTBS)	LbL	[101, 102]

(continued)

Table 8 (continued)

Materials	Categories	Molecules	Method	Refs.
		4,4'-Bis[2-di(b-hydroxyethyl) amino-4-(4-sulfophenylamino)-s-triazin-6-ylamino] stilbine-2,20-disulfonate (BBU)	LbL	[103]
		α-Naphthalene acetate (α-NAA) and β-naphthalene acetate (β-NAA)	Ions exchange	[104, 105]
		Fluorescein	Co-intercalation	[106, 107]
		2,2'-Azino-bis(3-ethylben-zothiazoline-6-sulfonate) (ABTS)	LbL	[108]
		Calcein	LbL	[109]
		Coumarin-3-carboxylate (C3C)	Coprecipitation	[110]
		(2,2'-(1,2-Ethenediyl)bis [5-[[4-(diethylamino)-6-[(2,5disulfophenyl) amino]-1,3,5-triazin-2-yl] amino]benzene sulfonate anion (BTZB)	Coprecipitation	[111]
		4-(4-Anilinophenylazo) benzenesulfonate (AO5)	Co-intercalation	[112]
		1,3,6,8-Pyrenetetrasulfonat acid tetrasodium salt (PTS)	LbL	[113]
		9-Fluorenone-2,7-dicarboxylate (FDC)	Co-intercalation	[114]
		2-Phenylbenzimidazole-5-sulfonate (PBS)	Co-intercalation	[115]
Inorganic	Quantum dots	CdTe	LbL	[102, 116–121]
		CdSe/ZnS	LbL	[118]
		InP/ZnS	Coprecipitation	[122]
		CdSe/CdS/ZnS	LbL	[123]
		CdS	Ions exchange	[124]
		ZnS	Ions exchange	[125]
	Polyoxometalates	$Na_9[EuW_{10}O_{36}]\cdot 32H_2O$ (EuW_{10})	LbL	[126, 127]
		$[EuW_{10}O_{36}]^{9-}$, $[Eu(BW_{11}O_{39})(H_2O)_3]^{6-}$ and $[Eu(PW_{11}O_{39})_2]^{11-}$	Ions exchange	[128]

The LbL assembly process of the (APPP/LDH)$_n$ UTFs was monitored by UV–Vis absorption spectroscopy (Fig. 22), in which the absorption intensity at about 207 nm and 344 nm attributed to $^1E_{1u}$ and π–π* transition of phenylene increases linearly along with the bilayer number n (Fig. 22a, inset), indicating a stepwise and uniform deposition procedure. Figure 22b shows the fluorescence emission spectra of (APPP/LDH)$_n$ UTFs, and the sharp peak at 415 nm shows a monotonic increase, consistent with the increase of bilayer number of (APPP/LDH) unit. The gradual

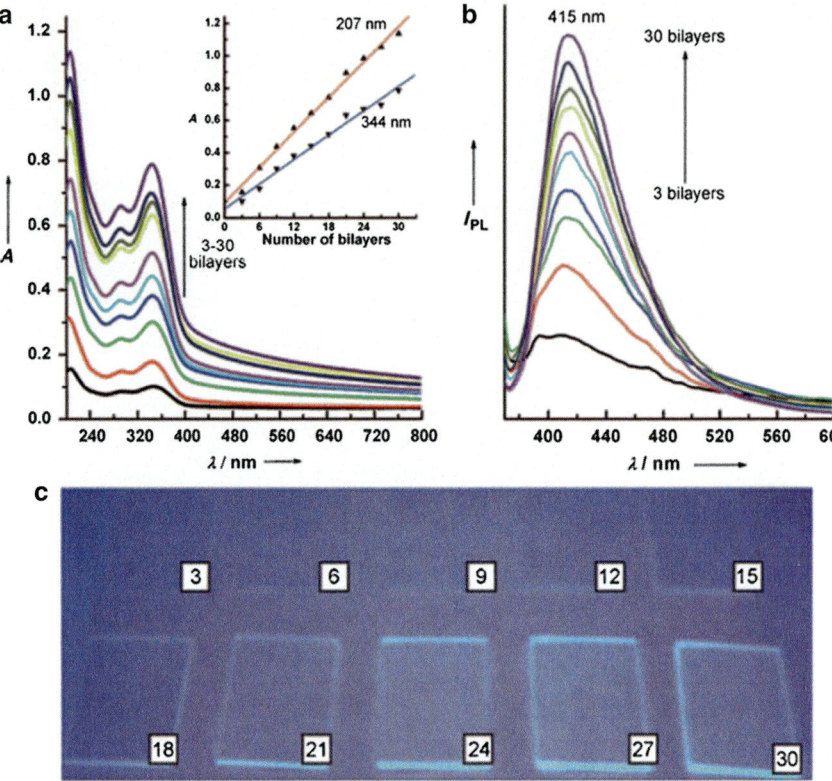

Fig. 22 Characterizations of (APPP/LDH)$_n$ ($n = 3$–30) UTFs: (**a**) UV–Vis absorption spectra (inset shows the linear correction of absorbance at 207 and 344 nm vs. n), (**b**) fluorescence spectra, and (**c**) photographs under 365 nm UV irradiation at different values of n [76]

change in luminescence can also be seen from photographs under UV light (Fig. 22c). Moreover, the UTFs with different n show no obvious shift or broaden in the absorption and fluorescence emission spectra, and this phenomenon suggests that no aggregation of APPP was formed during the assembly process.

The top-view SEM image (Fig. 23a) displays the homogeneity and uniformity of (APPP/LDH)$_9$ UTFs. Thickness of 20–23 nm for (APPP/LDH)$_9$ UTF was obtained from the side-view SEM image (Fig. 23b), with the average value about 2.5–2.8 nm for the basic unit of (APPP/LDH)$_1$. The AFM topographical image (3 μm × 3 μm) is illustrated in Fig. 23c with the root-mean-square roughness of 5.8 nm, indicating a smooth surface. Figure 23d shows the homogeneous distribution of the APPP chromophore throughout the film as dark blue color under the fluorescence microscope.

The fluorescence lifetime of (APPP/LDH)$_n$ UTF (18.63–21.05 ns) is largely prolonged compared with the pristine APPP solution (0.88 ns). This remarkable increase is a result of isolation of the adjacent polymer provided by rigid LDH

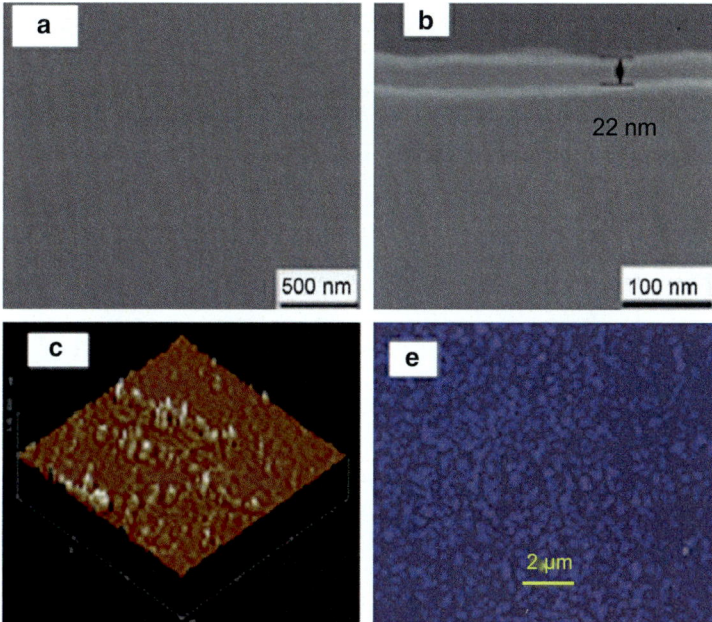

Fig. 23 The morphology of (APPP/LDH)$_9$ UTF: (**a**) top view of SEM image, (**b**) side view of SEM image, (**c**) tapping-mode AFM image, and (**d**) fluorescence microscope image [76]

nanosheets. Moreover, APPP molecules are confined within the interlayer environment of the LDH, and the inhibited thermal vibration of the APPP backbones may also contribute to the prolonged lifetime.

Photostability of (APPP/LDH)$_{27}$ UTF was performed under UV irradiation, and (APPP/PDDA)$_{27}$ (PDDA, polydimethyldiallylammonium chloride) was tested as a reference sample. As seen in Fig. 24, ~70 % of initial fluorescence intensity of (APPP/LDH)$_{27}$ UTF remained after continuous irradiation for 2 min, in contrast to ~50 % for (APPP/PDDA)$_{27}$ with the same measurement. Broadened emission band was observed for (APPP/PDDA)$_{27}$ UTF after 15 min, while no obvious band shift was found for (APPP/LDH)$_{27}$ in the whole test. It can be concluded that (APPP/LDH)$_{27}$ UTF possesses better UV light resistance ability compared with (APPP/PDDA)$_{27}$.

Assembly Based on Anionic Metal Complexes and Small Molecules

Transition-metal-based photoactive complexes have been extensively investigated in optoelectronics due to their high chemical stability and excellent luminescence properties [134]. Ruthenium-based complexes, tris(1,10-phenanthroline-4,7-diphenylsulfonate)ruthenium(II) (denoted as Ru(dpds)$_3$), have been studied by

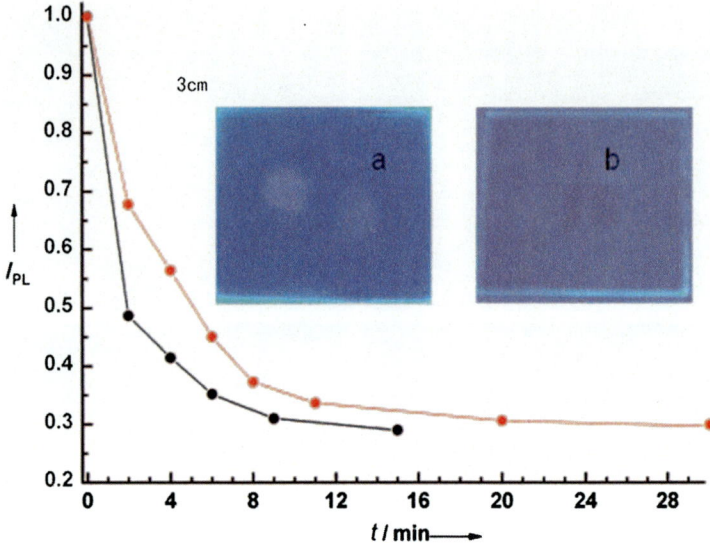

Fig. 24 Decay of normalized maximal PL intensity with time, normalized against the initial PL value; $\lambda_{ex} = 365$ nm to probe the UV irradiation resistance ability of (APPP/PDDA)$_{27}$ (*black data points*) and (APPP/LDH)$_{27}$ UTF (*red*) irradiated under 344 nm UV light. Insets: photographs under UV light of (**a**) (APPP/PDDA)$_{27}$ and (**b**) (APPP/LDH)$_{27}$ UTFs after the UV resistance experiment was finished [76]

assembly with LDH nanosheet to obtain orderly UTFs [83]. Figure 25 has illustrated the assembly process of (Ru(dpds)$_3$/LDH)$_n$ UTFs, and UV–Vis absorption and fluorescence emission spectra have been employed to monitor the process with the increased intensity when varying the number of bilayers. Except for the periodic long-range ordered structures, well-defined red photoluminescence for the UTFs, the ordered Ru(dpds)$_3$ has overcome symmetry due to interaction with LDHs, and polarized photoluminescence behavior was observed. Other small molecules such as sulfonated carbocyanine derivate (Scy) [100] can also be directly assembled with LDH nanosheet, and the fluorescence performance and stability can be largely improved for the assembled UTFs.

Assembly Based on Positive-Charged and Low-Charged Molecules

Due to the charge-balance rule, it has been known that only anionic species can be directly assembled into the galleries of the positively charged LDH layers. This largely restricts the development of LDH-based functional materials, and the assembly of LDH nanosheets with abundant functional cations remains a challenge. To meet this requirement, Yan et al. [95] have proposed a new assembly strategy for cation–LDH UTFs by employing a suitable polyanion as the carrier. Bis(*N*-methylacridinium) (BNMA, an important dye in the field of chemiluminescence)

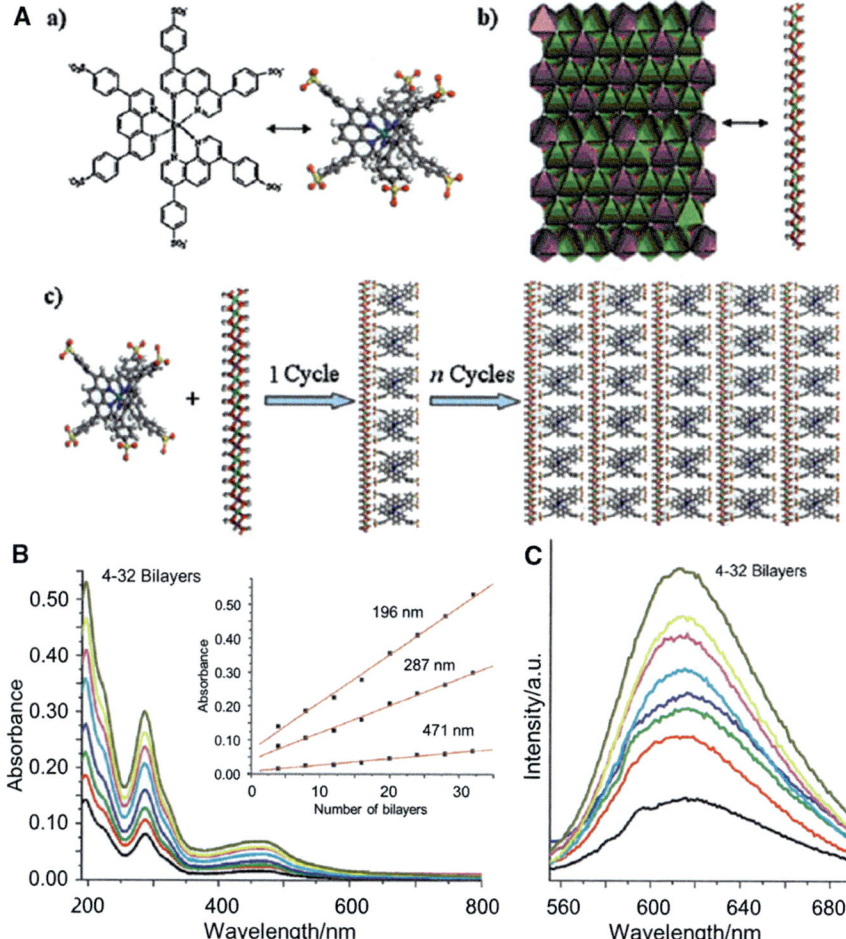

Fig. 25 (**A**) Schematic representation of assembly process of (Ru(dpds)$_3$/LDH)$_n$: (a) structure of [Ru(dpds)$_3$]$^{4-}$, (b) representation of a monolayer of MgAl-LDH (*dark pink*, Al(OH)$_6$ octahedra; *green*, Mg(OH)$_6$ octahedra), and (c) assembly process; (**B**) UV–Visible absorption spectra (inset shows plots of the absorbance at 196, 287 and 471 nm vs. *n*), and (**C**) fluorescence emission spectra [83]

and an optically inert polyanion polyvinylsulfinate (PVS) was taken as examples for fabrication of (BNMA@PVS/LDH)$_n$ UTFs, and the co-assembly process involves two steps. Firstly, BNMA is mixed with the main chain of the PVS, based on Coulombic interaction at the molecular level, and a cation–polyanion pair (BNMA@PVS) with negative charge and functional behavior of BNMA can be obtained. Subsequently, this negatively charged (BNMA@PVS) pair was assembled with LDH nanosheet via LbL process, whereby the cations and polyanion were stably coexisted in the gallery environment provided by LDH nanosheets.

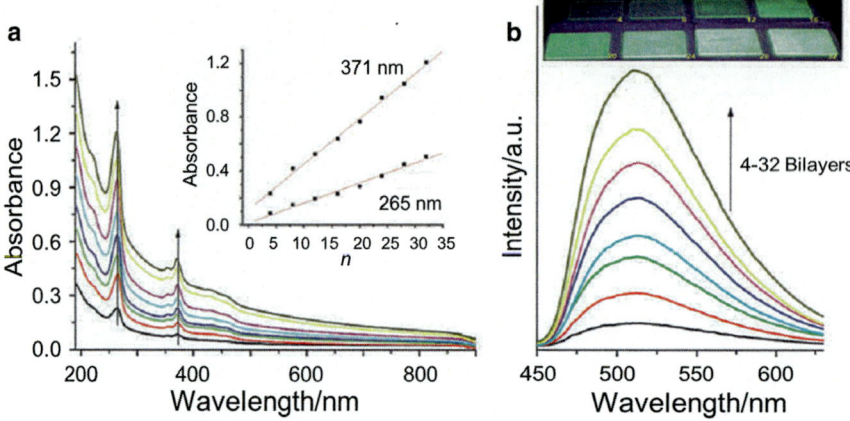

Fig. 26 (a) UV–Vis absorption spectra, (b) fluorescence spectra of the (BNMA@PVS/LDH)$_n$ ($n = 4$–32) UTFs. The insets in (a) and (b) show the plots of absorbance at 265 and 371 nm vs. n and optical photographs under 365 nm UV irradiation, respectively [95]

UV–visible absorption spectra were measured to monitor the assembly process for (BNMA@PVS/LDH)$_n$ UTFs. The linearly increased intensities at 265 and 371 nm which correspond to the characteristic absorption bands of BNMA demonstrated an ordered and regular film growth procedure (Fig. 26a). Fluorescence emission intensity at 510 nm also presents consistent enhancement with the increase of bilayer number n (Fig. 26b), which can be confirmed by increased luminescence brightness under UV light irradiation (the inset of Fig. 26b). These results demonstrated successful co-assembly of BNMA@PVS pairs into the interlayer of LDHs. Moreover, the absorption and fluorescence emission spectra show no obvious red or blue shift compared with the pristine BNMA solution, suggesting uniform distribution throughout the whole assembly processing.

Top-view SEM (Fig. 27a), AFM (Fig. 27c) and fluorescence microscopy (Fig. 27d) images show that the assembled UTFs exhibit a homogeneous and continuous surface, indicating that the BNMA chromophores are distributed uniformly throughout the whole film. Side-view SEM image (Fig. 27b) reveals an approximate 1.5 nm for per (BNMA@PVS/LDH)$_1$ bilayer, in accordance with the basal spacing obtained by XRD. This result further confirms a uniform and periodic layered structure of the multilayer film.

The molecular dynamic (MD) simulation was carried out to further understand the geometric structures of this new type UTF system. An idealized BNMA@PVS/LDH structural model was constructed, and the computational basal spacing is ca. 1.47 nm, in agreement with the experimental results. The simulation shows that the organic ions are found to adopt an ordered arrangement with a preferred orientation (15° between N-N axis to layers) in the as-prepared UTF, and the driving force for the UTF assembly is mainly dominated by the Coulomb interaction.

Moreover, several small low-charged molecules or molecules with poor assembled force can also be co-assembled into interlayer of LDHs by employing

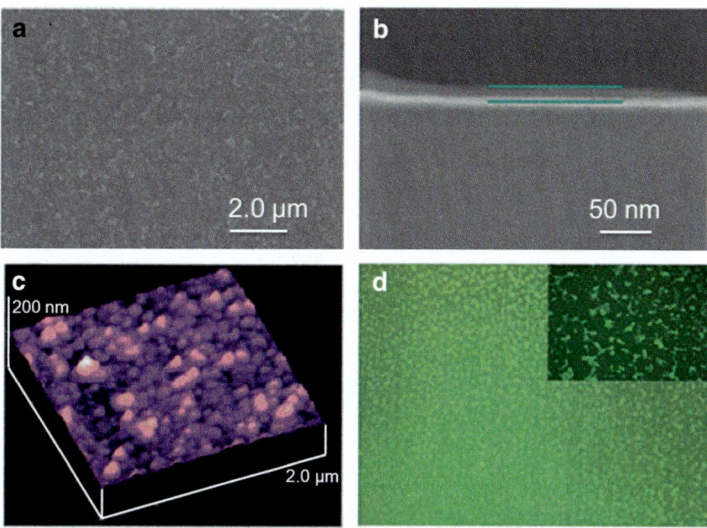

Fig. 27 The morphology of (BNMA@PVS/LDH)$_8$ UTF for (**a**) top-view SEM image, (**b**) side-view SEM image, (**c**) tapping-mode AFM image, and (**d**) fluorescence microscope image multiplied by 50-fold and 1,000-fold (the inset plot) [95]

polyanions. For example, Li et al. [84] have achieved the co-assembly of the tris (8-hydroxyquinolate-5-sulfonate) aluminum (AQS^{3-}) with three polyanions [polyanions poly(acrylic acid) (PAA), poly(styrene 4-sulfonate) (PSS), and poly [5-methoxy-2-(3-sulfopropoxy)-1,4-phenylene vinylene] (PPV)] to obtain the hybrid UTFs, denoted as (AQS@PAA/LDH)$_n$, (AQS@PSS/LDH)$_n$, and (AQS@PPV/LDH)$_n$ (Scheme 2). And similar work was also carried out by employing PSS and PVS to co-assemble with PTCB (perylene 3,4,9,10-tetracarboxylate) [97]. Therefore, the small complex anion–polyanion is anticipated to enhance the cohesion to assemble with LDH for UTFs. The luminous properties of such ordered UTFs can also be modulated by alternating the polyanion species and proportion.

Assembly Based on Electroneutral Molecules

A micellar assembly route for combination of neutral molecules and LDH nanosheets has been put forward in the absence of interaction between them. For example, neutral bis(8-hydroxyquinolate)zinc (Znq$_2$) can be encapsulated into the block copolymer poly(tert-butyl acrylate-co-ethyl acrylate-co-methacrylic acid) (PTBEM) micelles, and then the anionic block copolymer micelles Znq$_2$@PTBEM were assembled with the exfoliated LDH nanosheets to obtain hybrid (Znq$_2$@PTBEM/LDH)$_n$ UTFs (Fig. 28a) [87]. The as-prepared UTFs present a stepwise growth by varying number of bilayers, which can be observed in fluorescence emission spectra and corresponding photos under UV light (Fig. 28b).

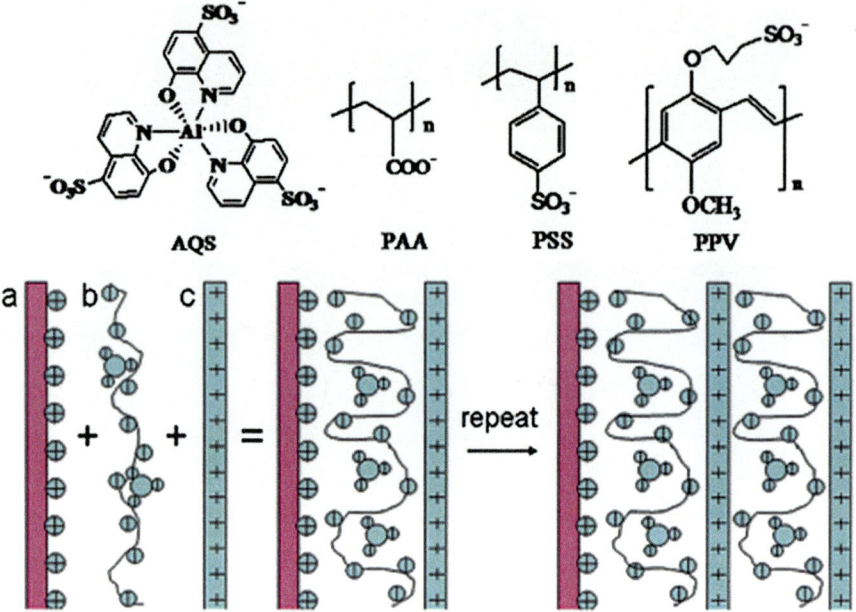

Scheme 2 Schematic representation of the process for the fabrication of (AQS@polyanion/LDH)$_n$ [84]

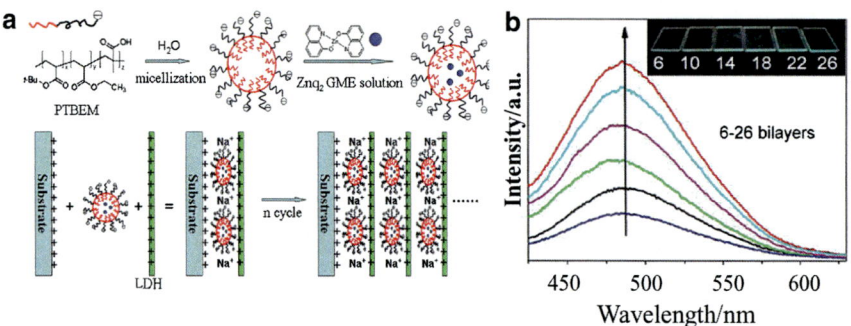

Fig. 28 (a) Schematic representation of the PTBEM, Znq$_2$, and MgAl-LDH nanosheet and the process for the fabrication of Znq$_2$@PTBEM micelle and (Znq$_2$@PTBEM/LDH)$_n$ film; (b) photoemission spectra of (Znq$_2$@PTBEM/LDH)$_n$ UTFs ($n = 6-26$) with the insets showing the photograph of UTFs under the 365 nm UV illumination [87]

Similar methods also appear for the neutral small molecules perylene [135] and tris [2-(4,6-difluorophenyl)pyridinato-C^2,N]iridium(III) (Ir(F$_2$ppy)$_3$) [93]. With poly (N-vinyl carbazole) (PVK) as the carrier, (PVK@perylene/LDH)$_n$ and (Ir(F$_2$ppy)$_3$@PVK/LDH)$_n$ UTFs can be obtained based on hydrogen-bond LbL assembly. The obtained UTFs exhibit decent brightness and reversible two-state switching of the fluorescence emission with external organic compounds.

6.2.2 Intercalation Assembly

Intercalation is one of the most common methods to incorporate the photoactive molecules into the interlayer of LDHs, and the powdered forms of layered organic–inorganic composites can be obtained. Various anionic molecules can be intercalated due to the diversity of the composition in layers and ion exchangeable properties of LDHs. Several intercalation methods have been applied and developed, such as direct intercalation (typical coprecipitation), ions exchange, and co-intercalation. The intercalation of chromophore molecules into LDHs may efficiently inhibit the quenching problems induced by molecular aggregation, and the corresponding fluorescence properties can be largely improved.

Direct Intercalation

By the use of coprecipitation method, the single-component chromophore molecules can be intercalated into the LDH galleries. As an example, 3,4,9,10-perylene tetracarboxylate (PTCB) anion has been intercalated into LDH layers by dropwisely adding the mixed solution of sodium hydroxide and PTCB to the solution containing magnesium nitrate and aluminium nitrate [96]. Figure 29 shows the XRD pattern of as-prepared PTCB/LDH composites, the main characteristic reflections appeared at 4.771° (003), 9.641° (006), and 14.331° (009) indicated the successful intercalation of PTCB molecules. Good multiple relationships can be observed for d_{003}, d_{006}, and d_{009} between the basal, second- and third-order reflections. The stronger reflection of 006 compared with 003 implied that PTCB molecules are located in the symmetric center of the half-way plane within the interlayer region. The photophysical properties, thermolysis behavior, and orientation arrangements of the PTCB/LDH have been studied to compare with the pure PTCB. It was found that the intercalation into LDH has weakened the

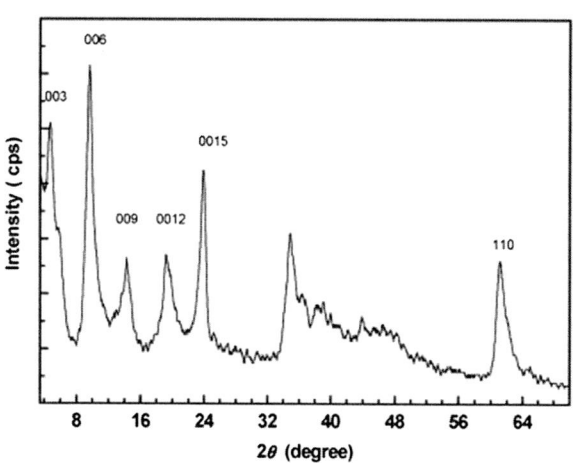

Fig. 29 Powder XRD pattern of PTCB/LDH [96]

strong π–π interaction between PTCB anions, which has paved a way for solid-state optoelectronic devices with high performance.

Ion Exchange

The ion-exchange method is often employed as an alternative of direct intercalation, particularly if the guest anions are unstable in aqueous solution with the divalent or trivalent metal cations. According to the different stability and affinity of the interlayer guests, nitrate-LDH is always applied as a precursor in the ion-exchange process. For example, Shi et al. [104] have fabricated α-naphthalene acetate (α-NAA) and β-naphthalene acetate (β-NAA) intercalated LDH materials through ion-exchange method. The XRD in Fig. 30 show the successful intercalation of these two compounds. The basal spacing of Zn_2Al-NO_3-LDH (Fig. 30a) powder is 0.87 nm, while 2.80 and 1.94 nm of basal spacing can be obtained when NO_3^- is replaced by α-NAA and β-NAA (Fig. 30b, c).

UV–Vis absorption spectra were used to investigate the status and arrangement of NAA in the interlayer of LDH. The pristine NAA in solution, physical mixture of NAA and LDH, and NAA intercalated LDH samples were measured. For the mixture of α-NAA and LDH, a broadened and red-shifted absorption band can be observed. The broaden of band is due to the existence of a continuous set of vibrational sublevels in each electronic state; red shift of 14 nm is the result of ordered and dense packing of molecules in the interlayer, and this may cause intermolecular interactions between NAA (Fig. 31A, line b). In contrast, the α-NAA intercalated LDH sample (curve c in Fig. 31A) exhibits a monomer form in the LDH galleries without any shift caused by intermolecular interactions. Similar results can be obtained for β-NAA samples (Fig. 31B). Moreover, other inorganic species (such as quantum dots [124, 125] and polyoxometalates (POMs) [128] can also be assembled into LDH through ion-exchange process, and the host–guest interaction can effectively affect the fluorescence properties of the guest molecules.

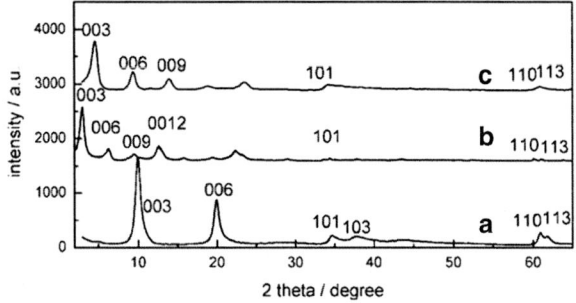

Fig. 30 XRD patterns for (a) Zn_2Al-NO_3-LDH powder, (b) α-NAA-LDH powder, and (c) β-NAA-LDH powder [104]

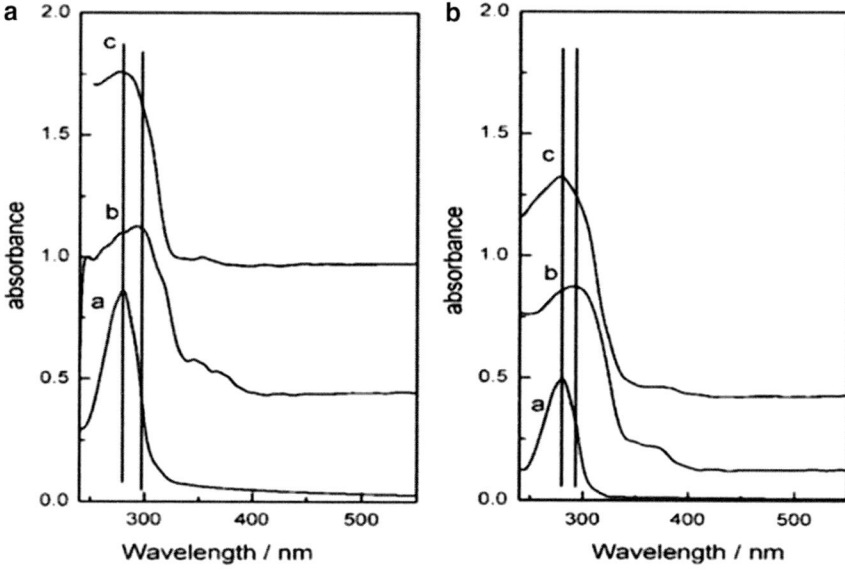

Fig. 31 UV–Vis absorption spectra of (**A**) α-NAA for (a) in solution, (b) mixed with LDH, and (c) intercalated in LDH; (**B**) β-NAA for (a) in solution, (b) mixed with LDH, and (c) intercalated in LDH [104]

Co-intercalation

Co-intercalation is an effective way if guest molecules are difficult to accommodate in the LDH layers. Frequently, surfactant, such as dodecyl sulfonate (DDS$^-$), 1-heptanesulfonic acid sodium (HES), and dodecylbenzene sulfonate (DBS) can be employed as the co-intercalation agents. The introduction of surfactant has provided an enlarged interlayer spacing, which facilitates the successful incorporation of guest. Moreover, a nonpolar and homogeneous microenvironment can be obtained due to surfactant, and thus, the adjustment of the orientation and aggregation status of photofunctional molecules can be achieved. The adjustment of ratio of fluorophore and surfactant within the LDH layer can effectively influence the fluorescence properties of guest molecules. Several fluorophores have been co-intercalated in the LDH layers, such as sulforhodamine B (SRB) [94], zinc phthalocyanines (ZnPc) [92], fluorescein (FLU) [106, 107, 136], ammonium 1-anilinonaphthalene-8-sulfonate (ANS) [137], and other similar systems (as shown in Table 8). The quantity of intercalated guest can be modulated through tuning the ratio of guest and surfactant, and the detailed information will be further introduced in Sect. 6.3.1.

6.3 Adjusting Photofunctionalities of LDH-Based Materials

6.3.1 Static Adjustment

Modulation Based on the Charge Density and Elemental Composition of Host Layer

By controlling the synthesis condition and mole ratio of raw materials, the charge density of the LDH layer can be varied in a wide range, which enabled the tunable photo-related properties of intercalated guest molecules. Luminescent benzocarbazole anions (BCZC), which possesses high quantum yield and good color purity, has been incorporated into LDH via coprecipitation with different layered charge densities of LDHs [98]. MgAl-LDHs with the Mg/Al ratio at 2 and 3.5 were employed to adjust the arrangement of the BCZC. Figure 32A shows the structural model for LDHs with different charge density. LDHs with host layer containing 12 Mg atoms and 6 Al atoms (Mg/Al ratio of 2, Fig. 32A-a) and 14 Mg atoms and 4 Al atoms (Mg/Al ratio of 3.5, Fig. 32A-b) have been adopted. The main characteristic peaks for BCZC/LDH samples present a good multiple relationship for the basal, second- and third-order reflections, with the basal spacing of 2.31 nm for Mg/Al ratio of 2 and 2.39 nm for Mg/Al ratio of 3.5 (Fig. 32B).

UV–Vis spectra of the pristine BCZC and intercalated BCZC have been studied. The absorption bands at 265, 289, and 320 nm can be attributed to the characteristic absorptions of carbazole-type dyes for BCZC in solution, while the solid sample of pristine BCZC and BCZC intercalated LDH composites possess a broadband at 295 and 410 nm due to the molecule stacking (Fig. 33A). Figure 33B shows the fluorescence emission spectra for these samples, and emissions in the range of 490–520 nm are displayed. It can be observed that the spectra of pristine BCZC in solution is similar with BCZC intercalated LDH with Mg/Al ratio at 2, while photoemission of the interlayer BCZC (Mg/Al ratio at 3.5) is close to that of its solid state. Therefore, it can be concluded that variation on the layered charge

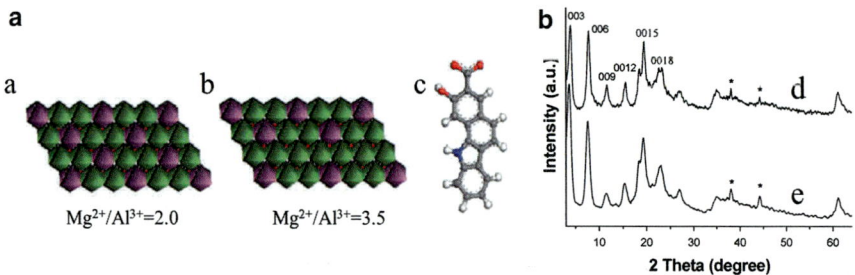

Fig. 32 (**A**) Structure model for the superlattice layer model for MgAl-LDH (*dark pink*: Al(OH)$_6$ octahedra; *green*: Mg(OH)$_6$ octahedra) for Mg/Al ratio of (a) 2 and (b) 3.5, and (c) structural model of BCZC; (**B**) XRD patterns of BCZC/MgAl-LDH (d, e stand for samples with Mg/Al ratio of 2 and 3.5, respectively; *Asterisk*: the Al substrate) [98]

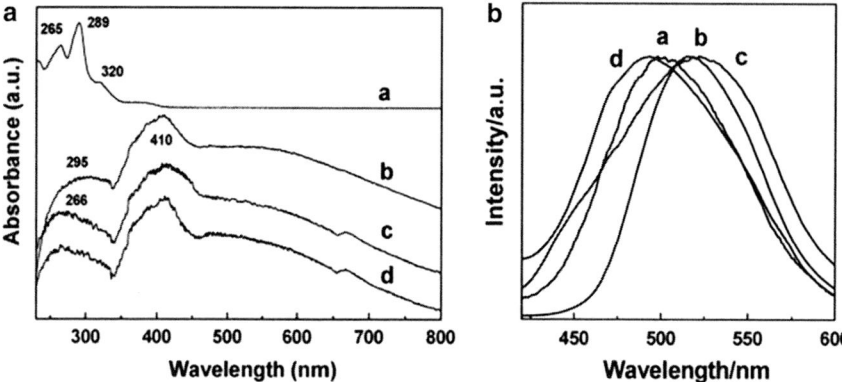

Fig. 33 (**A**) The UV–Vis absorption spectra and (**B**) normalized photoemission spectra for (a) pristine BCZC aqueous solution (10 mM), (b) solid BCZC, and BCZC/MgAl-LDH with the Mg/Al ratio of (c) 2 and (d) 3.5 [98]

density could effectively modify the luminescence performance of the guest dye molecules.

The chemical composition in the LDHs layer also serves as a parameter in modulating the luminescent behavior. For a sulfonated derivate of stilbene intercalated LDHs system [138], compared with the pure sample with the maximum emission wavelength (λ^{em}_{max}) located at 463 nm, the intercalation products exhibit tunable λ^{em}_{max} in the range from 456 nm (Co$_2$Al-LDH as the layer) to 484 nm (Mg$_2$Al-LDH as the layer), demonstrating that the fluorescent wavelength can be continuously tuned by changing the chemical composition of LDHs, as a result of different host–guest interactions between the host layer and the interlayer anions.

Modulation Based on the Quantity of Interlayer Molecules

The orientation and aggregation state of interlayer guest can be tuned by controlling the contents of the interlayer chromophores, and thus, the fluorescence performance can also be varied. For example, rhodamine is a widely used fluorescent dye due to its high quantum yields, but its fluorescence properties are severely affected by its concentration, and a high concentration usually leads to aggregation and subsequent quenching [139]. Yan et al. have achieved the co-intercalation of sulforhodamine B (SRB) and dodecylbenzene sulfonate (DBS) with different ratios into LDHs [94]. Figure 34A shows the XRD pattern of SRB-DBS/LDH with different interlayer concentration of SRB, and it was observed that when the ration of intercalated SRB increased from 0 to 16.67 %, the spacing increased from 28.87 Å to the maximum 30.96 Å (Fig. 34A, curve a–g). And the spacing value has a decreasing trend when the quantity of SRB increased from 16.67 % to the maximum 100 % (Fig. 34A, curve h–k). This variation of the interlayer spacing may be a result of changeable distribution and arrangement of interlayered SRB with different ratios.

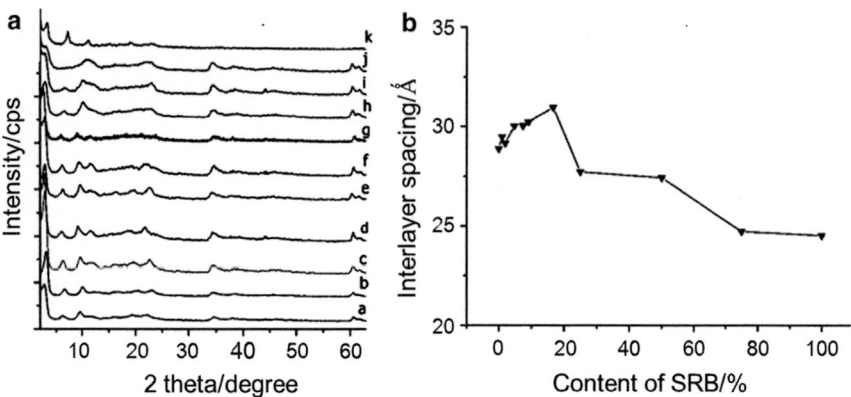

Fig. 34 (**A**) Powder XRD patterns of SRB(x%)-DBS/LDH (a, 0 %; b, 0.99 %; c, 1.96 %; d, 4.76 %; e, 7.41 %; f, 9.09 %; g, 16.67 %; h, 25.0 %; i, 50.0 %; j, 75.0 %; k, 100 %). (**B**) The plot of interlayer spacing vs. SRB concentration [94]

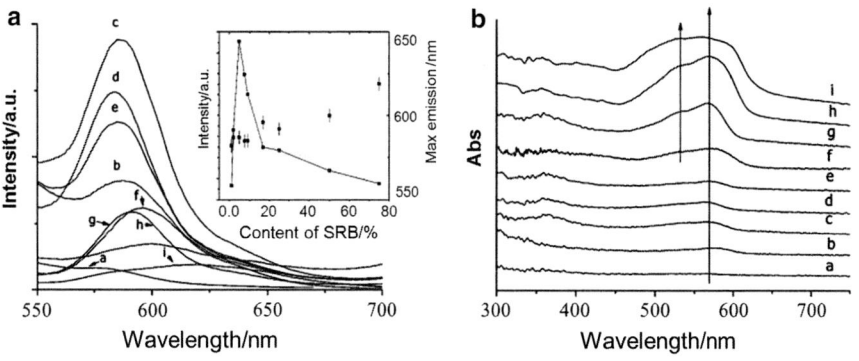

Fig. 35 Study of SRB(x%)-DBS/LDH with different SRB content: (a) 0.99 %, (b) 1.96 %, (c) 4.76 %, (d) 7.41 %, (e) 9.09 %, (f) 16.67 %, (g) 25.0 %, (h) 50.0 %, and (i) 75.0 % for (**A**) photoemission spectra (with the excitation wavelength of 360 nm and the inset shows the fluorescence intensity and the variation of maximum emission wavelength) and (**B**) solid UV–Vis absorption spectra [94]

The fluorescence emission behavior of SRB-DBS/LDHs with different SRB content has been studied, and the emission spectra are displayed in Fig. 35A. With the increase of SRB in SRB-DBS/LDHs, the intensity of fluorescence emission exhibits an increase to maximum (content of SRB is 4.76 %), followed by a continuous decrease. The emission peak varies from 587 to 619 nm as the SRB content increased to 75 %. This can be attributed to the variation of distribution status and arrangement of interlayer SRB, and high concentration may facilitate the formation of dye aggregation, accounting for the red shift and quenching in fluorescence emission. UV–Vis spectra (Fig. 35B) show that absorption peaks at

Table 9 The maximum wavelength of absorption and emission of the ZQS(x%)-DDS/LDH [86]

Sample	λ_{abs} (nm)	λ_{em} (nm)	PLQY (%)
DDS-ZQS(2 %)/Mg$_2$Al-LDH	258, 379	464	5.0
DDS-ZQS(5 %)/Mg$_2$Al-LDH	258, 381	479	24.4
DDS-ZQS(10 %)/Mg$_2$Al-LDH	256, 390	486	18.1
DDS-ZQS(20 %)/Mg$_2$Al-LDH	256, 395	487	17.3
DDS-ZQS(40 %)/Mg$_2$Al-LDH	256, 399	488	16.4
DDS-ZQS(60 %)/Mg$_2$Al-LDH	256, 396	491	9.6
DDS-ZQS(80 %)/Mg$_2$Al-LDH	256, 389	495	5.8
ZQS(100 %)/Mg$_2$Al-LDH	256, 385	497	2.3
DDS-ZQS(5 %)/Mg$_3$Al-LDH	258, 382	490	9.7
DDS-ZQS(5 %)/Zn$_2$Al-LDH	258, 377	491	27.4
DDS-ZQS(5 %)/Zn$_3$Al-LDH	258, 373	491	25.9
H$_2$ZQS crystal	258, 396	466	20.4
ZQS solution (5×10^{-5} mol L^{-1})	254, 368	516	–

360 and 570 nm increase as more SRB molecules were intercalated. An absorption shoulder at ca. 530 nm appears (SRB content of 16.67 %), indicating a formation of H-type aggregates.

For the fluorescence lifetime, compared with pristine SRB solution (2.38 ns), the fluorescence lifetime of SRB-DBS/LDH firstly increases with SRB content changed from 0.99 to 16.7 %, and the lifetime decreases when the content increased to 25 % (Table 9). The change in lifetime further indicates the aggregates of SRB, which is in accordance with emission and absorption spectra. Therefore, the absorption/emission intensity, wavelength, and fluorescence lifetime could be changed due to the variation of interlayer content.

As another example, Li et al. have prepared co-intercalated systems based on bis (8-hydroxyquinolate-5-sulfonate)zinc anion (ZQS^{2-}) and DDS, and the fluorescence properties can also be changed obviously through adjustment of intercalation content [86]. Table 9 shows the fluorescence emission information of ZQS-DDS/LDHs. For different ratio of ZQS in the interlayer and different host layer composition, the maximum emission wavelength and quantum yields can be finely controlled. Furthermore, the brightest luminous intensity, together with the longest fluorescence lifetime, was observed for DDS-ZQS (5 %)/Mg$_2$Al-LDH.

Liang et al. have investigated the dispersion and corresponding singlet oxygen production efficiency of zinc phthalocyanines (ZnPc) after intercalation [92]. XRD patterns showed the basal spacing varied in the range of 23.6–24.6 Å with the content of ZnPc changed from 1 to 10 %, and the co-intercalation of ZnPc and SDS can be further demonstrated by FT-IR spectra. Compared with the pristine form of ZnPc in water and ethanol, the monomer dispersed state with maximum absorption at 678 nm can be obtained via intercalation into the LDH galleries (Fig. 36A, curve c), indicating a low polarity microenvironment provided by LDH gallery for the formation of monomeric ZnPc. With the ZnPc content increased from 1 to 10 %, the absorption intensity exhibit a gradual increase at 678 nm, while a broad absorption

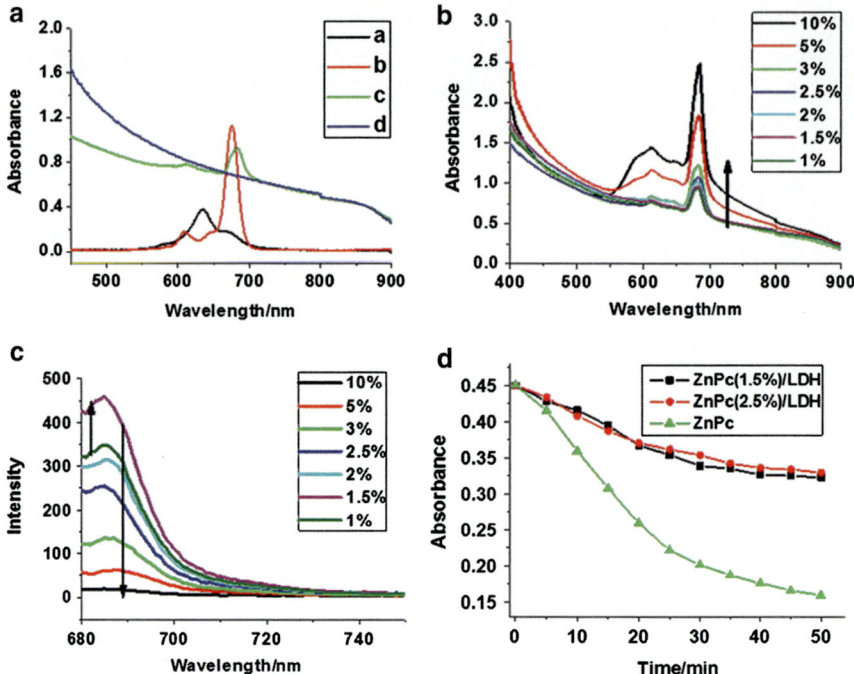

Fig. 36 (**A**) The UV–Vis absorption spectra: (a) pristine ZnPc (2×10^{-5} M) in aqueous solution, (b) pristine ZnPc (2×10^{-5} M) in 60 % ethanol solution, (c) ZnPc (1.5 %)/LDH (3×10^{-4} M) aqueous suspension, (d) pristine LDH (3×10^{-4} M) aqueous suspension. (**B**) The UV–Vis absorption spectra and (**C**) photoluminescence spectra of various ZnPc (x%)/LDH suspension (3×10^{-4} M) with x ranging in 1–10 %. (**D**) Decay curves of absorption at 678 nm for ZnPc (1.5 %)/LDH, ZnPc (2.5 %)/LDH, and 635 nm for pristine ZnPc, respectively, as a function of irradiation time (650 nm) [92]

band appeared in 580–610 nm when ZnPc content increased to 10 % (Fig. 36B). The fluorescence intensity displayed a first increase and then decrease with the optimal luminous intensity presents in 1.5 % ZnPc/LDH (Fig. 36C). This result indicates an aggregate-induced-quenching process. Moreover, better stability against UV irradiation can be obtained for ZnPc/LDH composites compared with the pristine ZnPc. In addition, the high dispersed ZnPc as monomeric state in the interlayer region of LDH induced by the host–guest and guest–guest interactions can make a big contribution to large singlet oxygen production efficiency in vitro and in vivo photodynamic therapy.

Modulation of the Orientation and Polarized Luminescence

As described above, due to the confined environment of LDHs, interlayer guest molecules possess ordered arrangement and preferred orientation, and thus, the

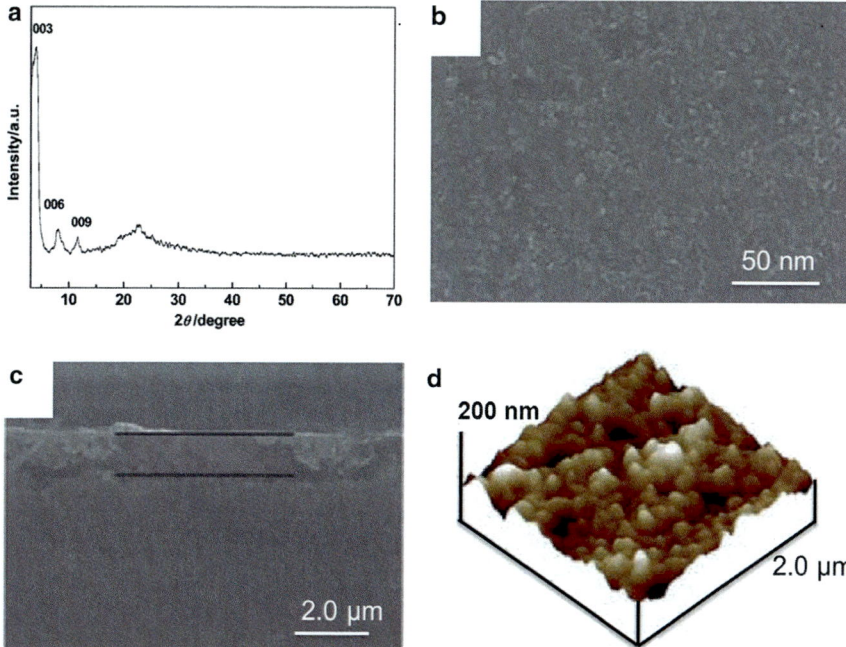

Fig. 37 (a) XRD pattern, (b) top-view SEM, (c) side-view SEM, and (d) tapping-mode AFM images for C3C-DDS/LDH (1.96 %) thin film [110]

obtained LDHs can present polarized luminescence properties. For example, coumarin-3-carboxylate (C3C) and DDS has been co-intercalated into LDH layer, and a C3C-DDS/LDH thin film can be subsequently obtained through solvent evaporation method [110]. Figure 37a shows the sole strong basal reflections (00l) for the thin films of C3C-DDS/LDH, indicating a well c-oriented assembly of LDH platelets. The uniform and smooth morphology of this film can also be confirmed by top SEM and AFM (Fig. 37b, d). The obtained thin film possesses continuous structure with the average thickness of 1.2 μm (Fig. 37c).

The polarized luminescence of the C3C–DDS/LDH thin films was investigated. Generally, polarized luminescence can be evaluated by determining the anisotropy value r, which can be expressed by the formula:

$$r = \frac{I_{\parallel} - I_{\perp}}{I_{\parallel} + 2I_{\perp}} \quad \text{or} \quad r = \frac{I_{VV} - GI_{VH}}{I_{VV} + 2GI_{VH}},$$

where I_{VH} stands for the PL intensity obtained with vertical polarized light excitation and horizontal polarization detection and I_{VV}, I_{HH}, I_{HV} are defined in a similar way; $G = I_{HV}/I_{HH}$ is determined from an aqueous solution of the C3C. Theoretically, the value of r is in the range from -0.2 (absorption and emission transition dipoles are perpendicular to one another) to 0.4 (two transition dipoles are parallel

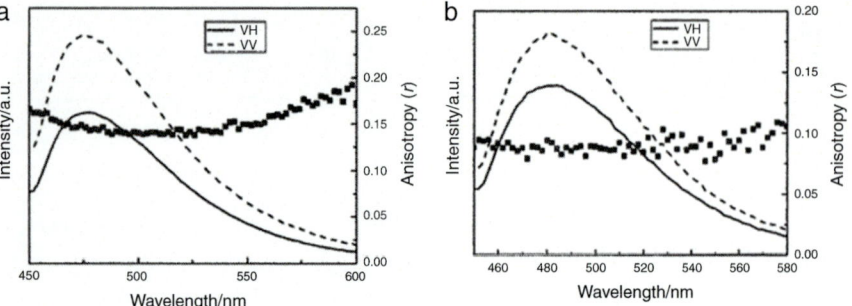

Fig. 38 Polarized fluorescence profiles in the VV, VH modes and anisotropic value (r) for the C3C-DDS/LDH (1.96 %) thin film sample. (**a**) and (**b**) correspond to excitation light with glancing and normal incidence geometry, respectively [110]

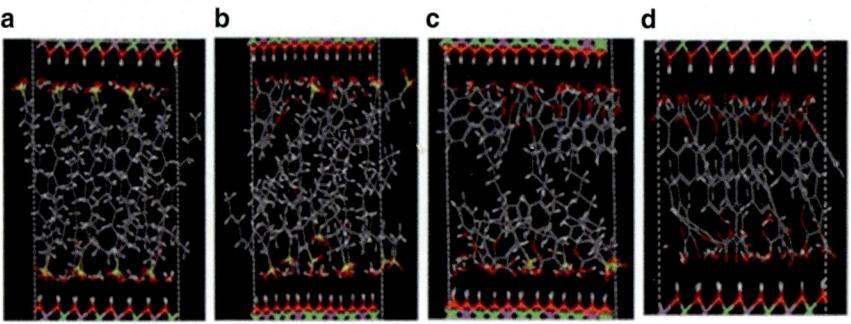

Fig. 39 Snapshots of the simulated equilibrium structures for C3C-DDS/LDHs (x%): (**a**) 0 %, (**b**) 25 %, (**c**) 75 %, (**d**) 100 % [110]

to each other). The polarized photoemission spectra of the C3C-DDS/LDH (optimal luminous properties was obtained when the content of C3C was 1.96 %) thin film are displayed in Fig. 38. Two typical measurement setups of polarized fluorescence were employed to determine the fluorescence anisotropic value r. It can be observed that the value r is about 0.15 for horizontal excitation (Fig. 38a) and 0.10 for vertical excitation (Fig. 38b), demonstrating highly dispersed and well-oriented arrangement of C3C in the gallery of LDH.

To further investigate variation of the basal spacing influenced by the ratio of C3C/DDS, molecular dynamic simulation was employed to study arrangement and orientation of the C3C anions in the LDH galleries with C3C content ranged from 0, 25, 50, 75, to 100 % (Fig. 39). The results show that the DDS anions under low and high concentration exhibit twisting and stretching state, which severely influence the basal spacing of the C3C-DDs/LDH composites.

Angle θ was defined as the orientational angle of the C3C plane with respect to the LDH layer. It was observed that θ varies between 29 and 55 ° with the most

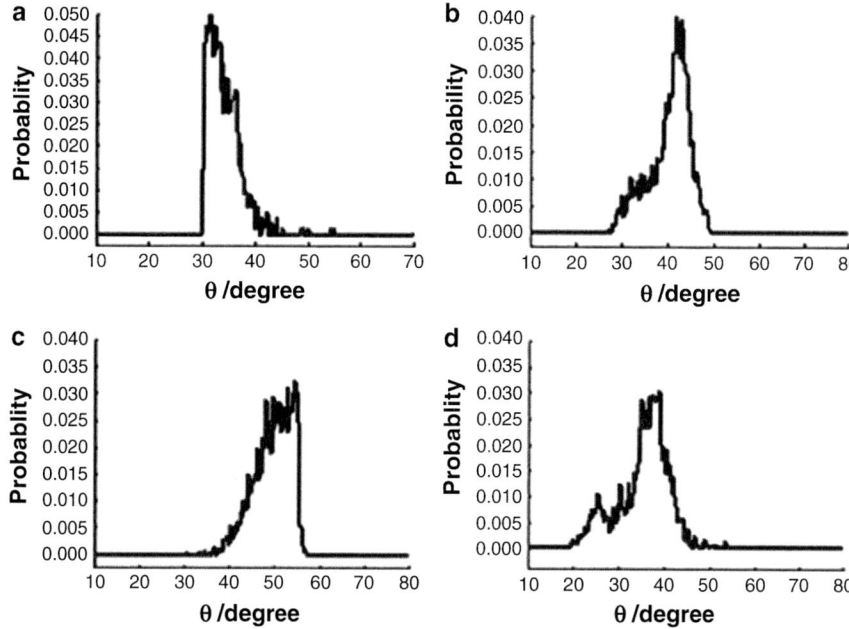

Fig. 40 The distributions of orientational angle θ (the plane of C3C with respect to the LDH layer) with different ratios of C3C to DDS for C3CDDS/LDH ($x\%$): (**a**) 100 %, (**b**) 75 %, (**c**) 50 %, and (**d**) 25 % [110]

probable angle of ca. 32° for the intercalated C3C ratio at 100 %. Stimulation from Fig. 40 shows that all the C3C molecules are inclining to the layers, attributed to the strong electrostatic interaction between the positive-charged host layers and the C3C anions. The optimal angles of 32°, 42°, and 54° can be obtained when the ratio of DDS and C3C varies from 0 to 1, with a tendency to change from a tilted to a vertical arrangement with respect to the layers. Moreover, the distribution of θ angle for the C3C-DDS/LDH (25 %) with high DDS content corresponds to the relative rigid environment imposed by the DDS which restricts the flexible space for the C3C guest anions. These results show that the changeable interlayer ratio of two components can highly influence the preferred orientation of the photoactive guests.

Several LDH-based LbL thin films also exhibit polarized luminescence properties. Due to the rigid-rod chain configuration or intrinsic one-dimensional structural anisotropy for these luminescent molecules, well-oriented thin films with a macroscopic polarized optical effect can be prepared. For example, sulfonated carbocyanine derivate (Scy) has been assembled with LDH nanosheets, which presents polarized near-infrared (NIR) emission luminescence [100]. The average anisotropy of pristine Scy solution is about 0.2 in the range from 770 to 820 nm, which is related to the random distribution of the dye molecules in the nonviscous liquid. NIR polarized fluorescence with the anisotropic value of ~0.8 can be

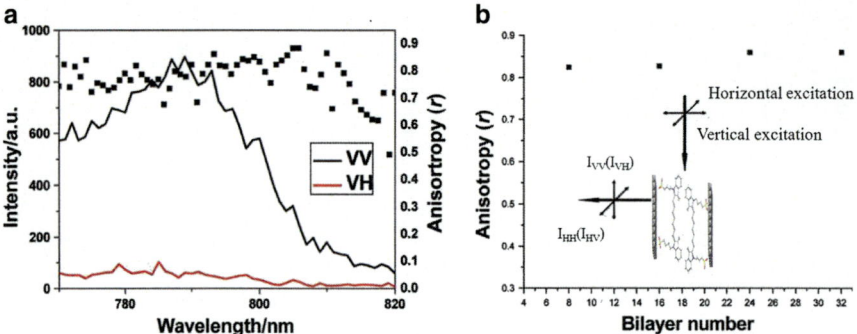

Fig. 41 (a) Polarized fluorescence profiles in the VV, VH modes and anisotropic value r for the (Scy/LDH)$_8$ UTF. (b) The correlation between fluorescence anisotropic values r of the Scy/LDH UTFs (averaged in the range 770–820 nm) and the bilayer number in the measurement mode with glancing incidence geometry [100]

observed for the Scy/LDH UTF (Fig. 41a), confirming the ordered arrangement of Scy between the LDH nanosheets. There were no obvious changes for UTFs with different number of bilayers (Fig. 41b), indicating the independent macroscopic polarized luminescence on assembly process.

Moreover, due to the regular assembly within the LDH nanosheets, some molecules with no luminescence polarization can also exhibit well-defined anisotropy: the as-mentioned Ru(dpds)$_3$/LDH UTF is such an example [83]. The pristine Ru(dpds)$_3$ in aqueous solution shows r at 0.02, while the r value for typical (Ru (dpds)$_3$/LDH)$_{32}$ achieved about 0.2–0.3 when comparing the parallel and perpendicular directions to the excitation polarization (I_{VV} vs. I_{VH}) for the in-plane polarized excitation light. Upon increasing the number of bilayers, the r value increases systematically, indicating enhancement of the orderly degree for Ru (dpds)$_3$ between LDH nanosheets.

Construction of Tunable Multicolor Emissive Film Materials

Materials with multicolor luminescence (especially white-light emission) have been considered as excellent candidates in full-color displays [140], light-emitting diodes (LEDs) [141], and optoelectronic devices [142]. Conventional strategies usually rely on combination of three primary color luminescence species, such as polymers [143], rare-earth compounds [144], metal complexes [145], and semiconductor nanocrystals [146]. However, the different luminous efficiency and intensity, complex fabrication process, as well as complicated interaction between these chromophores have severely limited the development of multicolor materials. Yan et al. [77] have introduced LDH nanosheet as building block to combine different chromophores; in this case, materials with a rigid and ordered microenvironment, finely tuning emission color, and polarized multicolor luminescence can be achieved.

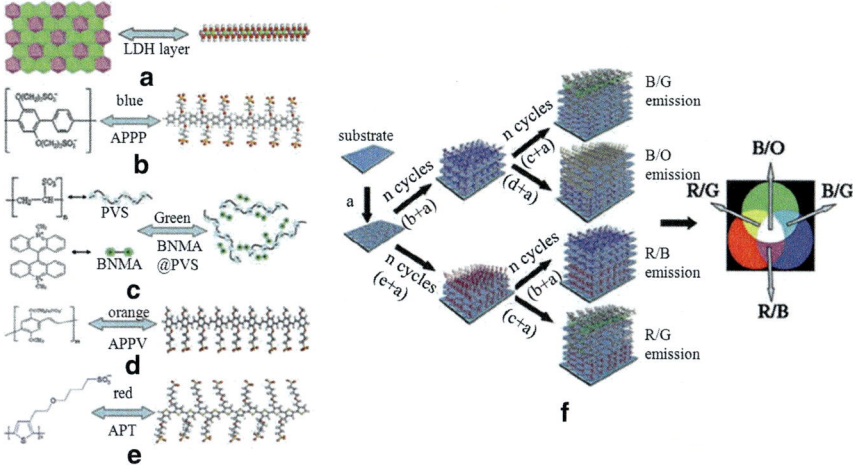

Fig. 42 (a) Representation of one monolayer of MgAl-LDH nanosheet; the chemical formulae of (b) APPP (*blue luminescence*), (c) BNMA@PVS (*green luminescence*), (d) APPV (*orange luminescence*), and (e) APT (*red luminescence*); (f) procedure for assembling two-color emitting UTFs with *blue/green* (B/G), *blue/orange* (B/O), *red/blue*(R/B), and *red/green* (R/G) luminescence [77]

Figure 42 shows the assembly process of multicolor emission materials employing APPP (blue) [76], APPV (orange) [78], SPT (red) [79], and BNMA (green) [95] as primary chromophores. The combination of these individual chromophores can obtain organic/inorganic UTFs with emission color throughout the whole visible region, even the white color emission.

Taking (APPP/LDH)$_{12}$/(BNMA@PVS/LDH)$_n$ ($n=0$–6) UTFs (B/G luminescence) as the example, the assembly process was carried out by assembling (BNMA@PVS/LDH)$_n$ ($n=1$–6) bilayers onto the as-prepared (APPP/LDH)$_{12}$ UTF. Figure 43a1 shows the fluorescence monotonic increase at 481 and 504 nm which can be attributed to the emission of BNMA@PVS, while the emission peak at 410 nm is assigned to the pristine (APPP/LDH)$_{12}$ UTF. The similar spectra between assembled UTFs and solutions indicate that the aggregation of BNMA can be avoided throughout the assembly process. The insets show the photographs taken under UV illumination, and high luminescence can be observed for these UTFs. The color coordinates in Fig. 43a2 demonstrated that the color of the UTFs can be tuned from blue (CIE 1931: (0.172, 0.149); $n=0$) to bluish green (CIE 1931: (0.208, 0.343); $n=6$) effectively. The multicolor emission of blue/orange, red/blue, and red/green can also be tuned by this method throughout the whole visible region (Fig. 43b–d). In addition, based on the periodic ordered structure of these UTFs, they exhibit well-defined multicolor polarized luminescence with high anisotropy, which may effectively extend its application in multicolor or white polarized photoemission devices.

For the inorganic building blocks, quantum dots (QDs) [147] have attracted great attention due to high quantum yields, good stability, and high color purity [148],

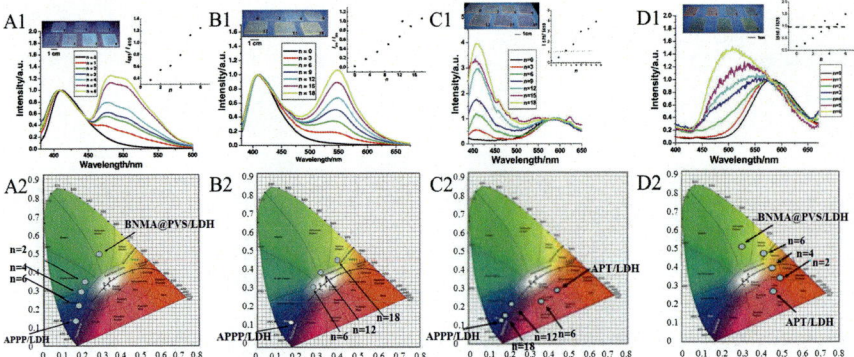

Fig. 43 Fluorescence spectra (the inset shows the ratio of two chromophores), photographs under UV light (365 nm), and the change in color coordinates with n for (**a**) (APPP/LDH)$_{12}$/(BNMA@PVS/LDH)$_n$ ($n=0$–6) UTFs (for B/G luminescence), (**b**) (APPP/LDH)$_{12}$/(APPV/LDH)$_n$ ($n=0$–18) UTFs (for B/O luminescence), (**c**) (APPP/LDH)$_{12}$/(APT/LDH)$_n$ ($n=0$–18) UTFs (for R/B luminescence), and (**d**) (APT/LDH)$_{12}$/(BNMA@PVS/LDH)$_n$ ($n=0$–6) UTFs (for R/G luminescence) [77]

and application of QDs in the multicolor emission materials has also been carefully investigated. Liang et al. [117] have assembled QDs with LDH nanosheets to obtain highly ordered structure and finely tunable fluorescence, and the tunable color in the red–green region could be achieved by adjusting the assembly cycle number and sequence of red- and green-emission QDs. In addition, Tian et al. [102] have introduced fabrication of precisely tuned white color emission UTFs employing red- and green-emission QDs and blue 2,2′-(1,2-ethenediyl) bis [5-[[4-(diethylamino)-6-[(2,5-disulfophenyl) amino]-1,3,5-triazin-2-yl] amino] benzene-5 sulfonic acid] hexasodium salt (BTBS). By adjusting the deposition sequence and relative ratio of the blue-, green- and red-emission units, luminescent UTFs with finely tunable photoemission in the white spectral region were obtained (Fig. 44B). Moreover, the color coordinates of the best white-light emission (denoted as f in Fig. 44C) could achieve (0.322, 0.324) which is very close to the standard coordinates of white light (0.333, 0.333).

The emission color of hybrid composites can be tuned in response to different excitation light, which has been recognized as an intelligent photoluminescence material. This composite UTF was illustrated as (BTBS/LDH)$_{12}$(QD-530/LDH)$_{20}$(QD-620/LDH)$_7$, composed of three chromophores: red-emission and green-emission QDs and blue-emission organic compound BTBS. Due to the different excitation and emission spectra of these three chromophores, it can be concluded that BTBS can be excited at 360 nm while photon energy above 500 and 600 nm can efficiently excite QD-530 and QD-620. Therefore, excited at 360, 460, and 560 nm, the UTFs display three different emission spectra as well as corresponding white, orange, and red color, respectively (Fig. 45). These hybrid materials responding to external excitation light via the alteration of emission color can be widely used in anti-forgery, colorimetric sensors, and display devices.

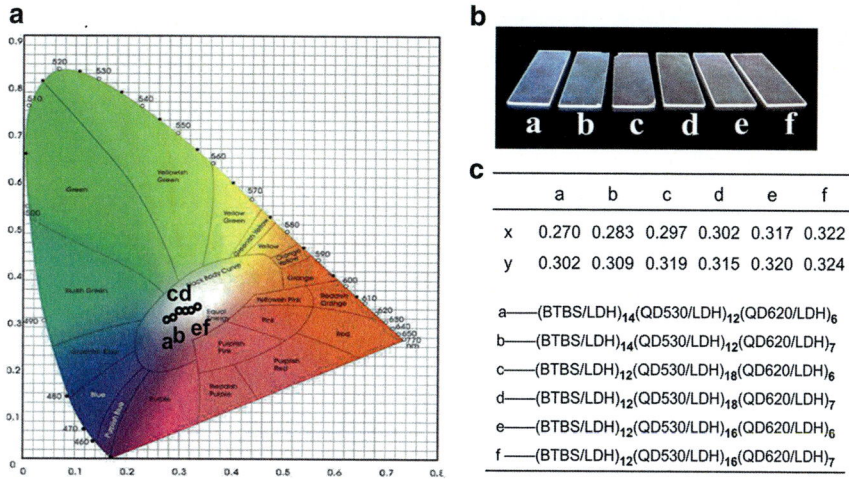

Fig. 44 Optical spectroscopy characterization of the (BTBS/LDH)$_m$(QD-530/LDH)$_n$(QD-620/LDH)$_p$ UTFs: (**A**) the change in color coordinates in the *white-light* region, (**B**) their photographs under UV light (365 nm), (**C**) the color coordinates and detailed compositions of these UTFs [102]

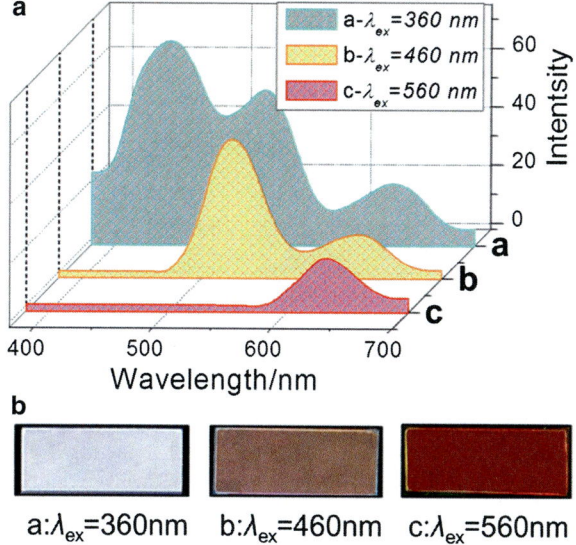

Fig. 45 (**A**) The emission spectra, (**B**) photographs of the (BTBS/LDH)$_{12}$(QD-530/LDH)$_{20}$(QD-620/LDH)$_7$ UTF with different excitation wavelength: (a) excited at 360 nm, (b) excited at 460 nm, (c) excited at 560 nm, respectively [102]

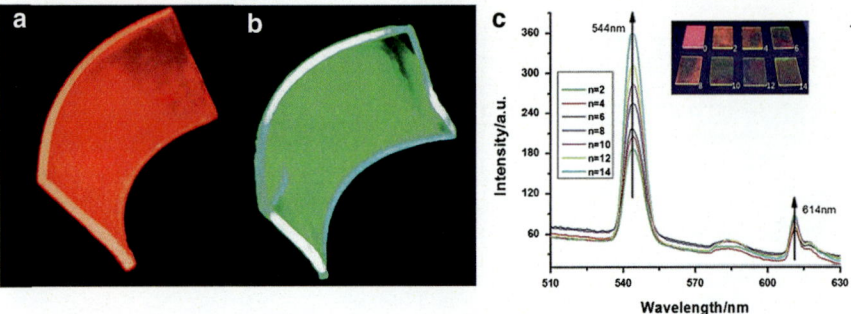

Fig. 46 Digital photos of flexible (**a**) (Eu(DBM)$_3$bath/LDH)$_8$ and (**b**) (Tb(acac)$_3$Tiron/LDH)$_8$ UTFs under UV irradiation; (**c**) fluorescence spectra of (Eu(DBM)$_3$bath/LDH)$_{10}$/(Tb(acac)$_3$Tiron/LDH)$_n$ ($n = 0$–14) UTFs, and insets show the photographs under UV light [149]

Moreover, sulfonated Eu- and Tb-based lanthanide complexes can also be employed in fabrication of multicolor emission materials [149]. After the synthesis of red-emissive Eu(DBM)$_3$bath and green-emissive Tb(acac)$_3$Tiron (DBM, dibenzoylmethane; bath, bathophenanthroline disulfonate disodium; acac, 2,4-pentanedione), these lanthanide complexes can be assembled with LDH to obtain highly emissive luminescence with tunable color on tuning of their relative ratio (Fig. 46). In addition, these UTFs exhibit well-defined one and two-color polarized fluorescence and obvious red and green up-conversion emission upon excitation by near-IR laser light.

6.3.2 Dynamic Tuning of the Emission

The photofunctional materials with smart-responsive properties can present dynamic changes by tuning the external stimuli. Such dynamic-responsive photofunctional materials should have rapid response time, high sensitivity, obvious signal change, and high stability for long-term use [150]. In order to achieve these requirements, LDHs have been employed as an effective inorganic matrix and building block for the ordered arrangement and improved stability of active chromophores in the solid state [151]. Recently, the smart-responsive LDH photofunctional materials responding to environmental stimuli (such as chemical substances, pH value, temperature, pressure, light irradiation, and electricity) have been studied, which present dynamic tuning of their photofunctionalities.

Detection of Chemical Substances

Chemo-sensor plays an important role in the detection of heavy metal ions (HMIs), biomolecules, environmental pollutants, and chemical warfare agents [152–156]. The basic mechanisms of fluorescence detection for most of hazardous substances rely on the quenching induced by the binding between analytes and fluorescence probe molecules. However, the commonly studied fluorescence probe molecules are in aqueous or powder states, and the complicated operation process and difficulty in recyclability have severely limited the application of such materials. Therefore, the development of suitable species and/or new approaches toward the facile fabrication of chemo-sensor with high sensitivity, fast response, and decent recyclability remains a stimulating challenge.

Fabrication of photofunctional UTFs based on the LbL assembly with LDH nanosheets can serve as an effective method to solve the above difficulties. A certain amount of fluorescence molecules have been applied (such as BTBS [101], calcein [109], and 1-amino-8-hydroxy-3,6-disulfonaphthalene (H-acid) [157]) as probes for the detection of HMIs (such as mercury ion). Taking BTBS/LDH system as an example, fluorescence emission spectrum shows a monotonic increase with n at ca. 445 nm, and the UTFs show well-defined blue luminescence under UV light (Fig. 47a and inset). The systematical and significant photoemission decrease of the $(BTBS/LDH)_{32}$ UTF can be observed in responding to the addition of Hg^{2+} (Fig. 47b), while different degrees of quenching for metal ions were presented in Fig. 47d. The strongest quenching by Hg^{2+} is related to the high thermodynamic affinity with typical O-chelate ligands and metal-to-ligand binding kinetics. Reproducibility was carried out for alternative dipping the UTFs into Hg^{2+} and EDTA, and the UTFs can recover to the pristine fluorescence intensity at least five cycles (Fig. 47c). The immobilization by LDH has guaranteed the repeatability of HMI detection, while the drop-casted sample is easy to fall off in solution.

Moreover, Jin et al. have proposed the detection of biomolecules dextran-40 by employing 8-amino-1,3,6-naphthalenetrisulfonates (ANTS) and LDH as building block [158]. The obtained $(ANTS/LDH)_n$ UTF can be adapted in the detection of Mg^{2+} which is important in stabilizing the structure of DNA and transferring of biological signals [159]. The detection limit of 2.37×10^{-7} M can be obtained. Trinitrotoluene (TNT), which is an important explosives commonly applied in both military and forensic science, can be specifically detected by $(BBU/LDH)_n$ UTFs, which are assembled by 4,4'-bis[2-di(b-hydroxyethyl)amino-4-(4-sulfophe-nylamino)-s-triazin-6-ylamino] stilbine-2,2'-disulfonate (BBU) with LDH nanosheets [103].

Li, Qin, and coworkers have efficiently detected the volatile organic compounds (VOCs) based on modulating the FRET process. The basal spacing and the interlayer distance between donor and acceptor can be varied with external stimuli (VOC vapor), and thus, the color change from blue to violet can be observed for the detection [93, 135].

pH Sensor

Optical pH sensor with stable lifetime and signal is of great importance in environmental research, toxicological assay, blood measurement, and biotechnology [160–163]. The red-emissive $(SPT/LDH)_{32}$ UTF described above exhibited reversible responses toward pH in alternative dipping in aqueous at pH = 4 and pH = 13 [79].

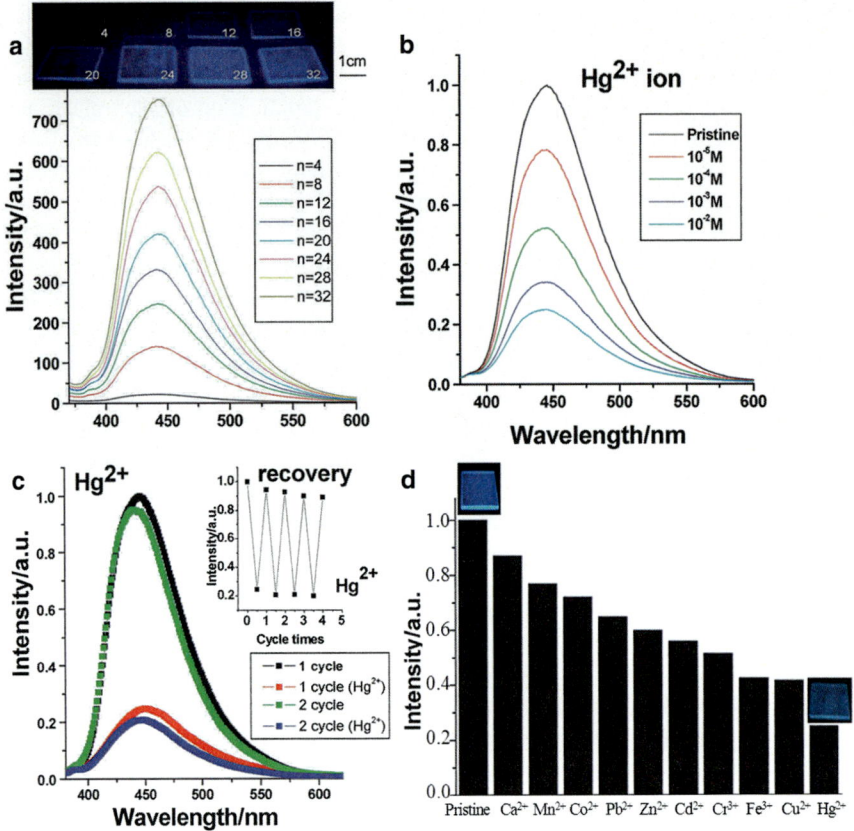

Fig. 47 Fluorescence spectra of the (BTBS/LDH)$_{32}$ UTFs for (**a**) variation in bilayer number (inset shows the photograph under UV light), (**b**) treated with Hg^{2+} at different concentration, and (**c**) detecting the Hg^{2+} after alternate treatment by metal ions and EDTA aqueous solutions over continuous reversible cycles (the inset shows the fluorescence intensities over five cycles); (**d**) comparison of the normalized luminescence intensity of the UTF responding to aqueous solutions of different metal ions (10^{-2} M) [101]

It has been found that dipping the (SPT/LDH)$_{32}$ UTFs in aqueous solutions with different pH values ranging from 4 to 14 for 10 s is sufficient for the proton to transfer into the UTFs, and the UTFs can display a variation of maximum emission wavelength (Fig. 48a). This result can be attributed to the occurrence of protonation and deprotonation of SPT between the LDH nanosheets. Reproducibility was also carried out in alternation between pH = 4 and pH = 13 for five cycles (Fig. 48b), and rapid change from bright red luminescence at pH = 13 to dark orange at pH = 4 can be observed under UV light (365 nm).

Shi et al. have fabricated another pH sensor based on the co-intercalation of the fluorescein (FLU) and 1-heptanesulfonic acid sodium (HES), and the obtained samples exhibit good photo- and storage stability as well as high repeatability

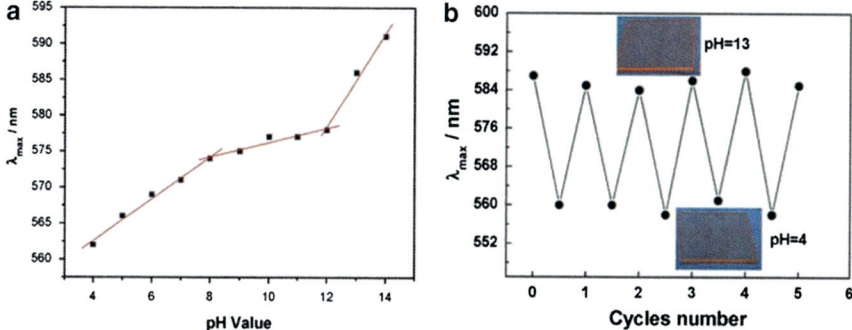

Fig. 48 (a) Variation of the maximum photoemission wavelength of the (SPT/LDH)$_{32}$ UTF with different pH values and (b) the reversible photoemission response upon alternation between pH = 4 and pH = 13 (the inset show the luminescence photographs of the UTFs at the two pH conditions) [79]

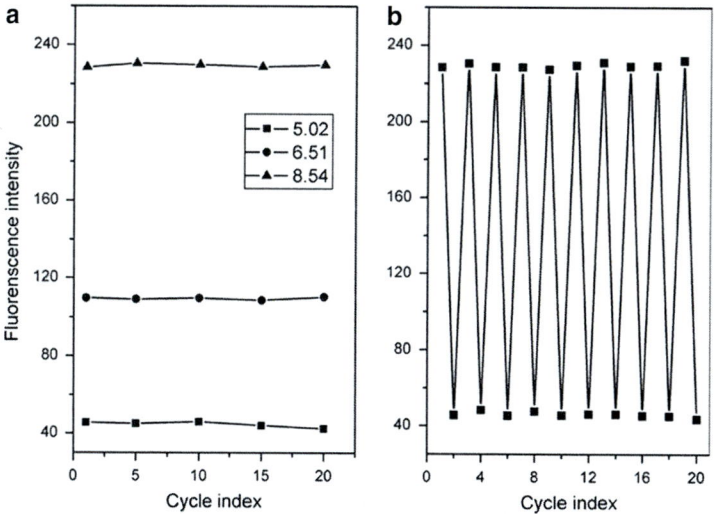

Fig. 49 (a) Fluorescence intensity of the pH sensor recorded after 1, 5, 10, 15, and 20 cycles at pH 5.02, 6.51, and 8.54 and (b) the reversibility recorded by alternate measurement in two solutions with pH 5.02 and 8.54 [106]

toward responses of pH [106]. Employing electrophoretic deposition method, a sensor with broad linear dynamic range in pH (5.02–8.54) can be obtained (Fig. 49a). Remarkable repeatability in 20 cycles can be achieved with the relative standard deviation (RSD) less than 1.5 % (Fig. 49b). The sensitive and rapid responses of SPT/LDH and FLU/LDH toward pH materials indicated their potential application as pH luminescence sensor.

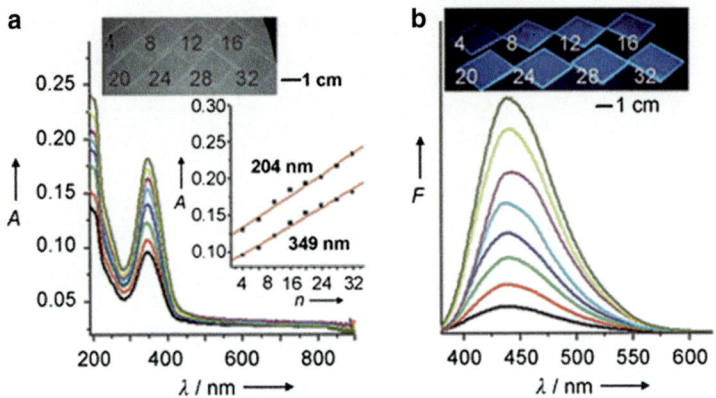

Fig. 50 Characterization of (BSB/LDH)$_n$ ($n = 4$–32) UTFs. (**a**) UV–Vis absorption spectra (inset: plots of the absorbance at 204 and 349 nm vs. the number of bilayers n) and (**b**) fluorescence spectra. The insets in (**a**) and (**b**) are photographs of UTFs with different n under daylight and UV light (365 nm), respectively [99]

Temperature Sensor

Temperature is one of the most common and natural environmental factor, and thermochromic materials can be potentially used in temperature sensor and devices [164–166]. Therefore, much attention has been paid on fabrication of thermochromic materials with good sensitivity and reversibility. Bis (2-sulfonatostyryl)biphenyl (BSB) was employed as thermochromic luminescence (TCL) molecule to assemble with LDH nanosheet to achieve an ordered supramolecular UTFs [99]. Figure 50 shows the assembly process of (BSB/LDH)$_n$ with the monotonic increase in absorption and fluorescence emission spectra. The maximum emission peak is located at 444 nm, and no broadening or shift can be observed in the whole assembly process, indicating no interaction between molecules occurred. The XRD data show the ordered and monolayer arrangement of BSB in the interlayer of LDHs.

The TCL behavior of the UTFs was studied on heating from 20 to 100 °C, and a gradual decrease in fluorescence intensity with a red shift from 444 to 473 nm can be observed (Fig. 51a). A visible change can be observed under UV light with the emission color changes from blue to bluish green upon heating. The recover in color and spectrum can be accomplished when the UTF cooled to 20 °C, and this heating–cooling process can be readily repeated (Fig. 51a, inset). Fluorescence lifetimes of these (BSB/LDH)$_n$ UTFs were also studied to obtain the excited-state information. The results show that lifetime of the UTF is longer at 100 °C (1.9 ns) than that at 20 °C (0.9 ns), and the recycle can also be achieved (Fig. 51b). Figure 51c shows the polarized fluorescence measurements of the UTFs, and the degree of structural order can be obtained by detecting the fluorescence anisotropy r. The r value decreases from 0.15 at 20 °C to 0.02 for UTFs at 100 °C, which exhibit a less ordered arrangement of molecules at high temperature.

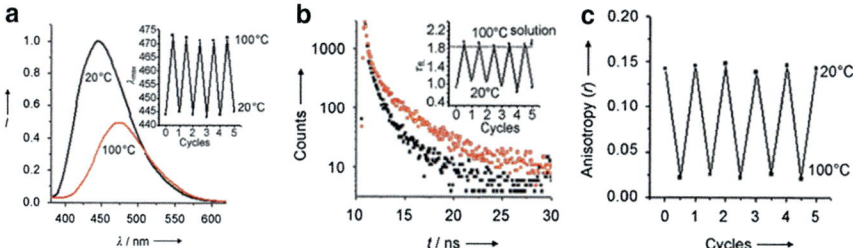

Fig. 51 TCL of the (BSB/LDH)$_8$ UTF at 20 and 100 °C. (**a**) Fluorescence emission spectra (inset: the reversible fluorescence response over five consecutive cycles); (**b**) typical fluorescence decay curve (inset: fluorescence lifetimes over five consecutive cycles, *black*: 20 °C, *red*: 100 °C; (**c**) fluorescence anisotropy over five consecutive cycles [99]

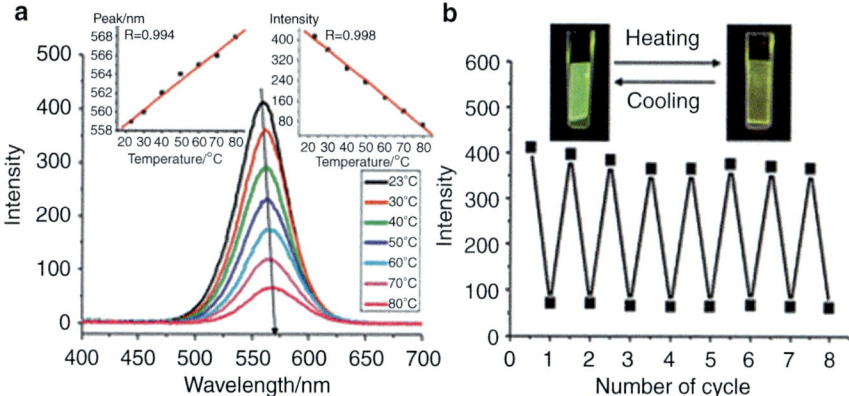

Fig. 52 (**a**) The luminescence spectra of the (CdTe QDs/LDH)$_{30}$ UTF in the temperature range 23–80 °C (inset: the emission position or intensity as a function of temperature, respectively); (**b**) the reversible fluorescence response of eight consecutive cycles (the inset shows the UTF photographs at 23 and 80 °C, respectively) [116]

Similar observations can also be obtained for (CdTe QDs/LDH)$_{30}$ UTF [116]; Fig. 52a shows the responses toward temperature upon heating, and an intensity decrease along with red shift can be observed. These inorganic–inorganic hybrid UTFs possess good sensitivity, and -1% per °C decrease and 0.1 nm °C^{-1} red shift can be obtained. The UTFs exhibit excellent repeatability for eight cycles in the heating–cooling process (Fig. 52b).

Aromatic azo dyes, such as 4-(4-anilinophenylazo)benzenesulfonate (AO5), have been widely used due to the easy transformation between the azo A-form and hydrazone H-form (Scheme 3). It was well known that this tautomerism is very sensitive to the surrounding environment; therefore, it has been widely used as thermal sensors and in molecular memory storage. In order to further modulate the environment in an inorganic matrix, AO5 was co-intercalated with sodium dodecylbenzene sulfonate (SDS) into the interlayer of LDH [112]. The absorption

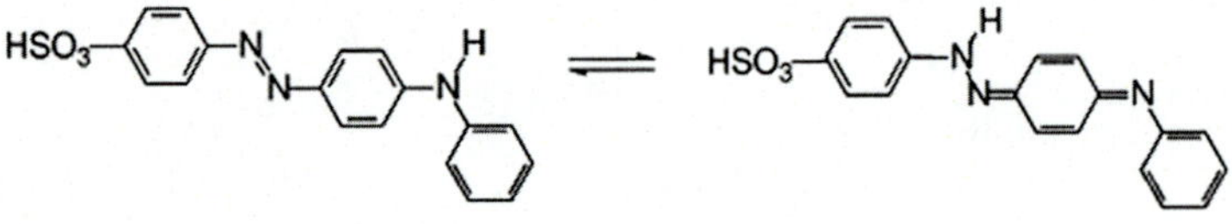

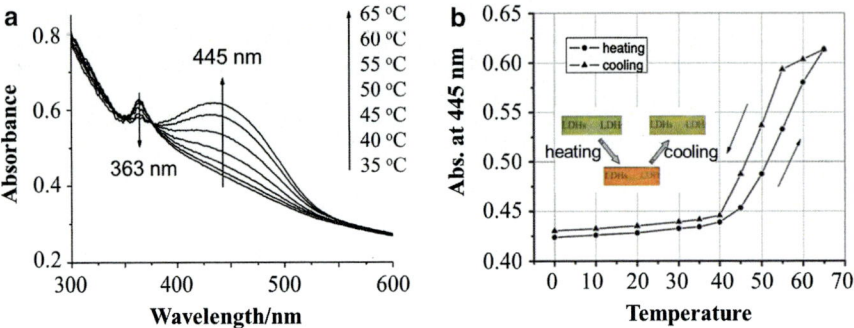

Scheme 3 Tautomeric equilibrium between the A-Form and H-Form of AO5 [112]

Fig. 53 (**a**) In situ absorption spectra of the 10 % AO5-LDH film during heating in the range 35–65 °C. (**b**) The correlation between the absorbance of the band at 445 nm and temperature over a complete heating–cooling cycle (inset shows the corresponding digital photographs) [112]

band at 363 nm at room temperature is attributed to the absorption of A-form, and the intensity increases with elevated temperature (Fig. 53a). The color of the composite film exhibits reversible change from light yellow to orange in the range of 35–65 °C, which is in accordance with absorption at 445 nm. Since the pristine AO5 shows no thermal-responsive behavior, this thermochromic phenomenon for the UTF can be a result of tautomerism of interlayer AO5 and changes in host–guest and guest–guest interactions.

In addition, thermal colorimetric and fluorescent response behaviors of polydiacetylene (PDA) in the LDH layers were also studied [81]. Diacetylene (DA) was firstly intercalated into the interlayer of LDHs, and subsequent PDA/LDH films were fabricated through UV-induced polymerization of intercalated DA. And the reversible changes under both visible light and UV light can be observed in the temperature range of 20–130 °C. This phenomenon can be attributed to the influence of hydrogen bond formed between LDH layers and PDA guests.

Pressure Sensor

Pressure-induced chromic, known as piezochromic, is an interesting phenomenon for luminescence materials responding to natural external force [167]. However, the research on piezochromic luminescent (PCL) materials remains relative seldom

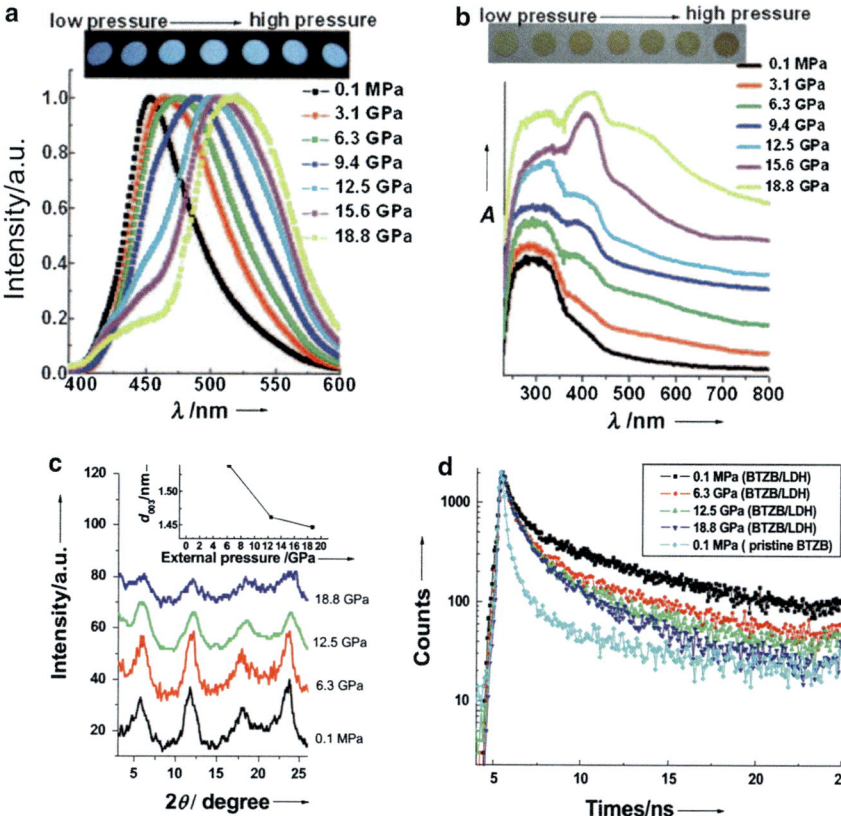

Fig. 54 (a) Fluorescence emission and (b) UV–Vis spectra of the BTZB/Mg$_2$Al-LDH at different pressures (inset shows the photographs of pressurized samples under daylight and UV light, respectively); (c) XRD pattern with the varying interlayer distance and (d) fluorescence lifetime of the BTZB/Mg$_2$Al-LDH at different pressures [111]

compared with other environment-responsive materials (such as temperature-, pH- and HMI-sensitive materials). Therefore, the investigation on PCL materials with excellent reversibility, repeatability, and rapid responsiveness is still a challenge [168, 169].

Yan et al. have studied the fluorescence properties as well as molecular packing mode of (2,2′-(1,2-ethenediyl)bis[5-[[4-(diethylamino)-6-[(2,5disulfophenyl) amino]-1,3,5-triazin-2-yl] amino]benzene sulfonate anion (BTZB) after intercalation into LDHs matrix [111]. The fluorescence of the pristine BTZB and BTZB/ LDH composites is shown at approximately 452 nm, and significant luminescent changes can be observed for the BTZB/LDHs treated with pressures while pristine BTZB remains unchanged. Figure 54a shows the changes of fluorescence spectra upon increasing the pressure from 0.1 MPa to 18.8 GPa; the emission wavelength shifts from 452 to 515 nm, which can be assigned to the formation of *J*-type aggregates in the LDH galleries. The change of this phenomenon can be obviously seen from photographs of the BTZB/LDH pellet under UV light (Fig. 54a, inset) with the color varies from blue to green with increasing pressure. UV–Vis spectra also exhibit an obvious change on varying the pressure (Fig. 54b). Except for the absorption band at 290 nm for the pristine BTZB/LDH, a new band, ranging from 350 to 450 nm, appears when the pressure is sustainable increased. A broad shoulder band at 450–550 nm can be observed when the pressure was increased to 15.6 GPa. The appearance of this new band is due to the formation of J-type aggregates of BTZB in the interlayer of LDH, which is in accordance with the results from fluorescence tests above. In order to take an insight into the influences of pressure on the supramolecular organization in this BTZB/LDH host–guest system, XRD was carried out. Figure 54c shows that upon increasing the pressure from 0.1 MPa to 18.8 GPa, the basal d_{003} spacing changes from 1.57 to 1.45 nm, maintaining the structure of LDHs. This may due to the contracted gallery height and subsequent rearrangement of intercalated guest molecules. And the fluorescence lifetime of BTZB/LDH is much shorter at high pressure (2.14 ns at 18.8 GPa and 4.63 ns at 0.1 MPa), further confirming the aggregation of the BTZB (Fig. 54d).

Reversibility and repeatability tests are also carried out by grinding and subsequent heating for the pellet sample, and a reversible performance for practical application of PCL materials is satisfactorily obtained. Moreover, the different composition of host layer has been investigated, and the results show that Mg_2Al-LDH possesses better PCL performances compared to Zn-Al-LDH or Mg_3Al-LDH. Therefore, these BTZB/LDH materials exhibit obvious changes in UV–Vis absorption, fluorescence emission, and lifetime upon the increase of pressure, and the color change from blue to green under UV light can be visually seen. Theoretical results show that the variations of the arrangement and aggregation status of BTZB in the interlayer of LDH cause the subsequent PCL phenomenon.

Mechanofluorochromism can be observed in the NFC-HPS (NFC, niflumic acid, a typical aggregation-induced emission molecule; HPS, heptanesulfonate) co-intercalated LDHs [170]. The results show that NFC-HPS/LDH (HFC: HPS = 5 %) possesses obvious change in fluorescence on mechanical force, while the pristine solid-state NFC sample remains nearly unchanged under pressure. This phenomenon can be attributed to the slipping of LDH sheets and reorientation of the interlayer surfactant; and the changed host–guest interactions can tune the molecular conformation and intermolecular interaction of NFC anions.

Light-Induced Sensor

Photochromism is a phenomenon that substance exhibits color change upon light irradiation, and it has been widely used in light switching, liquid-crystal alignment, information storage, and nonlinear devices [171–173]. It is well known that the photoisomerization process is very sensitive to the surrounding microenvironment, and the low isomerization efficiency and stability has limited the development of photochromic materials. Embedding the photo-responsive molecules into the inorganic matrix, such as LDH nanosheets, may effectively inhibit the aggregation of the molecules and thus enhance the photoisomerization free space for photochromism.

Wang et al. have studied the photoisomerization process of a kind of Schiff base (azomethine-H anions, AMH) assembled LDH system [174]. Generally, after the excitation of the initial enol (anti-enol) tautomer, the excited-state intramolecular proton transfer (ESIPT) occurs for AMH, leading to the excited keto tautomer (cis-keto or its zwitterionic form), exhibiting a characteristic Stokes fluorescence band. After the structural changes (rotation around $C=C$ and/or C–N bond) involving the cleavage of the intramolecular hydrogen bond, the red long-lived photochromic tautomer (trans-keto or its zwitterionic form) in the ground state is generated either by thermal treatment or by irradiation of visible light.

In order to obtain advanced performance along with good stability, AMH was co-intercalated with 1-pentanesulfonate (PS) into the interlayer of LDH, and the corresponding thin films of AMH/PS-LDH were obtained by the solvent evaporation. Firstly, the photoisomerization process of the pristine AMH solution has been studied, and the transient absorption spectra were illustrated in Fig. 55a. The bands at 470 nm are attributed to the absorption from the S_0 state of the trans-keto tautomer, and the band with a maximum at ~370 nm is possibly related to the long-lived syn-enol form. Upon irradiation of laser, a dramatic decrease can be observed. For AMH/PS-LDH films, an excited-state intramolecular proton transfer to the trans-keto tautomer was occurred when the initial AMH (as an enol tautomer) receives UV excitation, and the relaxed back isomerization to the ground state was observed through transient UV–Vis spectroscopy (Fig. 55b). The band located at 465 nm is due to the S_0 state of the trans-keto tautomer, and a 5 nm blue shift due to the changed weak polar microenvironment can be observed. The absence of syn-enol form indicates the inhibition of the syn-enol/anti-enol isomerization in the

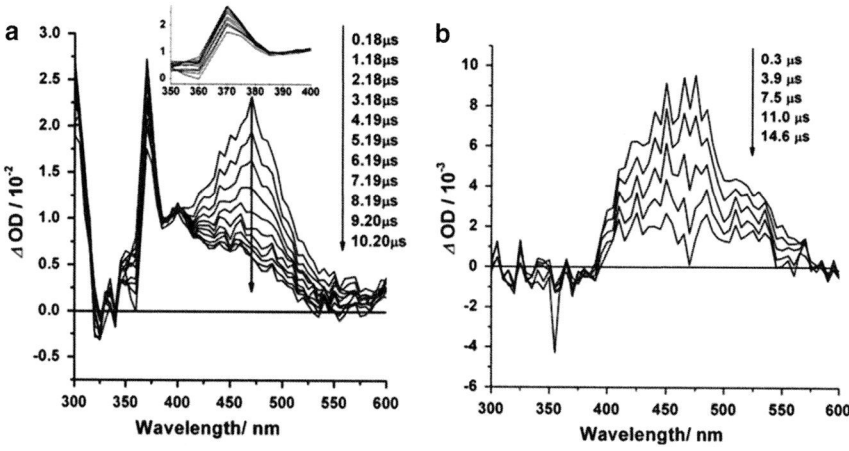

Fig. 55 Transient absorption spectra obtained by laser photolysis for (**a**) nitrogen-saturated solution of AMH (5.0×10^{-5} M) in water (inset shows the magnified absorption from 350 to 400 nm), and (**b**) the 2 % AMH-LDH film [174]

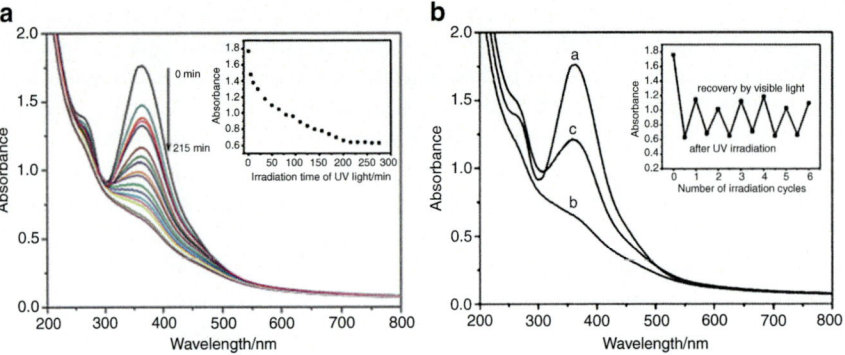

Fig. 56 UV–Vis absorption spectra of photoisomerization for the (LDH/PAZO)$_{50}$ film: (**A**) the absorption intensity changes with time; (**B**) repeatability tests for (a) the trans isomer, (b) the same sample after irradiation with UV light, and (c) the recovered trans isomer after irradiation with visible light. The inset shows the change in the absorbance at the maximum of 376 nm [82]

2D confined environment. As a result, the trans-keto tautomer of AMH is less stable in the 2D confined environment, and relaxed back isomerization is more easily to take place.

Moreover, photoactive azobenzene polymer PAZO was also applied in the assembly with LDH to obtain a photochromism material [82]. Upon UV irradiation, the absorption at 376 nm decreases as a function of time (Fig. 56A) due to the transformation of *trans* to *cis* of PAZO. A complete isomerization can be achieved (Fig. 56B, line a–b) after sufficient irradiation time. The π–π* absorption band increases again (Fig. 56B, line c) when irradiated by visible light ($\lambda > 450$ nm), and the back isomerization for the *cis–trans* can be achieved. The extent of 51 % is obtained for the *cis* to *trans* back isomerization, and repeatability for several cycles can be accomplished by alternative irradiation of UV and visible light. The results show that the intertwisting of polymer chains and π–π stacking of the azobenzene chromophores can be efficiently reduced, and sufficient free space is available for the isomerization of the azobenzene moiety due to the employment of LDH nanosheets.

Light-triggered materials containing porphyrin and phthalocyanine molecules for biological and medical application were also largely investigated. Through the intercalation, the inherent properties of these porphyrin and phthalocyanine molecules may be altered [175]. [Tetrakis(4-carboxyphenyl)-porphyrinato]zinc (II) (ZnTPPC) [88], Pd(II)-5,10,15,20-tetrakis(4-carboxyphenyl) porphyrin (PdTPPC) [89], 5,10,15,20-tetrakis(4-sulfonatophenyl) porphyrin (TPPS), and its palladium complex (PdTPPS) [89–91] intercalated LDH-based hybrid materials have been studied. Figure 57 shows the relaxation of PdTPPS phosphorescence (triplet states) and luminescence signals of $O_2(^1\Delta_g)$ in the assembled LDH-DS/PdTPPS films. The proximity of the porphyrin triplet states and their interaction with $O_2(^1\Delta_g)$ contribute to the repopulation of the fluorescent excited singlet states of TPPS. Moreover, Liang et al. have also studied the singlet oxygen production

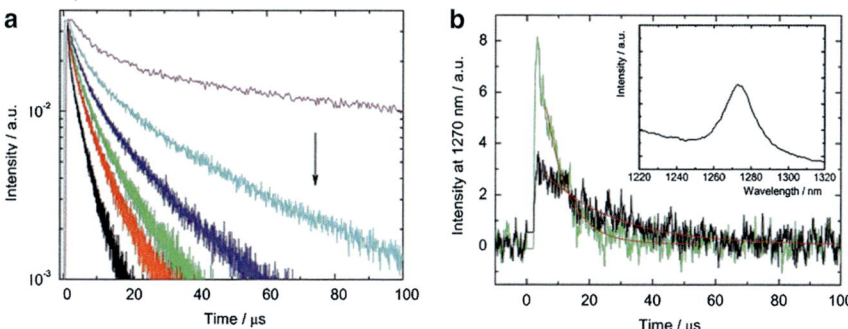

Fig. 57 (a) Quenching of the PdTPPS triplet states in the Mg$_2$Al-LDH-DS/PdTPPS film at different oxygen pressures (excitation at 425 nm, phosphorescence recorded at 690 nm); the *arrow* shows increasing O$_2$ pressure (vacuum, 100, 210, 300, 400, 760 Torr). (b) Time dependencies of the O$_2$($^1\Delta_g$) luminescence signals in air (*black line*) and oxygen (*green line*) after the excitation of TPPS in the Mg$_2$Al-LDH-DS/TPPS film. The *red lines* are least-squares monoexponential fits (inset shows emission band of O$_2$ ($^1\Delta_g$) in air) [91]

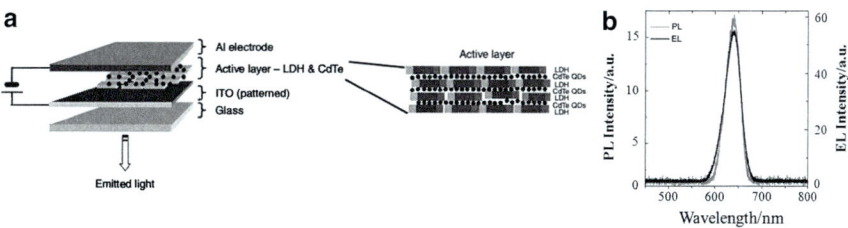

Fig. 58 (a) Schematic of the architecture of the device, which is based on a "sandwich" structure, and (b) photoluminescence peak of aqueous CdTe QDs (*gray curve*) compared to the electroluminescence peak of the device (*black curve*) [119]

efficiency of zinc phthalocyanines intercalated LDHs, and the composite materials exhibit excellent anticancer behavior when used for photodynamic therapy [92].

Electroluminescence, Chemiluminescence, and Electrochemiluminescence

Electroluminescence (EL) is the luminescent behavior induced by external electric field, which requires saturated emission colors, high thermal stability, and long operational lifetimes of chromophores [176]. However, traditional EL devices often suffered from low efficiency, short lifetimes, and complicated fabrication technique under specially controlled environment. A method for fabricating high efficiency and thermally stable light-emitting device (LED) involving simple technique is quite needed. Bendall et al. [119] have designed an all-inorganic EL device constructed by assembly highly luminescent CdTe QDs with LDH nanosheet (Fig. 58a). EL emission peak, which can be observed at 637 nm, is almost the

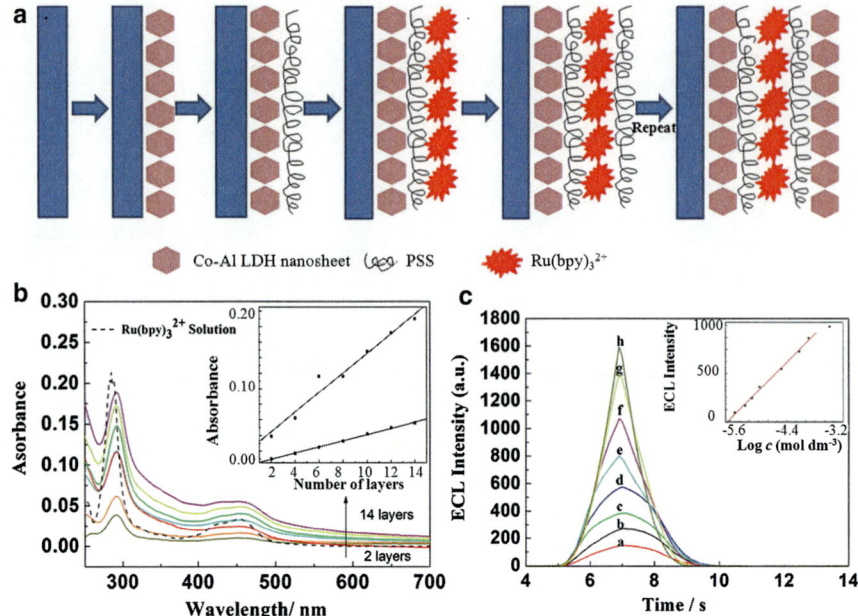

Fig. 59 Assembly and properties of (LDH/PSS/Ru(bpy)$_3^{2+}$/PSS)$_n$ multilayer films: (**a**) schematic representation, (**b**) UV–Vis absorption spectra along with different tetra-layer number n (the inset shows the absorbance at 291 and 457 nm vs. n), and (**c**) ECL spectra in the presence of NADH with various concentrations (from a to h: 3.16×10^{-6} to 3.23×10^{-4} M, scan rate = 100 mV s^{-1}, and the inset shows plot of ECL intensity vs. NADH concentration) [179]

same as the photoluminescence (PL) emission peak of the bare QD without any shift or broadening occurs. Such device maintains the requirements for the common organic LEDs, and possesses better thermal stability (can be operated over 80 °C) due to the employment of LDHs. Thus, a bright, monochrome, and stable EL device with excellent thermal stability, brightness, and simple fabrication process is achieved.

Chemiluminescence (CL) stands for the luminescence derived from a chemical reaction. Much attention has been paid on enhanced CL for detection in biology and sensor fields [177]. Lu et al. have investigated the CL phenomenon based on nanocomposite material constructed by the assembly of CdTe QDs upon the surface of DBS-LDHs in an ordered manner. The obtained material exhibited a remarkable CL amplification of the luminol–H$_2$O$_2$ system due to ordered arrangement of QDs and host–guest interaction between LDH and CdTe QDs [121].

Electrochemiluminescence (ECL) is a combined process of EL and CL, which is caused by energetic electron transfer (redox) reactions of electrogenerated species [178]. Recently, ECL based on a solid-state sensor has also been investigated employing ruthenium(II) tris(bipyridine) and poly(sodium 4-styrene sulfonate) to assemble with LDH nanosheets [179]. Figure 59a shows the assembly process for this (LDH/PSS/Ru(bpy)$_3^{2+}$/PSS)$_{10}$ multilayer films, and UV–Vis spectra has been used to monitor the stepwise growth of the UTFs (Fig. 59b). A linear response to dihydronicotinamide adenine dinucleotide (NADH) in the range 3.16×10^{-6} to 3.23×10^{-4} M can be obtained, and the detection limit can reach as low as 0.023 μM (Fig. 59c).

Therefore, the application of LDH-based materials can be extended to EL, CL, and ECL devices due to the uniform dispersion of photoactive molecules in 2D confined microenvironment supported by the superlattice structure of the multilayer films.

7 Conclusion and Outlook

In conclusion, LDH-based compounds have become promising photofunctional materials, which can be used in infrared radiation absorption, ultraviolet shielding, and tunable color/luminescence with both static and dynamic manners. Due to their inherent high stability, ion-exchangeable property, and tailorable photo-related performance, LDHs can serve as an effective building block to incorporate photoactive molecules for the fabrication of new hybrid materials with diverse potential applications. Through the adjustment of the elemental composition, charge density, and particle size of LDH precursors, the quantity and distribution of confined guests can be tuned, which enable the fabrication of photofunctional materials with preferred orientations. Host–guest interaction and interlayer confinement effect can further influence the arrangement and orientation of intercalated guest molecules, and hybrid materials with superior optical properties compared with the pristine form can be achieved. For example, the LDH-based luminescent materials with precisely tuned emission in the whole visible range are obtained. And the hybrid materials also present repeatable and reversible responses toward external stimuli (such as light, pH, pressure, temperature, electricity, and chemicals), which can be utilized in the fields of photochromic, piezochromic, thermochromic, and polarized luminescence, chemical sensors, and switches. Moreover, some of LDHs systems (IR absorption and UV-shielding materials) have already been used as functional addictives due to the significant improvement of IR/UV properties for polymers and asphalt.

From both academic and engineering perspectives, two key scientific problems still need to be further understood for these LDH-based photofunctional materials: (1) How the fine structures in both the inner layer and interlayer of LDHs (such as interface, surface, and defect) as well as their size effect can influence the macroscopical photo-related properties; (2) how the host–guest interactions between the LDHs layers and photoactive guests can influence the arrangement, orientation, and aggregation of the interlayer species. In addition, to meet the need of practical application, further explorations and interests can be concentrated on: (1) Intelligent materials: the development of new types of stimuli-responsive luminescent materials with fast response, high on/off ratio, and good repeatability are highly desirable. As well, the multi-stimuli-responsive materials remain in an early stage and have been very seldom studied. (2) Large-scale preparation of LDHs photofunctional materials: up to now, most of these materials are still at the laboratory level, and how to scale up the LDHs materials for industrial applications needs further exploration from both process engineering and product engineering perspectives. (3) Combination of the photo-related performance with other functionalities, such as medical, pharmaceutical, environment, and energy fields, would be an important strategy and direction to broaden LDH-based multifunctional materials. Researches in these directions are continuing by us and many others worldwide.

Acknowledgment We would like to thank all the coworkers cited in the references for their invaluable contributions to the work described here. This work was supported by the 973 Program (Grant No. 2014CB932103), the National Natural Science Foundation of China (NSFC), Beijing Municipal Natural Science Foundation (Grant No. 2152016), the 111 Project (Grant B07004), and Program for Changjiang Scholars and the Innovative Research Team in University (PCSIRT: IRT1205).

References

1. Newman SP, Jones W (1998) New J Chem 22:105
2. Funnell NP, Wang Q, Connor L, Tucker MG, O'Hare D, Goodwin AL (2014) Nanoscale 6:8032
3. Jellicoe TC, Fogg AM (2012) J Phys Chem Solids 73:1496
4. Fogg AM, Dunn JS, Shyu SG, Cary DR, O'Hare D (1998) Chem Mater 10:351
5. Wang Q, O'Hare D (2012) Chem Rev 112:4124
6. Ma R, Liu Z, Li L, Iyi N, Sasaki T (2006) J Mater Chem 16:3809
7. He J, Wei M, Li B, Kang Y, Evans DG, Duan X (2006) Struct Bond 119:89
8. Kim JA, Hwang SJ, Choy JH (2008) J Nanosci Nanotechnol 8:5172
9. Guo X, Zhang F, Evans DG, Duan X (2010) Chem Commun 46:5197
10. Liu Z, Ma R, Osada M, Iyi N, Ebina Y, Takada K, Sasaki T (2006) J Am Chem Soc 128:4872
11. Evans DG, Slade RCT (2006) Struct Bond 119:1
12. Yan D, Lu J, Duan X (2013) Scientia Sinica Chimica 43:1
13. Li C, Dou Y, Liu J, Chen Y, He S, Wei M, Evans DG, Duan X (2013) Chem Commun 49:9992
14. Shao M, Ning F, Zhao J, Wei M, Evans DG, Duan X (2012) J Am Chem Soc 134:1071
15. Vialat P, Leroux F, Taviot-Gueho C, Villemure G, Mousty C (2013) Electrochim Acta 107:599
16. Choi SJ, Choy JH (2011) Nanomedicine 6:803
17. Yan D, Lu J, Wei M, Evans DG, Duan X (2011) J Mater Chem 21:13128
18. Gursky JA, Blough SD, Luna C, Gomez C, Luevano AN, Gardner EA (2006) J Am Chem Soc 128:8376
19. Gardner E, Huntoon KM, Pinnavaia TJ (2001) Adv Mater 13:1263
20. Li C, Wang L, Wei M, Evans DG, Duan X (2008) J Mater Chem 18:2666
21. Wang L, Li C, Liu M, Evans DG, Duan X (2007) Chem Commun 2:123
22. Zhang F, Sun M, Xu S, Zhao L, Zhang B (2008) Chem Eng J 141:362
23. Han J, Dou Y, Wei M, Evans DG, Duan X (2010) Angew Chem Int Ed 49:2171
24. Dou Y, Han J, Wang T, Wei M, Evans DG, Duan X (2012) J Mater Chem 22:14001

25. Zhang Z, Chen G, Xu K (2013) Ind Eng Chem Res 52:11045
26. Xu K, Zhang Z, Chen G, Shen J (2014) RSC Adv 4:19218
27. Posati T, Costantino F, Latterini L, Nocchetti M, Paolantoni M, Tarpani L (2012) Inorg Chem 51:13229
28. Zhao Y, Li JG, Fang F, Chu N, Ma H, Yang X (2012) Dalton Trans 41:12175
29. Gunawan P, Xu R (2009) J Phys Chem C 113:17206
30. Gao X, Hu M, Lei L, O'Hare D, Markland C, Sun Y, Faulkner S (2011) Chem Commun 47:2104
31. Gao X, Lei L, Kang L, Wang Y, Lian Y, Jiang K (2014) J Alloys Compd 585:703
32. Zhang Z, Chen G, Liu J (2014) RSC Adv 4:7991
33. He Q, Yin S, Sato T (2004) J Phys Chem Solids 65:395
34. Shi W, Lin Y, Zhang S, Tian R, Liang R, Wei M, Evans DG, Duan X (2013) Phys Chem Chem Phys 15:18217
35. Tuo Z (2006) Beijing University of Chemical Technology, China
36. Chai H (2008) Beijing University of Chemical Technology, China
37. Yu S, Zheng Y, Du J, Liu J, Shang H, Liu L (2005) China Surf Det Cosm 352:48
38. Xing Y (2003) Beijing University of Chemical Technology, China
39. Xing Y, Li D, Guo C, Evans DG, Duan X (2003) Fine Chem 20:1
40. Wang G, Rao D, Li K, Lin Y (2014) Ind Eng Chem Res 53:4165
41. Xing Y, Li D, Reng L, Evans DG, Duan X (2003) Acta Chim Sin 61:267
42. Zhang L, Lin Y, Tuo Z, Evans DG, Li D (2007) J Solid State Chem 180:1230
43. Cui G, Evans DG, Li D (2010) Polym Degrad Stab 95:2082
44. Cui GJ, Xu XY, Lin YJ, Evans DG, Li DQ (2010) Ind Eng Chem Res 49:448
45. Feng Y, Li D, Wang Y, Evans DG, Duan X (2006) Polym Degrad Stab 91:789
46. Li D, Tuo Z, Evans DG, Duan X (2006) J Solid State Chem 179:3114
47. Sun W, He Q, Luo Y (2007) Mater Lett 61:1881
48. Lin Y, Tuo Z, Chai H, Evans DG, Li D (2006) Chin J Inorg Chem 22:1431
49. Chai H, Lin Y, Evans DG, Li D (2008) Ind Eng Chem Res 47:2855
50. Chai H, Xu X, Lin Y, Evans DG, Li D (2009) Polym Degrad Stab 94:744
51. Zhu H, Feng Y, Tang P, Cui G, Evans DG, Li D, Duan X (2011) Ind Eng Chem Res 50:13299
52. Wang J (2012) Beijing University of Chemical Technology, China
53. Wang L, Xu X, Evans DG, Li D (2010) Ind Eng Chem Res 49:5339
54. Liu Y (2002) Plastics Sci Technol 2:22
55. Wang L, Wang L, Feng Y, Feng J, Li D (2011) Appl Clay Sci 53:592
56. Gao J (1999) China Plast 13:66
57. Ding S (1998) Chin Synth Res Plast 15:32
58. Tian Y (2004) China Plast 18:1
59. Zhu H, Tang P, Feng Y, Wang L, Li D (2012) Mater Res Bull 47:532
60. Wang S, Zhao X, Li SL (2003) Plast Sci Technol 3:29
61. Kunihiko K (2003) China Plast 17:1
62. Wang L, Xu X, Evans DG, Duan X, Li D (2010) J Solid State Chem 183:1114
63. Jiao Q, Zhao Y, Xie H, Evans DG, Duan X (2002) Chin J Appl Chem 19:1011
64. Xie H (2000) Beijing University of Chemical Technology, China
65. Liu J, Lian S, Zhu A, Li Q, Liu L, Zeng L (2007) Chin J Lumin 28:67
66. Duan X, Lu J (2013) Nano science and technology. The two-dimensional nano composite hydroxide: structure, assembly and function. Science Press, China
67. Chen Z, Kang J (2006) China Plast 20:13
68. Wang L, Cui H, Jiao H (2007) Spectrosc Spect Anal 27:259
69. Wang L, Xu X, Evans DG, Li D (2010) Chin J Inorg Chem 26:970
70. Wang L (2011) Beijing University of Chemical Technology, China
71. Badreddinea M, Legrouri A, Barroug A, DeRoy A, Besse JP (1999) Mater Lett 38:391
72. Xu G, Guo C, Duan X (1999) Chin J Appl Chem 16:45

73. Haurie L, Fernández AI, Velasco JI, Chimenos JM, Lopez Cuesta JM, Espiell F (2006) Polym Degrad Stab 91:989
74. Roy PK, Surekha P, Rajagopal C, Chatterjee SN, Choudhary V (2007) Polym Degrad Stab 92:1151
75. Singh R, Samra KS, Kumar R, Singh L (2008) Radiat Phys Chem 77:53
76. Yan D, Lu J, Wei M, Han J, Ma J, Li F, Evans DG, Duan X (2009) Angew Chem Int Ed 48:3073
77. Yan D, Lu J, Wei M, Qin S, Chen L, Zhang S, Evans DG, Duan X (2011) Adv Funct Mater 21:2497
78. Yan D, Lu J, Ma J, Wei M, Wang X, Evans DG, Duan X (2010) Langmuir 26:7007
79. Yan D, Lu J, Ma J, Wei M, Evans DG, Duan X (2011) AIChE J 57:1926
80. Gao R, Lei X, Chen M, Yan D, Wei M (2013) New J Chem 37:4110
81. Shi W, Lin Y, He S, Zhao Y, Li C, Wei M, Evans DG, Duan X (2011) J Mater Chem 21:11116
82. Han J, Yan D, Shi W, Ma J, Yan H, Wei M, Evans DG, Duan X (2010) J Phys Chem B 114:5678
83. Yan D, Lu J, Wei M, Ma J, Evans DG, Duan X (2009) Chem Commun 42:6358
84. Li S, Lu J, Ma H, Xu J, Yan D, Wei M, Evans DG, Duan X (2011) Langmuir 27:11501
85. Li S, Lu J, Wei M, Evans DG, Duan X (2010) Adv Funct Mater 20:2848
86. Li S, Lu J, Xu J, Dang S, Evans DG, Duan X (2010) J Mater Chem 20:9718
87. Li S, Lu J, Ma H, Yan D, Li Z, Qin S, Evans DG, Duan X (2012) J Phys Chem C 116:12836
88. Daniel SR, Prabir KD (1996) Langmuir 12:402
89. Eva K, Kamil L, Pavel K, Mariana K, Jiří M, Miroslav Š, Anne-Lise TT, Fabrice L, Vincent V, Christine TG (2010) J Mater Chem 20:9423
90. Eva K, Christine TG, Petr B, Mariana K, Petr K, Pavel K, Jiří M, Miroslav P, Kamil L (2010) Chem Mater 22:2481
91. Marie J, Jan D, Pavel K, Jiří H, František K, Kamil L (2011) J Phys Chem C 115:21700
92. Liang R, Tian R, Ma L, Zhang L, Hu Y, Wang J, Wei M, Yan D, Evans DG, Duan X (2014) Adv Funct Mater 24:3144
93. Qin Y, Lu J, Li S, Li Z, Zheng S (2014) J Phys Chem C 118:20538
94. Yan D, Lu J, Wei M, Evans DG, Duan X (2009) J Phys Chem B 113:1381
95. Yan D, Lu J, Chen L, Qin S, Ma J, Wei M, Evans DG, Duan X (2010) Chem Commun 46:5912
96. Yan D, Lu J, Wei M, Ma J, Evans DG, Duan X (2009) Phys Chem Chem Phys 11:9200
97. Zhang Y, Song M, Yun R, Meng Q, Yan D (2014) Chin J Chem 32:859
98. Yan D, Lu J, Ma J, Wei M, Evans DG, Duan X (2010) Phys Chem Chem Phys 12:15085
99. Yan D, Lu J, Ma J, Wei M, Evans DG, Duan X (2011) Angew Chem Int Ed 50:720
100. Yan D, Lu J, Ma J, Wei M, Li S, Evans DG, Duan X (2011) J Phys Chem C 115:7939
101. Yan D, Lu J, Wei M, Li S, Evans DG, Duan X (2012) Phys Chem Chem Phys 14:8591
102. Tian R, Liang R, Yan D, Shi W, Yu X, Wei M, Li LS, Evans DG, Duan X (2013) J Mater Chem C 1:5654
103. Ma H, Gao R, Yan D, Zhao J, Wei M (2013) J Mater Chem C 1:4128
104. Shi W, Wei M, Lu J, Li F, He J, Evans DG, Duan X (2008) J Phys Chem C 112:19886
105. Shi W, Wei M, Lu J, Evans DG, Duan X (2009) J Phys Chem C 113:12888
106. Shi W, He S, Wei M, Evans DG, Duan X (2010) Adv Funct Mater 20:3856
107. Shi W, Sun Z, Wei M, Evans DG, Duan X (2010) J Phys Chem C 114:21070
108. Shi W, Ji X, Zhang S, Wei M, Evans DG, Duan X (2011) J Phys Chem C 115:20433
109. Shi W, Ji X, Wei M, Evans DG, Duan X (2012) Langmuir 28:7119
110. Yan D, Lu J, Ma J, Wei M, Qin S, Chen L, Evans DG, Duan X (2010) J Mater Chem 20:5016
111. Yan D, Lu J, Ma J, Qin S, Wei M, Evans DG, Duan X (2011) Angew Chem Int Ed 50:7037
112. Wang X, Lu J, Shi W, Li F, Wei M, Evans DG, Duan X (2010) Langmuir 26:1247
113. Shi W, Lin Y, Kong X, Zhang S, Jia Y, Wei M, Evans DG, Duan X (2011) J Mater Chem 21:6088

114. Yan D, Zhao Y, Wei M, Liang R, Lu J, Evans DG, Duan X (2013) RSC Adv 3:4303
115. Zheng S, Lu J, Li W, Qin Y, Yan D, Evans DG, Duan X (2014) J Mater Chem C 2:5161
116. Liang R, Tian R, Shi W, Liu Z, Yan D, Wei M, Evans DG, Duan X (2013) Chem Commun 49:969
117. Liang R, Xu S, Yan D, Shi W, Tian R, Yan H, Wei M, Evans DG, Duan X (2012) Adv Funct Mater 22:4940
118. Liang R, Yan D, Tian R, Yu X, Shi W, Li C, Wei M, Evans DG, Duan X (2014) Chem Mater 26:2595
119. Bendall JS, Paderi M, Ghigliotti F, Li Pira N, Lambertini V, Lesnyak V, Gaponik N, Visimberga G, Eychmüller A, Torres CMS, Welland ME, Gieck C, Marchese L (2010) Adv Funct Mater 20:3298
120. Dong S, Guan W, Lu C (2013) Sens Actuators B Chem 188:597
121. Dong S, Liu F, Lu C (2013) Anal Chem 85:3363
122. Cho S, Jung S, Jeong S, Bang J, Park J, Park Y, Kim S (2013) Langmuir 29:441
123. Cho S, Kwag J, Jeong S, Baek Y, Kim S (2013) Chem Mater 25:1071
124. Xu X, Zhang F, Xu S, He J, Wang L, Evans DG, Duan X (2009) Chem Commun 48:7533
125. Wu G, Wang L, Evans DG, Duan X (2006) Eur J Inorg Chem 2006:3185
126. Han Z, Guo Y, Tsunashima R, Song YF (2013) Eur J Inorg Chem 2013:1475
127. Omwomaa S, Chena W, Tsunashimab R, Song Y (2014) Coord Chem Rev 258–259:58–71
128. Sousa FL, Pillinger M, Sá Ferreira RA, Granadeiro CM, Cavaleiro AMV, Rocha J, Carlos LD, Trindade T, Nogueira HIS (2006) Eur J Inorg Chem 2006:726
129. Li L, Ma R, Ebina Y, Iyi N, Sasaki T (2005) Chem Mater 17:4386
130. Ma RZ, Sasaki T (2012) Recent Pat Nanotechnol 6:159
131. Takakazu Y, Kiyoshi S, Takashi K, Tetsuji I, Takaki K (1996) J Am Chem Soc 118:3930
132. Sirringhaus H (2005) Adv Mater 17:2411
133. Igor AL, Jinsang K, Timothy MS (1999) J Am Chem Soc 121:1466
134. Gao ZQ, Mi BX, Tam HL, Cheah KW, Chen CH, Wong MS, Lee ST, Lee CS (2008) Adv Mater 20:774
135. Li Z, Lu J, Li S, Qin S, Qin Y (2012) Adv Mater 24:6053
136. Shi W, Wei M, Evans DG, Duan X (2010) J Mater Chem 20:3901
137. Sun Z, Jin L, Shi W, Wei M, Evans DG, Duan X (2011) Langmuir 27:7113
138. Yan D, Lu J, Wei M, Evans DG, Duan X (2013) Chem Eng J 225:216
139. Ray K, Nakahara H (2002) J Phys Chem B 106:92
140. Wood V, Panzer MJ, Chen J, Bradley MS, Halpert JE, Bawendi MG, Bulovic V (2009) Adv Mater 21:1
141. Kim S, Kim T, Kang M, Kwak SK, Yoo TW, Park LS, Yang I, Hwang S, Lee JE, Kim SK, Kim SW (2012) J Am Chem Soc 134:3804
142. Lee M, Yang R, Li C, Wang ZL (2010) J Phys Chem Lett 1:2929
143. Chiang CL, Tseng SM, Chen CT, Hsu CP, Shu CF (2008) Adv Funct Mater 18:248
144. Shang M, Geng D, Kang X, Yang D, Zhang Y, Lin J (2012) Inorg Chem 51:11106
145. Wang M, Guo S, Li Y, Cai L, Zou J, Xu G, Zhou W, Zheng F, Guo G (2009) J Am Chem Soc 131:13572
146. Ki W, Li J (2008) J Am Chem Soc 130:8114
147. Kwak J, Bae WK, Lee D, Park I, Lim J, Park M, Cho H, Woo H, Yoondo Y, Char K, Lee S, Lee C (2012) Nano Lett 12:2362
148. Rosson TE, Claiborne SM, McBride JR, Stratton BS, Rosenthal SJ (2012) J Am Chem Soc 134:8006
149. Gao R, Zhao M, Guan Y, Fang X, Li X, Yan D (2014) J Mater Chem C 2:9579
150. Kim HN, Guo Z, Zhu W, Yoon J, Tian H (2011) Chem Soc Rev 40:79
151. Li W, Yan D, Gao R, Lu J, Wei M, Duan X (2013) J Nanomater 2013:437082
152. Tomchenko AA, Harmer GP, Marquis BT (2005) Sens Actuators B 108:41
153. Davis AP, Wareham RS (1999) Angew Chem Int Ed 38:2978

154. Zampolli S, Elmi I, Ahmed F, Passini M, Cardinali GC, Nicoletti S, Dori L (2005) Sens Actuators B 105:400
155. Yoon S, Miller EW, He Q, Do PH, Chang CJ (2007) Angew Chem Int Ed 46:6658
156. Rose A, Hu Z, Madigan CF, Swager TM, Bulovic V (2005) Nature 434:876
157. Sun Z, Jin L, Zhang S, Shi W, Pu M, Wei M, Evans DG, Duan X (2011) Anal Chim Acta 702:95
158. Jin L, Guo Z, Wang T, Wei M (2013) Sens Actuators B 177:145
159. Jin L, Guo Z, Sun Z, Li A, Jin Q, Wei M (2012) Sens Actuators B 161:714
160. Michelle M, Ward M, Asher SA (2008) Adv Funct Mater 18:1186
161. Schroeder CR, Weidgans BM, Klimant I (2005) Analyst 130:907
162. Mattu J, Johansson T, Holdcroft S, Leach GW (2006) J Phys Chem B 110:15328
163. Roy I, Gupta MN (2003) Chem Biol 10:1161
164. Chandrasekharan N, Kelly LA (2001) J Am Chem Soc 123:9898
165. Allison SW, Gillies GT (1997) Rev Sci Instrum 68:2615
166. Tian L, He F, Zhang H (2007) Angew Chem Int Ed 46:3245
167. Kunzelman J, Kinami M, Crenshaw BR, Protasiewicz JD, Weder C (2008) Adv Mater 20:119
168. Zhang X, Chi Z, Li H (2011) Chemistry 6:808
169. Chi Z, Zhang X, Xu B (2012) Chem Soc Rev 41:3878
170. Zhao Y, Lin H, Chen M, Yan D (2014) Ind Eng Chem Res 53:3140
171. Durr H (1989) Angew Chem Int Ed 28:413
172. Zhang J, Zou Q, Tian H (2012) Adv Mater 25:378
173. Fukaminato T, Irie M (2006) Adv Mater 18:3225
174. Wang X, Lu J, Yan D, Wei M, Evans DG, Duan X (2010) Chem Phys Lett 493:333
175. Jan D, Kamil L (2012) Eur J Inorg Chem 2012:5154
176. Schreuder MA, Xiao K, Ivanov IN, Weiss SM, Rosenthal SJ (2010) Nano Lett 10:573
177. Gill SK, Brice LK (1984) J Chem Educ 61:713
178. Richter MM (2004) Chem Rev 104:3003
179. Zhang B, Shi S, Shi W, Sun Z, Kong X, Wei M, Duan X (2012) Electrochim Acta 67:133

Layered Rare Earth Hydroxides: Structural Aspects and Photoluminescence Properties

Jianbo Liang, Renzhi Ma, and Takayoshi Sasaki

Contents

1 Introduction 70
2 Background Survey 71
3 Synthetic Strategies 72
   3.1 Transformation from Oxide Solid Precursor 72
   3.2 Precipitation of Aqueous Rare Earth Ions 73
   3.3 Synthesis via Solvothermal Reaction 74
4 Structural Features 75
   4.1 LREH-I Category 75
   4.2 LREH-II Category 81
5 Topological Features 84
6 Structural Evolution Associated with "Lanthanide Contraction" 88
   6.1 A Series of $RE_2(OH)_5Cl \cdot nH_2O$ 88
   6.2 A Series of $RE_2(OH)_5NO_3 \cdot nH_2O$ 90
   6.3 A Series of $RE_2(OH)_4SO_4 \cdot 2H_2O$ 91
7 Photoluminescence Properties and Potential Applications 92
   7.1 Photoluminescence Features of LREH Compounds 93
   7.2 Oxide Phosphors Derived from LREH Compounds 97
   7.3 Phosphors Based on Exfoliated Nanosheet Crystallites 100
8 Summary and Outlook 102
References 102

Abstract Layered rare earth hydroxides (LREHs), a special class of layered solids featuring cationic host layers of rare earth (RE) hydroxides, have become recognized as novel multifunctional materials in which the intercalation reactivity and host–guest interaction are coupled with the appealing physicochemical properties of RE elements. This chapter presents a background survey and an up-to-date overview on the development of LREH materials in terms of their synthesis, structural characterization, and photoluminescence properties. We first summarize the synthetic strategies to produce LREHs in various forms. In the following

---

J. Liang • R. Ma • T. Sasaki (✉)
International Center for Materials Nanoarchitectonics (MANA), National Institute for Materials Science (NIMS), 1-1 Namiki, Tsukuba, Ibaraki 305-0044, Japan
e-mail: Sasaki.takayoshi@nims.go.jp

section, the basic structural features of LREH compounds are illustrated for typical anionic forms, and we highlight the critical importance of this knowledge in interpreting their fundamental properties and functionality hunting. Then, the photoluminescence properties of LREH compounds are discussed. Various phosphors with tunable or enhanced performance, including forms of oriented films, exfoliated nanosheet crystallites, and hybrid nanocomposites, are designed based on the structural features of LREH compounds. We describe the major contributions to this topic from studies conducted before 2015.

Keywords Anion exchange • Crystal structure • Intercalation • Layered rare earth hydroxides • Phosphor • Photoluminescence • Pillared layered framework

Abbreviations

AQS   2,6-Anthraquinonedisulfonate
DS   Dodecyl sulfate
HMT   Hexamethylenetetramine
LDHs   Layered double hydroxides
NDS   2,6-Naphthalenedisulfonate
PL   Photoluminescence
RH   Relative humidity
SAED   Selected area electron diffraction
SEM   Scanning electron microscopy
TEM   Transmission electron microscopy
XRD   X-ray diffraction

1 Introduction

The term of layered rare earth hydroxides (LREHs) is used to designate a special class of lamellar solids built up from cationic host layers of rare earth hydroxides and counter anionic species, which are stacked in an alternating sequence [1–3]. The compounds have recently emerged as new cationic layered hosts besides the well-known layered double hydroxides (LDHs). Rare earth hydroxide layers, which can be described as two-dimensional corrugated sheets, are constituted by RE^{3+}-centered polyhedral units interconnected via hydroxyl groups and water vertexes. The counter anions, either inorganic or organic species, or even negatively charged metal clusters, are accommodated in the hydroxide galleries either by electrostatic interaction or covalent bonding as well as hydrogen bond networks. LREH compounds can be synthesized with variable compositions. The general chemical formula is expressed as $RE_2(OH)_{6-m}(A^{n-})_{m/n} \cdot xH_2O$ ($1 \leq m \leq 2$, A^{n-} represents the n-valence anion), in which the ideal phase of $m=1$ and $m=2$ represents two distinctive types of host layer topologies. Reports of LREH compounds in the literature can be traced back to the 1950s [4], when a compound of the $m=2$ phase was isolated. The lamellar structure of this phase was revealed in the following years. Though compounds of the $m=1$ phase were reported in the 1970s, it was not until the middle of the 2000s that their structural features were clearly unveiled. On the solid basis of structural knowledge, the compounds have been identified as a new family of layered solids, in which the intercalation chemistry and host–guest interaction are coupled with the attractive physiochemical properties of RE elements. This chapter provides an overview of the efforts that have led to the development of LREH compounds as multifunctional phosphors.

2 Background Survey

Reports on LREH compounds date back to the 1950s, when Aksel'rud et al. [4–6] observed the formation of anion-containing rare earth hydroxides when they studied the precipitation of aqueous RE^{3+} ions. These precipitates were obtained as amorphous powders, and the compositions were determined as $RE(OH)_2Cl$ by chemical analysis. Crystalline samples of $RE(OH)_2Cl$ were later obtained by other researchers. Klevtsov and coworkers [7] isolated single crystals of Y$(OH)_2Cl$ as a side product in the hydrothermal growth of rare earth ferrite–garnets and solved the crystal structure. The compounds crystallize in two polymorphs [8, 9], monoclinic and orthorhombic systems, and comprise identical cationic layers of $\{[Y_2(OH)_4]^{2+}\}_\infty$ stacking in a different sequence. In addition to $Y(OH)_2Cl$, isostructural compounds containing other RE members such as La, Pr, Sm, Gd, and Nd [10–13] or nitrate as a counter anion were reported. In some papers, these compounds are referred as basic rare earth salts [12], and they are the first category of LREH compounds to be structurally established.

It is known that rare earth oxide solids can be transformed into a hydroxide form under proper hydrothermal conditions. Haschke and his coworkers [14–21] studied the hydrothermal equilibria of RE_2O_3–REX_3–H_2O (X represents the anions of halides, nitrate, sulfate, or carbonate) systems and obtained a number of hydroxide compounds incorporated with inorganic anions. In addition to the phases mentioned above, they identified some non-stoichiometric compounds such as $RE(OH)_{3-x}(NO_3)_x$ [14, 15] and $RE(OH)_{3-x}Cl_x$ [18] with x values close or equal to 0.5. Many details of their studies, including the synthesis conditions, the thermal stability of the products, and some structural insights were described. The synthetic method employed by Haschke and his coworkers required high temperature and pressure. In an alternative route, compounds of similar composition could be obtained under mild conditions based on precipitation of aqueous RE^{3+} ions [20]. For a long period, a full understanding of the structural features of these non-stoichiometric compounds was not available, and little attention was paid to their potent intercalation properties. In 2006, Monge's group [3] demonstrated a new type of host layer, $\{[RE_2(OH)_5(H_2O)_2]^+\}_\infty$, from compounds of

$Yb_4(OH)_{10}[C_{14}H_6O_2(SO_3)_2]\cdot 4H_2O$ and $Y_4(OH)_{10}[C_{10}H_6(SO_3)_2]\cdot 4H_2O$. This is believed to be the first time that the layer topology and intercalation chemistry of LREH compounds have come to be compared and discussed with those of the LDH compounds. In separate studies, Fogg's group and our group successfully synthesized compounds of $Ln_2(OH)_5NO_3\cdot 1.5H_2O$ [22] and $Ln_2(OH)_5Cl\cdot nH_2O$ [23], respectively, and reported their anion-exchange properties, opening the avenue to their rich intercalation chemistry relating to their dehydration/rehydration, swelling, and exfoliation behaviors. Then, a new class of anion-exchangeable lamellar hosts has been recognized, which inspired a range of studies to explore the novel multifunctionality based on the unique properties of RE elements and the interesting intercalation reactivity.

The LREH family has become rich and diverse. The composition covers the whole lanthanide series including their congener Y, and the compounds can accommodate inorganic, organic, and polyoxometalate species. The empirical formula of LREH compounds is generally described as $RE_2(OH)_{6-m}(A^{n-})_{m/n}\cdot xH_2O$ ($1 \leq x \leq 2$). Here, we denote the phase of $m = 1$ and 2 as LREH-I and LREH-II category, respectively. Compounds in these two separate categories crystallize in different structures and exhibit different photoluminescence properties.

3 Synthetic Strategies

In general, LREH compounds are synthesized through solution processes. Two major routes have been developed relating to starting reactants. One method starts from solid precursors, transforming rare earth oxides into hydroxide forms via hydrothermal treatment, and the other method is based on the precipitation of aqueous RE^{3+} ions through hydrolysis via titration or homogeneous alkalization. Details of these routes are described in this section.

3.1 Transformation from Oxide Solid Precursor

Rare earth oxides can undergo hydrolysis to corresponding hydroxide compounds:

$$RE_2O_3 + 3H_2O \rightarrow 2RE(OH)_3$$

Because of the low solubility of RE_2O_3 solids, hydrothermal treatment at high temperature and pressure near or at the supercritical conditions is required to promote the reaction. This reaction was modified and employed to synthesize LREH compounds. The products are often obtained as mixed phases in the form of a single crystal or a polycrystalline powder. Klevtsov and coworkers [7–10] isolated crystals of $RE(OH)_2Cl$ from a system of $FeCl_3$–NH_4Cl solution/Y_2O_3–Fe_2O_3 solids. Haschke and coworkers [14–21] systematically investigated the phase equilibria of RE_2O_3–REX_3–H_2O systems and obtained a number of LREH compounds as well as hydroxide compounds in other structures (X: anions of Cl^-, Br^-, I^-, NO_3^-, and SO_4^{2-}). This method can produce LREH compounds containing robust inorganic anions but is not suitable to prepare compounds containing organic species due to the low stability even at critical conditions.

3.2 Precipitation of Aqueous Rare Earth Ions

3.2.1 Precipitation by Titration with an Alkali

Aksel'rud et al. [4–6] reported the formation of $RE(OH)_2Cl$ as an amorphous precipitate when adding NaOH into a $RECl_3$ solution at room temperature. Fogg and coworkers [22, 24–28] reported that crystalline compounds of $RE_2(OH)_5NO_3 \cdot H_2O$ were derived from such precipitates by drying at 80 °C. To acquire samples of an improved quality, hydrothermal treatments were further employed. In a typical procedure, a solution containing RE^{3+} ions was titrated by an alkali solution to a certain pH value. The as-yielded slurry was transferred into an autoclave, tightly sealed, and heated at an elevated temperature (usually not higher than 250 °C) to produce a crystalline sample. In addition to inorganic anions such as the halides and nitrate [22, 24–28], compounds containing specific organic anions, for example, dodecyl sulfate (DS^-, $C_{12}H_{25}OSO_3^-$) [29], 2,6-naphthalenedisulfonate (NDS^{2-}), or 2,6-anthraquinonedisulfonate ($AQDS^{2-}$) [3, 30], could be obtained by adding an alkali into a solution containing RE ions and target anions. The conditions for precipitating RE^{3+} ions are restricted compared with the preparation of LDH compounds. Carefully controlling the pH value in a narrow range is essential to avoiding the formation of $RE(OH)_3$ as well as other RE-related compounds. In addition, the alkali employed in the titration has a significant impact on the products. Thus, careful selection of the alkali is needed. Samples obtained by this method are mostly in the form of a polycrystalline powder. But under optimal conditions, single crystals suitable for structure determination can be obtained [3, 9, 24, 27, 28].

3.2.2 By Homogeneous Alkalization

Homogeneous alkalization, which employs the hydrolysis of a special reagent to tune the pH value, is a special route to obtain highly crystalline LREH samples. Louër et al. [31] reported the synthesis of $Nd(OH)_2NO_3 \cdot H_2O$ by refluxing a concentrated solution of $Nd(NO_3)_3$ and NH_4Cl. The product was composed of well-developed microplatelets, and the crystal structure could be revealed from corresponding powder diffraction data. This method is simple, yet has not been extended to other LREH compounds. In an alternate strategy, our group found that hexamethylenetetramine [$(CH_2)_6N_4$, HMT] was useful in the synthesis of LREH compounds [23, 32–40]. The hydrolysis of HMT is expressed as follows:

$$(CH_2)_6N_4 + 6H_2O \rightarrow 6HCHO + 4NH_3$$
$$NH_3 + H_2O \rightarrow NH_4^+ + OH^-$$

In a typical synthesis, a solution containing RE^{3+} ions, HMT, and a source of target anions is prepared and refluxed under N_2 gas flow. Upon refluxing, the HMT reagent undergoes hydrolysis, progressively releasing OH^- to precipitate RE^{3+} ions. Employing a large excess of the anion source can effectively suppress the formation of $RE(OH)_3$ and promote sample crystallinity. Compared with the precipitation by adding an alkali solution, homogeneous alkalization can avoid the complicated steps needed to accurately control the pH value. Compounds of various RE members including the solid solution phase can be conveniently synthesized.

Urea is a well-known reagent to promote the precipitation of transition metal ions, and it has been effectively applied to synthesize a wide range of LDH compounds.

$$(NH_2)_2CO + H_2O \rightarrow CO_2 + 2NH_4^+ + 2OH^-$$
$$NH_3 + H_2O \rightarrow NH_4^+ + OH^-$$
$$CO_2 + 2OH^- \rightarrow CO_3^{2-} + 2H_2O$$

Due to the release of CO_2, employing urea often results in the precipitation of $RE_2(CO_3)_3$ and other phases other than LREH compounds. The formation of DS^- intercalated LREH compounds coexisting with $RE_2(CO_3)_3$ impurities by this method was reported [41].

3.3 Synthesis via Solvothermal Reaction

It is difficult to obtain a compound of $La_2(OH)_5NO_3 \cdot nH_2O$ using the methods described above. Byeon and coworkers [42] developed a non-aqueous solvothermal route to acquire the target sample. In their strategy, ethanol was selected as a solvent to dissolve the reactants of $La(NO_3)_3 \cdot 6H_2O$ and AOH (A = K, Rb, and Cs). Though AOH was only partially dissolved in ethanol at room temperature, complete dissolution could be attained under solvothermal conditions. The mixture was sealed in a Teflon-lined autoclave, heated to 150–160 °C, and aged for several hours, which produced a layered compound mixed with ANO_3 salt. The ANO_3 impurity was removed by washing with water. The final product was determined as $La_2(OH)_5NO_3 \cdot 1.5H_2O$ by chemical analysis.

4 Structural Features

Elucidating the crystal structure of LREH compounds is of critical importance to understanding their physicochemical properties. The structural information has been derived from either single crystals or polycrystalline powders. Well-crystallized single crystals were successfully obtained for $Yb_4(OH)_{10}[C_{14}H_6O_2(SO_3)_2]\cdot 4H_2O$ and $Y_4(OH)_{10}[C_{10}H_6(SO_3)_2]\cdot 4H_2O$ [3]. Using high-flux X-ray synchrotron radiation, the structural information of $Yb_2(OH)_5Cl\cdot 1.5H_2O$ [24] was acquired from a crystal of small size. On the other hand, for $Yb_2(OH)_5NO_3\cdot H_2O$ and $Yb_2(OH)_5NO_3\cdot 2H_2O$ [27], only the structure of the host layer was determined. The homogeneous precipitation route has enabled to yield well-crystallized polycrystalline samples suitable for structural analysis. On the basis of high-resolution synchrotron or laboratory powder XRD data, the structures of several anionic series were successfully determined by the direct method coupled with the Rietveld refinement [23, 32, 34, 40]. The structural features of the LREH compounds will be described below according to the anionic form in each category.

4.1 LREH-I Category

4.1.1 Aromatic Disulfonate Form

$Yb_4(OH)_{10}[C_{14}H_6O_2(SO_3)_2]\cdot 4H_2O$ (Yb-AQDS) and $Y_4(OH)_{10}[C_{10}H_6(SO_3)_2]\cdot 4H_2O$ (Y-NDS) are the first examples, showing the crystal structure of the LREH-I category [3, 30]. Single crystals were produced by a hydrothermal reaction using $(CH_3CH_2)_3N$ as an alkalization reagent. The two compounds crystallize in an orthorhombic system with the same space group of *I*bam [cell parameters: Y-NDS: $a = 1.2639(1)$, $b = 3.0525(2)$, $c = 0.71348(6)$ nm; Yb-AQDS: $a = 1.25401(6)$, $b = 3.5652(2)$, $c = 0.70347(4)$ nm], and crystallographic data show that the compounds comprise identical inorganic layers. Two types of RE coordination sites are observed: one is surrounded by eight hydroxy groups and one water molecule, forming a monocapped square antiprism of $\{RE(OH)_8(H_2O)\}$, while the other is enclosed with seven hydroxy groups and one water molecule, showing a dodecahedron of $\{RE(OH)_7(H_2O)\}$. RE^{3+} ions of the same type are packed along the [001] direction to produce a linear row. The two types of rows alternately combine with each other to form a cationic layer of $\{[RE_2(OH)_5(H_2O)_2]^+\}_\infty$. The anions of $AQDS^{2-}$ or NDS^{2-} are sandwiched between the inorganic layers, adopting an alternating configuration along the *b* axis. The aromatic rings are parallel to the *ab* plane. For the $AQDS^{2-}$ ions, the ordered arrangement gives rise to π–π stacking interactions between the aromatic rings and O–π stacking interactions among oxygen atoms from the quinine groups and the lateral aromatic rings. In the case of Y-NDS, only C–π stacking interactions

Table 1 Bond lengths (Å) and angles (degree) of hydrogen bonds in compounds of Y-NDS and Yb-AQDS (in italics)

D–H···A	d(D–H)	d(H···A)	d(D···A)	Angle (D H A)
O_{W1}–H···O (×2)	0.82	2.04	2.81	156
	0.85	1.94	2.77	167
O_{W2}–H···O (×2)				
	0.85	2.12	2.85	144
O_H–H···O (×2)	0.98	2.00	2.939	163
	1.12	2.32	3.023	119
O_H–H···O	0.98	2.14	3.032	151
	0.92	2.60	3.097	115

The hydrogen bonds are between oxygen atoms from the ligand sulfonate ions and water molecules (O_w) or hydroxyl groups (O_H) around the rare earth ions in the inorganic layers. D represents the donor and A represents the acceptor. Adapted with permission from [3]. Copyright 2006 Wiley-VCH Verlag GmbH & Co. KGaA

among NDS^{2-} anions are possible. The organic anions are bonded to the hydroxide layers via hydrogen bonds other than a covalent bond to the RE^{3+} ions. The hydrogen bonds are generated between oxygen atoms from the sulfonate ions and water molecules or hydroxy groups. Selected bond lengths and angles are listed in Table 1. The well-ordered position of the anions in the interlayer space renders the compound a porous structure with channels running along the c direction.

One noticeable characteristic is that the RE^{3+} ions in the $\{[RE_2(OH)_5(H_2O)_2]^+\}_\infty$ layer are arranged in a quasi-hexagonal order. The structure is distorted from an ideal hexagonal lattice to a larger orthorhombic one, producing two independent RE^{3+} coordinations and somewhat irregular interatomic distances. The crystallographic relation between such a LREH-I host lattice and that of the brucite-like layer of LDHs was addressed by Monge and coworkers (Fig. 1).

4.1.2 Chloride Form

Highly crystalline samples with a chemical formula of $RE_2(OH)_5Cl \cdot nH_2O$ (RE, from Nd to Er, $n \sim 1.6$) were synthesized via the homogeneous alkalization route by refluxing a solution containing $RECl_3$, HMT, and NaCl. These compounds isostructurally crystallize in an orthorhombic system. The crystal structures were solved from synchrotron X-ray powder diffraction data and found to comprise host layers of $\{[RE_2(OH)_5(H_2O)_{2-x}]^+\}_\infty$ [23, 32], close to those described by Monge and coworkers. In a typical example of $Eu_2(OH)_5Cl \cdot 1.6H_2O$ [space group $P2_12_12$, cell parameters: $a = 1.29152(3)$, $b = 0.73761(1)$, and $c = 0.87016(3)$ nm, $Z = 4$], two types of Eu^{3+} are found in the unit cell: one is surrounded by seven hydroxy groups and a fully occupied water molecule, leading to dodecahedron of $\{Eu(OH)_7(H_2O)\}$, and the other is bonded to eight hydroxy groups and one partially occupied water molecule, forming a monocapped square antiprism of

Fig. 1 Crystal structure of Y-NDS and Yb-AQDS. Reprinted with permission from [3]. Copyright 2004 Wiley-VCH Verlag GmbH & Co. KGaA

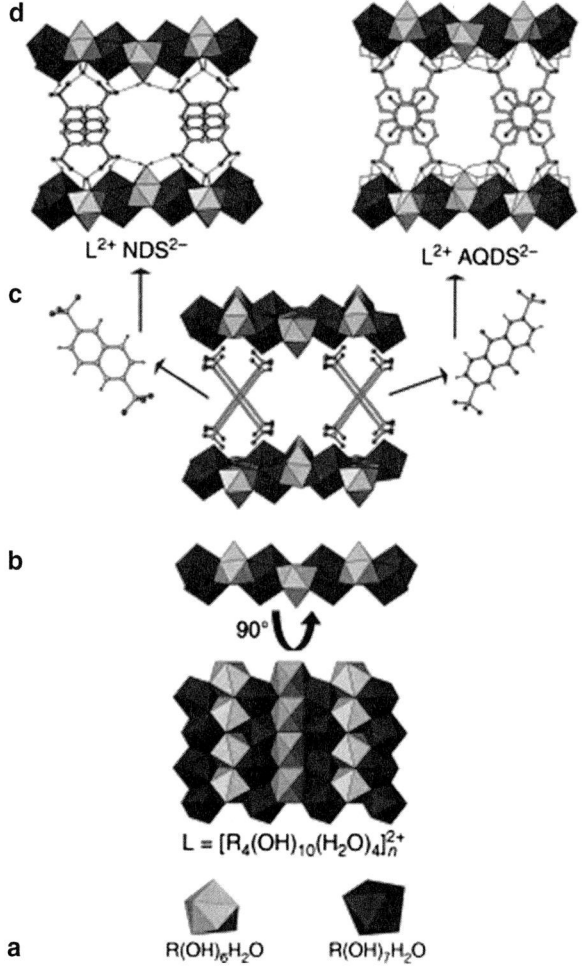

[Eu(OH)$_8$(H$_2$O)]. Polyhedra of the same type are connected in rows along the b axis, and the rows are alternatingly packed via OH–OH edges to form infinite layers of $\{[Eu_2(OH)_5(H_2O)_{2-x}]^+\}_\infty$. The 3D electron density distribution map indicates that the chloride ions are located in well-defined sites in the interlayer galleries, a feature differing from the heavily disordered arrangement in LDH compounds. The interatomic distances between the chloride ions and the closest Eu^{3+} ions are in the range of 4.5–4.9 Å, beyond the scale of a covalent bonding. The charge density of the $\{[Eu_2(OH)_5(H_2O)_{2-x}]^+\}_\infty$ layer is approximately 4.8×10^{-2} $q/Å^2$, which is comparable to that of $Mg_2Al(OH)_6(CO_3^{2-})_{0.5} \cdot 2H_2O$ (Fig. 2).

Fogg's group synthesized $RE_2(OH)_5Cl \cdot 1.5H_2O$ (Ln = Y, Dy, Er, Yb) via the titration method followed by hydrothermal treatment [24]. The powder XRD pattern of the Yb sample showed two sets of basal reflections, 0.84 and 0.80 nm.

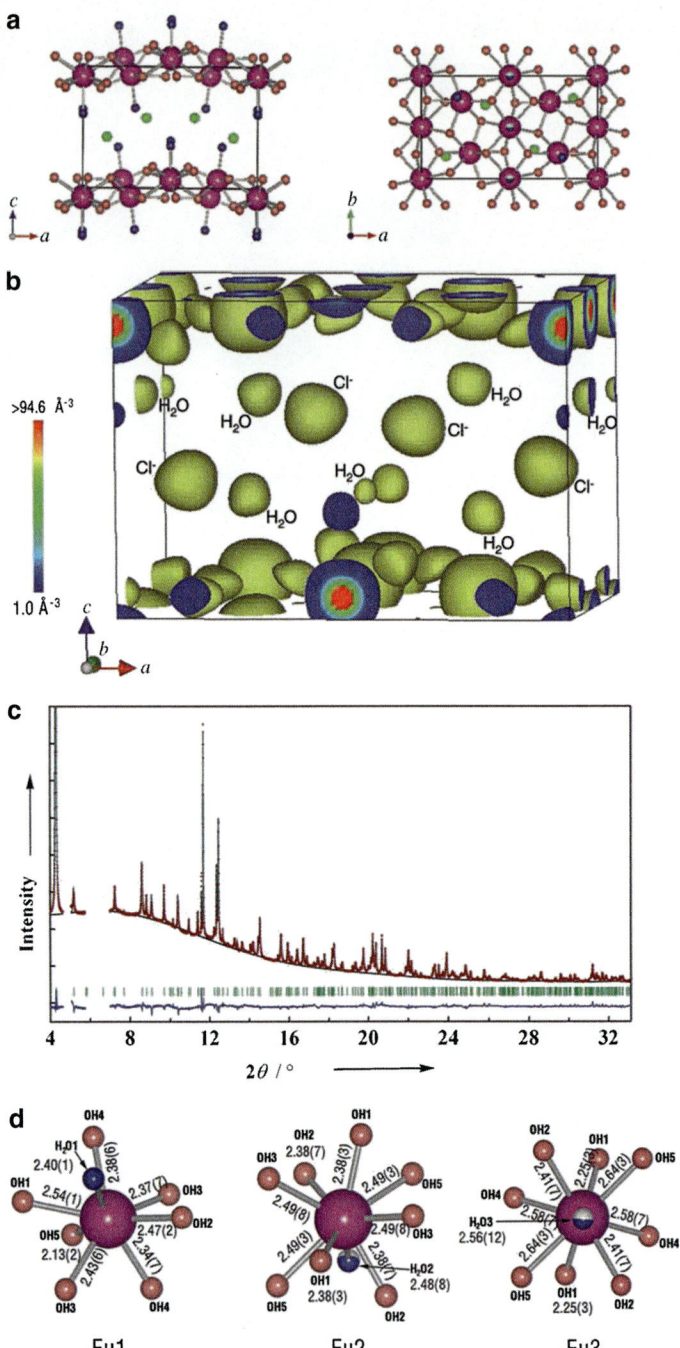

Fig. 2 (**a**) Crystal structure of $Eu(OH)_{2.5}Cl_{0.5} \cdot 0.8H_2O$; (**b**) 3D electron density distribution of the structure; (**c**) Rietveld refinement profiles; (**d**) coordination of the Eu ions in the cationic host layers. Reprinted with permission from [23]. Copyright 2008 Wiley-VCH Verlag GmbH & Co. KGaA

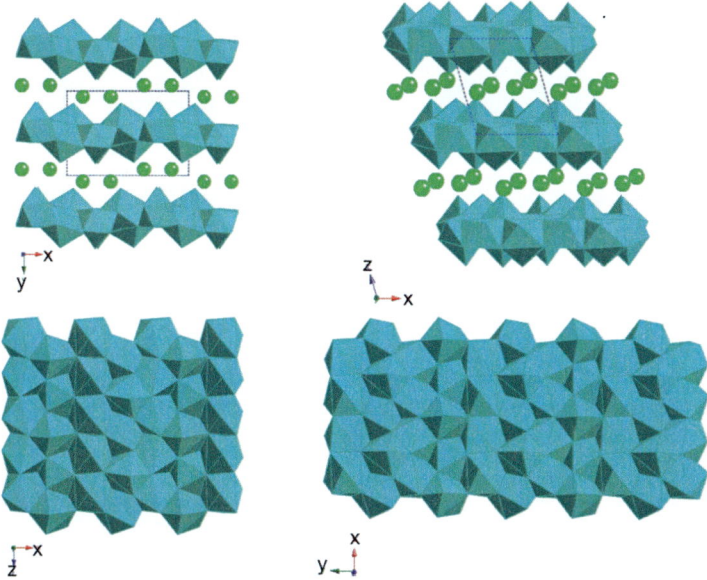

Fig. 3 Crystal structure of 0.84 nm phase (*left*) and 0.80 nm phase (*right*) $Yb_2(OH)_5Cl \cdot 1.5H_2O$ samples. Adapted with permission from [24]. Copyright 2008 American Chemical Society

Single crystals of each phase were isolated to solve the crystal structure. Briefly, the 0.84 nm phase crystallizes in an orthorhombic system (space group $Pca2_1$) with $\{[Yb_2(OH)_5(H_2O)_{1.5}]^+\}_\infty$ layers similar to those described above. In contrast, the 0.80 nm phase was less hydrated and crystallizes in a monoclinic system (space group of $P2_1$). The $\{[Yb_2(OH)_5(H_2O)_{1.5}]^+\}_\infty$ layers in the 0.8 nm phase differed from those of 0.84 nm phase in that a portion of the $\{Yb(OH)_8(H_2O)\}$ in the latter was transformed into $\{Yb(OH)_7(H_2O)\}$ by removing the capping water molecule, therefore lowering the symmetry of the 2D lattice. Both of these phases showed anion-exchange properties and could be converted into other anion forms (Fig. 3).

4.1.3 Nitrate Form

Fogg and coworkers [22] synthesized samples of $RE_2(OH)_5NO_3 \cdot 1.5H_2O$ (RE: Y, and Gd to Lu) by titrating a $RE(NO_3)_3$ solution with NaOH followed by hydrothermal treatment. The samples were polycrystalline, and their diffraction peaks showed heavy overlaps, preventing the determination of detailed structural information. By a combination of XRD and in-plane SAED data, the Er sample was determined as the monoclinic phase with cell parameters of $a \sim 0.71$, $b \sim 1.27$, $c \sim 1.88$ nm, and $\beta \sim 92.5°$. For Yb, two samples in different hydration states, $Yb_2(OH)_5NO_3 \cdot H_2O$ and $Yb_2(OH)_5NO_3 \cdot 2H_2O$ [27], were isolated by controlling the hydrothermal treatment temperature. Both of these two phases crystallize in an orthorhombic system. $Yb_2(OH)_5NO_3 \cdot 2H_2O$ adopts a space group of *Cmcm* (No. 63)

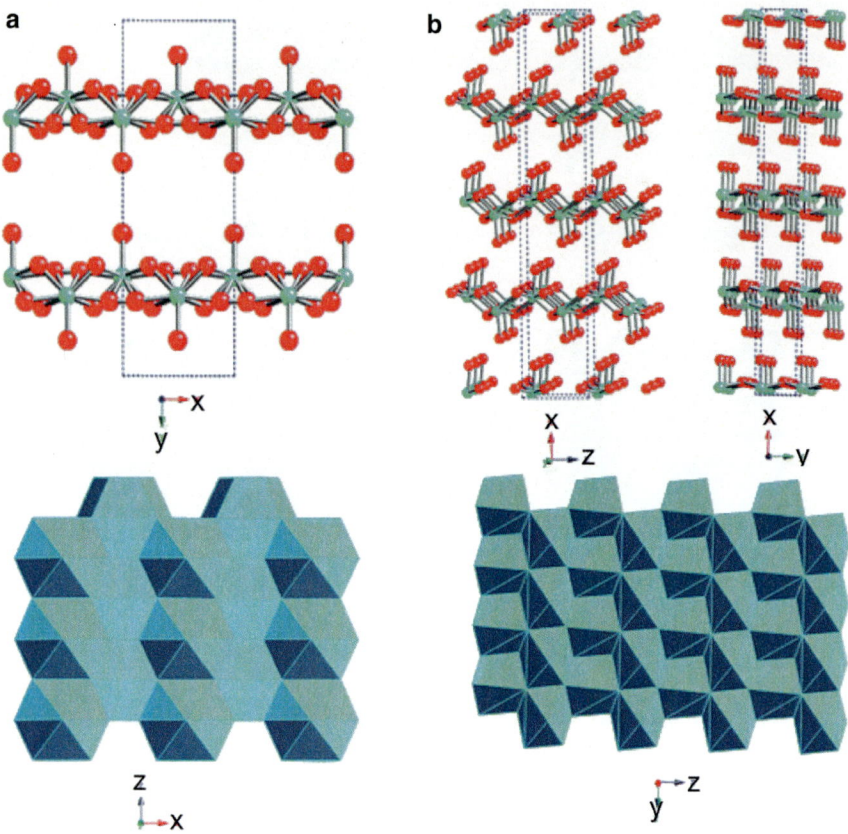

Fig. 4 Crystal structure of $Yb_2(OH)_5NO_3 \cdot 2H_2O$ (**a**) and $Yb_2(OH)_5NO_3 \cdot H_2O$ (**b**). Adapted with permission from [27]. Copyright 2010 American Chemical Society

with cell parameters of $a = 0.60000(12)$, $b = 1.8589(4)$, and $c = 0.37555(8)$ nm, and $Yb_2(OH)_5NO_3 \cdot H_2O$ is in a space group of $Pna2_1$ (No. 33) and the cell parameters are $a = 3.3866(5)$, $b = 0.37305(6)$, and $c = 0.59417(10)$ nm. It should be mentioned that the in-plane cell parameters of these two phases differ from those of $Yb_2(OH)_5NO_3 \cdot 1.5H_2O$, indicating different layer structures. The crystal structure was explored from corresponding single-crystal samples. The host layer was preliminarily determined, but the location of the interlayer species is yet undetermined. In contrast to $RE_2(OH)_5NO_3 \cdot 1.5H_2O$, nitrate ions in these two phases cannot be exchanged at room temperature. The nitrate anions may be chemically bonded to the Yb ions (Fig. 4).

Our group synthesized samples of $RE_2(OH)_5NO_3 \cdot nH_2O$ (RE: from Sm to Tm, and Y, $1.0 < n < 2.0$) employing the homogeneous alkalization protocol [33]. The products were composed of microcrystals with well-defined edges, showing high crystallinity. Based on XRD and SAED data, these compounds were identified as the monoclinic phase with in-plane unit cells nearly identical to the corresponding members of $RE_2(OH)_5Cl \cdot nH_2O$. It is reasonable to consider that the host layers in this series are identical to the chloride forms. Parameter c along the layer stacking direction is double that of the basal spacing, suggesting the gliding of the host layers with respect to each other. The possible space group was among Pc, $P2/c$, or $P2_1/c$. Due to the severe overlapping of diffraction peaks, determination of the exact crystal structures was unsuccessful.

4.2 LREH-II Category

4.2.1 Nitrate and Halide Forms

A polycrystalline powder of $Nd(OH)_2NO_3 \cdot H_2O$ was prepared by refluxing $Nd(NO_3)_3$ and NH_4NO_3 [31]. The compound crystallizes in a monoclinic system (space group of $C2/m$, $Z=4$) with unit cell dimensions of $a = 2.0996(4)$, $b = 0.3875(2)$, $c = 0.6282(2)$ nm, and $\beta = 113.73(3)°$. The structure solved from the powder XRD pattern clearly shows layers stacking along the [100] direction. Each hydroxy group is connected to three Nd^{3+} ions with distances ranging from 0.240 to 0.256 nm, while a water molecule is bonded to one Nd^{3+} ion with a distance of 0.26 nm. The Nd–O–Nd angles vary from 105° to 108°, implying the H position at the vertex of the tetrahedron. One Nd^{3+} is coordinated by nine oxygen atoms: six from hydroxy groups, two from two separated nitrate groups, and one from a water molecule. The subunits of $\{NdO_9\}$ are interconnected via the hydroxy groups to form infinite two-dimensional layers. The nitrate ions adopt two orientations that are statistically distributed in the layers (Fig. 5).

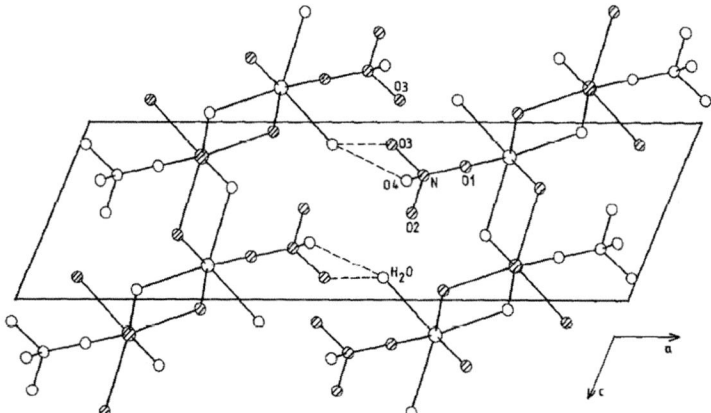

Fig. 5 Lattice description of $Nd(OH)_2NO_3 \cdot H_2O$. Reprinted with permission from [31]. Copyright 1987 Elsevier

Similar layered structures are found in compounds incorporated by halide ions, for example, La(OH)$_2$I·H$_2$O [43]. A single crystal of the compound was obtained by the hydrothermal treatment of a solution prepared by dissolving La$_2$(CO$_3$)$_3$·xH$_2$O into concentrated HI and adjusting the pH value to 5. The compound is monoclinic (space group $C2/m$, $Z=4$) with lattice parameters of $a=1.9691(3)$, $b=0.4136(1)$, $c=0.6286(1)$ nm, and $\beta=108.45(1)°$. Each La^{3+} ion is coordinated by six hydroxy groups, one water molecule, and two iodide ions forming two-dimensional host layers similar to those in Nd(OH)$_2$NO$_3$·H$_2$O.

4.2.2 Sulfate Form

RE$_2$(OH)$_4$SO$_4$·2H$_2$O [34, 39] (Ln = Pr, Nd, Sm ~ Tb) were synthesized by refluxing a solution containing RE$_2$(SO$_4$)$_3$, HMT, and Na$_2$SO$_4$. These compounds isostructurally crystallize in the monoclinic phase with cell parameters $a \sim 0.63$, $b \sim 0.37$, $c \sim 0.84$ nm, and $90.07 < \beta < 90.39°$. β is approximately equal to 90°, and the unit cell is subtly distorted from the orthorhombic one. The compounds adopt a face-centered space group of $A2/m$ (No. 12).

The crystal structure of a representative sample, Tb$_2$(OH)$_4$SO$_4$·2H$_2$O, was revealed by the Rietveld analysis [34]. In the unit cell, there is one crystallographically distinct site for Tb^{3+}, two distinct sites for hydroxy groups, two for oxygen atoms of SO$_4^{2-}$, and one unique site for S and oxygen in the water molecule. Each Tb^{3+} ion is surrounded by nine oxygen atoms: six from hydroxy groups, two from water molecules, and one from a SO$_4^{2-}$ ion. Each {TbO$_9$} polyhedron is connected with six surrounding neighbors by sharing the μ_3-OH groups and μ_2-H$_2$O molecules, forming an infinite layer of $\{[\text{Tb(OH)}_2\text{H}_2\text{O}]^{2+}\}_\infty$ parallel to the ab plane. Viewed along the c axis, the Tb^{3+} ions arrange themselves into a quasi-hexagonal network. The {TbO$_9$} polyhedra can be described as a distorted monocapped square antiprism with one oxygen atom of a sulfate group as the cap. The oxygen caps of two {TbO$_9$} polyhedra from neighboring layers that point to each other form a tetrahedron for a sulfate along with two other terminal oxygen atoms in the gallery. The Tb–O (OH$^-$) distances range from 2.322(4) to 2.446(4) Å, while the bond lengths of Tb–O (SO$_4^{2-}$) and Tb–O (H$_2$O) are 2.472(4) Å and 2.600(3) Å, respectively. The SO$_4^{2-}$ ion is covalently bonded to the rare earth ions in a *trans*-bidentate configuration, as supported by the FT-IR spectra, bridging the $\{[\text{Tb(OH)}_2(\text{H}_2\text{O})]^+\}_\infty$ layers to form a rigid pillared structure. The distance between the S and O atoms coordinated to the Tb^{3+} ions (1.414(6)/1.438(6) Å) is shorter than that to the other two terminal oxygen atoms [1.499(6) Å]. The O–S–O angles are 118.8(3)°, 108.6(3)°, 107.0(3)°, and 107.6(3)° with an average of 109.4°, which suggests that the SO$_4^{2-}$ ion is distorted into a quasi-C_{2v} symmetry. The strong chemical bonding is responsible for the distortion of the SO$_4^{2-}$ ions. In the unit cell, SO$_4^{2-}$ ions take two orientations, resulting in an alternative ordered structure supported by the superspots in the electron diffraction data. This alternate alignment of SO$_4^{2-}$ ions produces a void space between them, although it is not

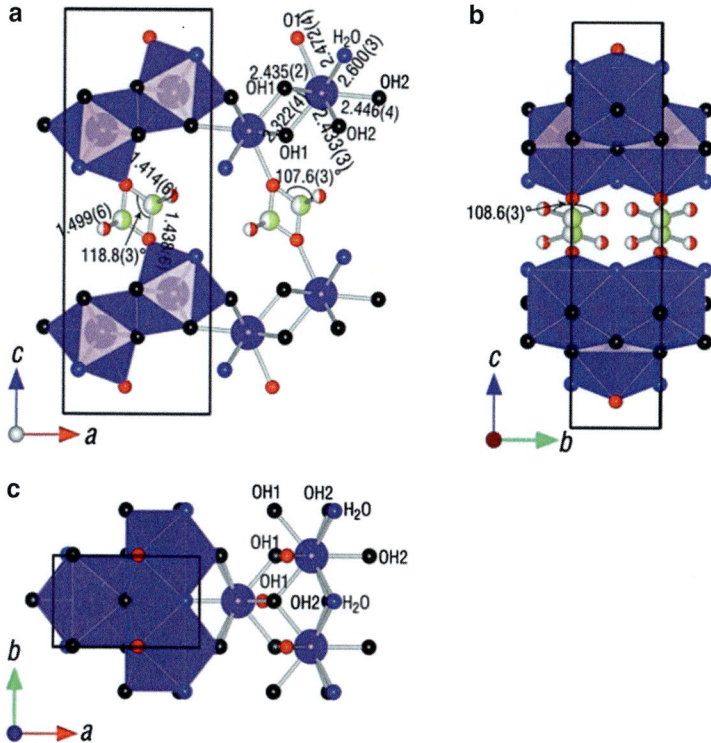

Fig. 6 Crystal structure of $Tb_2(OH)_4SO_4 \cdot 2.0H_2O$. Adapted with permission from [34]. Copyright 2011 American Chemical Society

accessible due to the very narrow open windows. This structure can be identified as a crystalline framework constructed entirely of inorganic components (Fig. 6).

4.2.3 Linear Alkyl Disulfonate Form

Inspired by the structural features of $RE_2(OH)_4SO_4 \cdot 2H_2O$ compounds, if the sulfate pillars can be replaced by some other bidentate ligands, a new type of frameworks may be created that provides new opportunities to engineer the gallery spaces. Screening synthesis found that linear organo-disulfonate salts, $Na_2[O_3S(CH_2)_nSO_3]$, were the right choice. A new series of $Ln_2(OH)_4[O_3S(CH_2)_nSO_3] \cdot 2H_2O$ (Ln: La, Ce, Pr, Nd, Sm, $n = 3, 4$) [40] was obtained by the homogeneous alkalization route. These samples isostructurally crystallize in a monoclinic system with a space group of P $2_1/n$ (No. 14). The in-plane unit cell dimensions are almost identical to those of the sulfate series, while the basal spacing is largely expanded from 0.84 nm for the $RE_2(OH)_4SO_4 \cdot 2H_2O$ to ~1.31 nm ($n = 3$) and ~1.40 nm ($n = 4$),

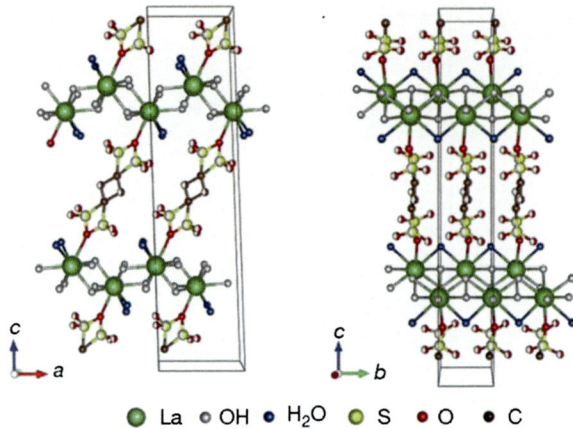

Fig. 7 Crystal structure of La$_2$(OH)$_4$[O$_3$S(CH$_2$)$_3$SO$_3$] ·2.2H$_2$O. Adapted with permission from [40]. Copyright 2013 American Chemical Society

reflecting the successful incorporation of organic moieties to yield inorganic-organic hybrids.

The structural features of these hybrids are illustrated by a typical sample of La$_2$(OH)$_4$[O$_3$S(CH$_2$)$_3$SO$_3$]·2.2H$_2$O. This structure can be described as a hybrid framework consisting of cationic host layers of $\{[La(OH)_2(H_2O)]^+\}_\infty$ bridged by α,ω-alkanedisulfonate pillars. Briefly, lanthanide ions are surrounded by nine oxygen atoms, among which six are from OH groups and two from H$_2$O molecules and the last one stems from a SO$_3$ group of the organic ligand. The $\{[La(OH)_2(H_2O)]^+\}_\infty$ layers are almost identical to those in the sulfate series. The La–O(SO$_3$) bond distance is 0.258 nm, close to that for the OH group and a H$_2$O molecule (0.252–0.277 nm), indicating a covalent bonding nature. Each α,ω-alkanedisulfonate ligand is coordinated to two lanthanide ions in the two neighboring layers, bridging the hydroxide layers to yield the hybrid 3D crystalline framework. In the gallery, the relevant positions for O and S are split in two, and thus the linear alkyl disulfonate ions present equivalent configurations. The alkyl chains are nearly parallel to the *ac* plane but tilted toward the *bc* plane at an angle of ~38.8°. Slot-like hydrophobic voids are noticed among these alkyl chains (Fig. 7).

5 Topological Features

Monge and coworkers [3, 30] addressed the topology features of Yb$_4$(OH)$_{10}$[C$_{14}$H$_6$O$_2$(SO$_3$)$_2$]·4H$_2$O (Yb-AQDS). From a crystallographic point of view, a direct relationship is observed between the LDH hexagonal in-plane lattice and that of the LERH compounds. Given that in the latter there are two independent Yb atoms in the asymmetric unit, the ideal hexagonal superlattice would have an "*a'*" parameter twice that of the LDHs. Because the distances among the Yb atoms are not uniform, the real structure involves some distortion from the hexagonal system to the orthorhombic. Vectors of the LREH unit cell can be deduced from those of the LDH by applying a transformation matrix of $2\bar{2}0, 00\bar{1}, 220$.

For the LREH-II compounds, the ninefold RE^{3+} coordination and the μ_3-OH connectivity are reminiscent of the UCl_3-type $RE(OH)_3$. In UCl_3-type $RE(OH)_3$ (hexagonal system, $P\ 6_3/m$), the $[RE(OH)_9]$ polyhedron is a regular tricapped trigonal prism with threefold symmetry. The polyhedra are connected into a linear column via sharing faces, and each column is laterally joined with three neighboring ones to form tunnel structure. In the case of $La_2(OH)_4(O_3S(CH_2)_3SO_3)\cdot 2.2H_2O$ [40], interatomic distances and bond angles in the polyhedra are very close to those in $La(OH)_3$. This aspect indicates that the $[RE(OH)_6(H_2O)_2A]$ subunit can be considered as a derivative of the $[RE(OH)_9]$, in which one third of the OH groups in the latter are replaced by two H_2O molecules and one oxygen from the guest anion. Such an enrolling of the SO_3 ligand and the H_2O molecules in lanthanide coordination is the key to yielding structural anisotropy. Accordingly, the columns are laterally linked to two neighbors, thereby extending in a planar manner to produce the 2D host layer. In this sense, the LREH-II compounds can be regarded as a substitutional derivative of UCl_3-type $RE(OH)_3$. The substituting transformation from $Ln(OH)_3$ with the highly symmetrical UCl_3-type structure to the LREH-II-type LREH compounds corresponds to a unique example of 3D to 2D topology engineering (Fig. 8).

In case of the LREH-I category, the $\{[RE_2(OH)_5(H_2O)_{2-x}]^+\}_\infty$ layer [23, 40] represents a different type of lamellar topology, which is built up from two different polyhedra, $\{RE(OH)_8(H_2O)\}$ and $\{RE(OH)_7(H_2O)_{1-x}\}$. The $\{RE(OH)_8(H_2O)\}$ polyhedra are arranged into a column in the same fashion as the LREH-II series, whereas the $\{RE(OH)_7(H_2O)\}$ polyhedra are linked into a linear chain by sharing edges. As discussed above, the $\{[RE_2(OH)_5(H_2O)_{2-x}]^+\}_\infty$ layer is generated by combining the columns and chains in an alternating sequence. The $\{RE(OH)_7(H_2O)\}$ subunits can be considered as "nodes," which connect the $\{RE(OH)_8(H_2O)_{1-x}\}$ polyhedra via both edge and face sharing. Such an arrangement produces a quasi-hexagonal arrangement of the nearest neighboring lanthanide ions, and thereby the in-plane unit cell dimensions are double those in the LREH-II compounds. This unique topology feature is evidenced by the "superspots" in the in-plane SAED pattern (Fig. 9).

The quasi-hexagonal polyhedral arrangements of the LREH compounds are compared with the LDH structure. In the host layer of LDH compounds, the metal hydroxide octahedra are connected via hydroxide edges yielding a hexagonal two-dimensional lattice. For the LREH compounds, because of the nonuniform

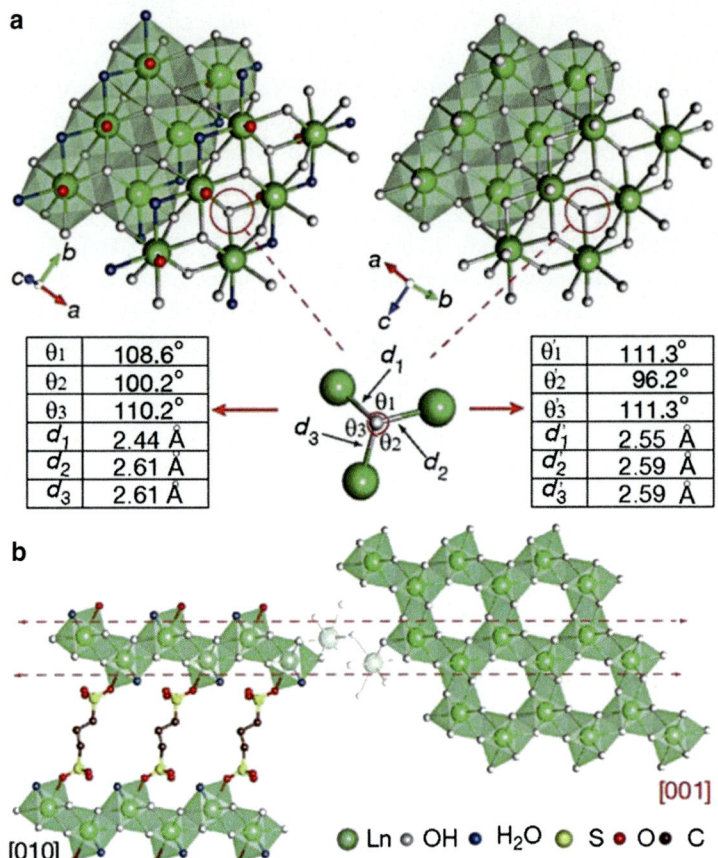

Fig. 8 Topology description of $La_2(OH)_4[O_3S(CH_2)_3SO_3]\cdot 2.2H_2O$ and UCl_3-type $La(OH)_3$. Reprinted with permission from [40]. Copyright 2013 American Chemical Society

distance between the neighboring polyhedra, the rare earth hydroxide layers present rectangular 2D lattices with different dimensions. The LREH-II type of host layer shows a larger dimension than that in the LREH-I category due to the presence of two types of polyhedra (Fig. 10).

Fig. 9 The in-plane SAED pattern of Eu $(OH)_{2.5}Cl_{0.5} \cdot 0.8H_2O$ and an illustration of the relationship between the fundamental cell and the superlattice cell. Labels with *asterisks* are the corresponding reciprocal axes. Adapted with permission from [23]. Copyright 2008 Wiley-VCH Verlag GmbH & Co. KGaA

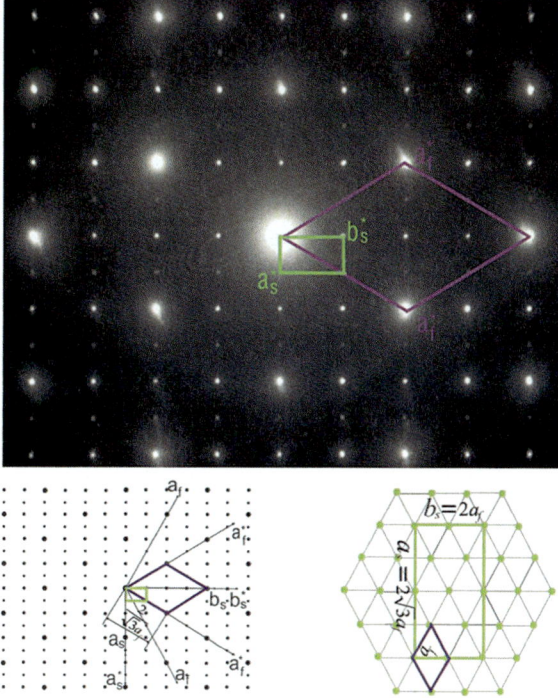

Fig. 10 Illustration of the quasi-hexagonal polyhedral arrangement in the LREH-I compounds (*upper right*) and LREH-II compounds (*lower right*) in comparison with LDH structure. The *solid lines* represent the unit cell and the *dotted lines* donate the assumed hexagonal lattice

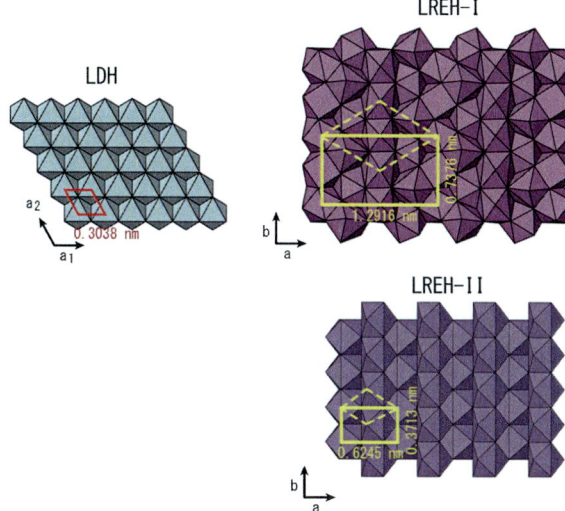

6 Structural Evolution Associated with "Lanthanide Contraction"

The homogeneous alkalization method successfully isolated a number of LREH series. The compounds show isostructural features across the RE series, clearly manifesting the structural trend arising from "lanthanide contraction" in a two-dimensional hydroxide lattice. This structural knowledge can provide insights into the nature of the host–guest interactions as well as the intercalation reactivities.

6.1 A Series of $RE_2(OH)_5Cl \cdot nH_2O$

The series of $RE_2(OH)_5Cl \cdot nH_2O$ [32] show a clear trend associated with "lanthanide contraction." In the orthorhombic unit cell, parameters such as the in-plane lattice dimensions, average Ln–O interatomic distances, and the charge density are monotonically or even linearly decreasing with RE^{3+} radius. The member of Gd shows a large derivation from the expected value, which may be related to the shielding effect of the half-filled 4f electron shells (the "Gd break" effect). Unlike the in-plane cell parameters, the interlayer distances are not regularly evolved but rather show two distinct sets of value, ~0.87 nm for members with a larger ionic radius (Nd, Sm, Eu, and Gd) and ~0.845 nm for Tb, Dy, Ho, Er, and Tm, which have a smaller radius (Fig. 11).

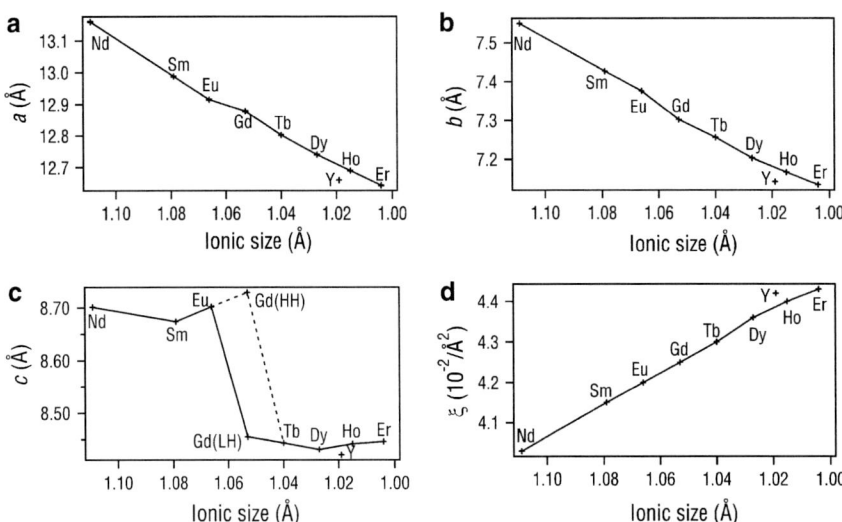

Fig. 11 Lattice parameters of $RE_2(OH)_5Cl \cdot nH_2O$. Reprinted with permission from [32]. Copyright 2008 American Chemical Society

The interlayer distance of the compounds is dependent on the relative humidity. Based on XRD and gravimetric measurements, the basal spacing difference is attributed to the differing water content in the layered solids. Accordingly, the two sets of basal spacing are assigned as the high hydration (HH) state and the low hydration (LH) state. Rietveld refinements indicate that the interlayer spacing difference originates from the occupancy of water molecules or, specifically, water molecules bonded to the ninefold coordinated RE^{3+} ions. In an ideal simplified structure where all the sites of H_2O molecules are occupied, the total hydration number per chemical formula is 8 (4 from the eightfold coordination and 4 from the ninefold coordination). In the Gd sample where both the HH and LH phases are isolated, the refined hydration numbers are 6.6 and 6.0, respectively. The basal spacing shrinks by approximately 0.02 nm as the sample transforms from the HH to the LH state. Across the series, the hydration numbers are 7.4, 6.3, and 7.2 for samples adopting the HH state (Nd, Sm, and Eu) and 5.8, 5.6, 5.4, and 4.9 for Tb, Dy, Sm, and Er in the low hydration state. The hydration number tends to decrease as RE^{3+} ions shrink in radius. In the crowded RE coordination sphere, the occupancy of water molecules is suppressed due to the repulsion between the hydroxyl groups. Therefore, compounds of small REs favor LH structure. Note that for Er, the right end of this series, the occupancy of water molecules bonded to the ninefold coordinated RE^{3+} ions is only 0.2. Water molecules play a crucial role in the stability of the hydroxide layer, and Er might represent the lower limit of the series. The dehydration/rehydration behaviors of this series are greatly influenced by such structural features. In situ XRD and gravimetric characterizations show that Nd adopted a HH state even at a RH of 5 %. For the intermediate members of Sm, Eu, and Gd, the phase transition between the HH and LH states occurred at RH values of 10 %, 20 %, and 50 %, respectively. For the members from Tb to Tm, the LH state was favored in the whole RH range from 5 % to 95 % (Table 2) (Fig. 12).

Table 2 Site occupancy factors and the number of water molecules from the structure refinement and chemical analysis

Ln	Occupancy, $H_2O(2)$/ $H_2O(3)$	Water number from refinement	Water number from chemical analysis
Nd	0.98/0.72	7.4	–
Sm	0.58/0.55	6.3	6.3
Eu	0.91/0.69	7.2	6.2
Gd	HH:73/0.58	HH:6.6	7.4
	LH:0.61/0.37	LH:6.0	–
Tb	0.36/0.54	5.8	7.0
Dy	0.50/0.30	5.6	6.7
Ho	0.33/0.36	5.4	6.3
Er	0.24/0.21	4.9	6.0
Y	0.47/0.71	6.4	6.3

Adapted with permission from [32]. Copyright 2008 American Chemical Society

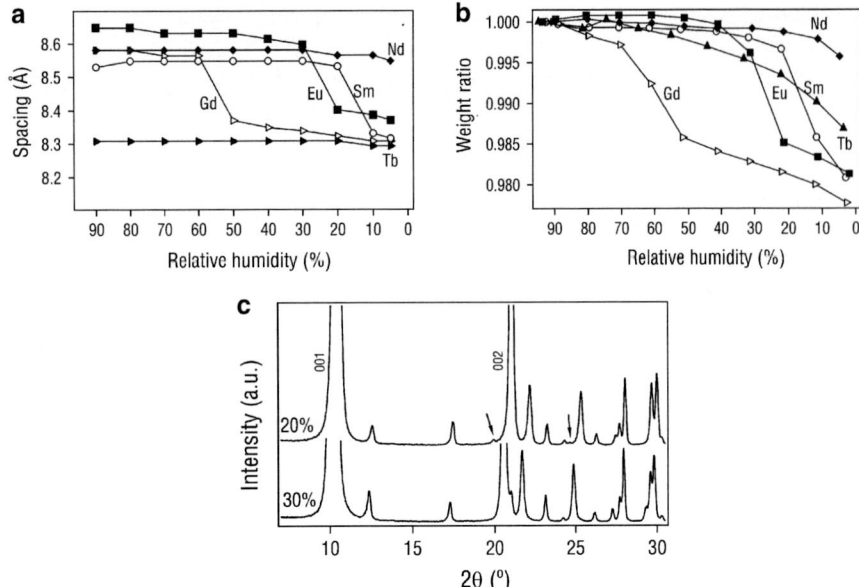

Fig. 12 Hydration/dehydration profiles of $RE_2(OH)_5Cl \cdot nH_2O$ series. (**a**) Basal spacing and (**b**) gravimetry as a function of the relative humidity. (**c**) XRD pattern of the Eu sample at RH of 20 % and 30 %, respectively. The *arrows* indicate the emergence of new peaks when the sample is subjected at RH of 20 %. Reprinted with permission from [32]. Copyright 2008 American Chemical Society

6.2 A Series of $RE_2(OH)_5NO_3 \cdot nH_2O$

In addition to the repulsion between ligands in RE^{3+} coordination, the hydrogen bonding networks between the host layers and the guest anions play an important role in the dehydration/rehydration processes. The interaction is more obvious in the case of anions having nonspherical geometry, for example, nitrate, in which the central nitrogen atom is surrounded by three identically bonded oxygen atoms in a triangular plane arrangement. Compounds of $RE_2(OH)_5NO_3 \cdot nH_2O$ [33] synthesized via the homogeneous alkalization route have almost identical in-plane lattice parameters to the corresponding chloride member and therefore are believed to consist of the same type of host layer. However, their basal spacing measured at normal conditions showed opposite trends to the chloride series, that is, the Sm sample was in the LH state while the Tm member adopted the HH structure. Moreover, in situ XRD and gravimetric measurements revealed that their dehydration/rehydration behaviors are different from the chloride members in three aspects. First, the basal spacing difference between the HH and the LH phases reached 0.06 nm, much larger than the 0.02 nm in the chloride series. Second, the phase transition between the HH and the LH states initiated at a critical RH value, and only partial transformation occurs, as indicated by the coexistence of the two sets of

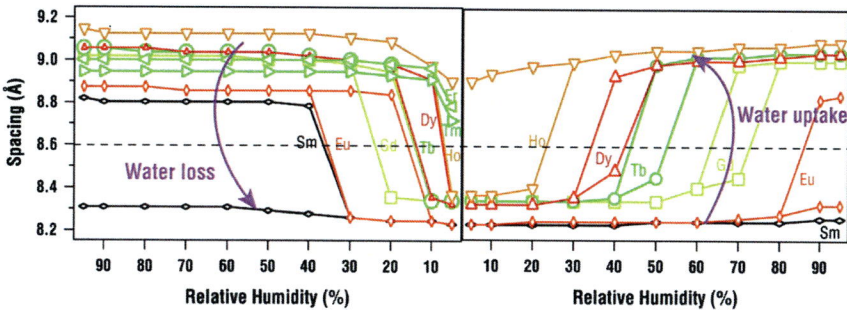

Fig. 13 Hydration/dehydration profile of $RE_2(OH)_5NO_3 \cdot nH_2O$. Reprinted with permission from [33]. Copyright 2009 American Chemical Society

basal series. Extending the aging time had little effect in prompting the phase transition. Third, unlike the promptly reversible transformation in the chloride series, hysteresis was observed. These differences strongly suggest that additional factors affect the phase transition in the dehydration/rehydration processes. The nitrate anions may adopt a perpendicular or tilted orientation in the HH state while prefer a lying-flat configuration in the LH states. Remarkably, in the nitrate series, the critical RH value for the phase transition decreased, and the RH hysteresis loop became narrow with the increasing atomic number of RE^{3+}, suggesting an increased stability of the HH states over the chloride series (Fig. 13).

6.3 A Series of $RE_2(OH)_4SO_4 \cdot 2H_2O$

The series of $RE_2(OH)_4SO_4 \cdot 2H_2O$ [34, 39] consists of typical LREH-II-type compounds crystallized in a monoclinic system in which the host layers of $\{[RE(OH)_2(H_2O)]^+\}_\infty$ are pillared by bidentate SO_4^{2-} to form a framework structure. The in-plane lattice parameters progressively decrease with the increasing RE atomic number, and the β angle also progressively decreases, varying from 90.387(1)° for Pr to 90.0718(3)° for Tb. The unit cell is only subtly distorted from the orthorhombic, especially for the right end member of Tb. The gallery height decreases from Pr to Tb, but in a moderate trend compared with the in-plane parameters, which is due to the rigid pillaring of the hydroxide layers with SO_4^{2-}. Due to the strong host–guest interactions, geometry distortion is observed for the SO_4^{2-} pillars, as revealed by the crystallographic data. Spectral features such as FT-IR data also confirm that the SO_4^{2-} ion is distorted from the ideal tetrahedral shape. The degree of distortion becomes less from Pr to Tb, as noted from the S–O distances and the O–S–O dihedral angles (Table 3) (Fig. 14).

Table 3 The S–O interatomic distances (Å) of sulfate ions in $RE_2(OH)_4SO_4 \cdot 2H_2O$

Ln	S–O1	S–O2	Difference between two S–O1 distances
Pr	1.375(8)/1.467(9)	1.477(7)	0.092
Nd	1.387(9)/1.473(9)	1.481(8)	0.086
Sm	1.402(8)/1.451(9)	1.484(8)	0.049
Eu	1.431(8)/1.408(9)	1.476(8)	0.023
Gd	1.415(9)/1.397(11)	1.468(9)	0.018
Tb	1.414(6)/1.438(6)	1.499(6)	0.024

Adapted with permission from [34]. Copyright 2011 American Chemical Society

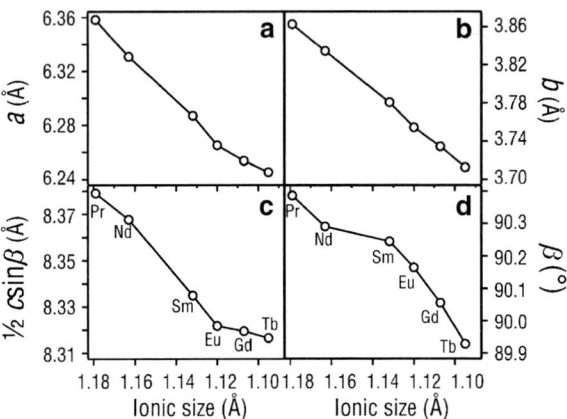

Fig. 14 Unit cell parameters of $RE_2(OH)_4SO_4 \cdot 2H_2O$ versus the ion size of RE^{3+}. Reprinted with permission from [34]. Copyright 2011 American Chemical Society

7 Photoluminescence Properties and Potential Applications

The photoluminescence properties of rare earth compounds are useful for their sharp line-like emissions, which generates light of a high color purity. The LREH compounds can accommodate various RE luminescent centers, for example, Eu^{3+} or Tb^{3+}, and display corresponding characteristic emissions at room temperature. The LREH-I compounds are anion exchangeable, which enables incorporation of anionic sensitizers or quenchers to tune the photoluminescence behavior. In the LREH-II compounds where anionic species are directly connected to the luminescent centers through robust covalent bonds, the photoluminescence behaviors can be finely tuned by selecting a proper ligand. On the other hand, it is known that water molecules and hydroxyl groups have quenching effects on the emission, showing a disadvantage of hydroxide materials for PL applications. High-performance phosphors can be acquired by posttreatment of the LREH compounds to oxides or another form. This section concerns the photoluminescence behaviors of LREH compounds and their derivatives.

7.1 Photoluminescence Features of LREH Compounds

Eu- and Tb-based LREH-I compounds show typical red and green emission, respectively, at room temperature. In a typical example of $Eu(OH)_{2.5}Cl_{0.5}\cdot 0.8H_2O$ [23], the excitation spectrum monitored at the $^5D_0-^7F_2$ lines (612 nm) showed a series of sharp lines ascribable to intra-$4f^6$ transitions within the $4f^6$ electronic configuration of the Eu^{3+} ions. The emission spectrum displayed typical $^5D_0-^7F_J$ ($J = 0-4$) transitions at 578, 595, 612, 649, and 697 nm. The emission from higher excited levels such as $^5D_1$ was not detected, indicating that the nonradiative relaxation to the $^5D_0$ level was efficient (Fig. 15).

It is known that the $^5D_0-^7F_J$ emission lines are sensitive to the local environment of the Eu^{3+} ions. The relative intensity of each line and the splitting of the emission peaks are often used to probe the local symmetry of the crystal field. Theoretically, if a Eu^{3+} ion is located in an inversion symmetric site, the magnetic dipole transition of $^5D_0-^7F_1$ (approximately at 590 nm) will be dominant. However, if the electric dipole transition $^5D_0-^7F_2$ (approximately in 610–620 nm) is prominent, the ion should be located in a site without inversion symmetry. The intense emission at 612 nm suggests that the Eu ions are accommodated in a site without inversion symmetry. Moreover, the $^5D_0-^7F_1$ and $^5D_0-^7F_2$ emissions show three- and twofold

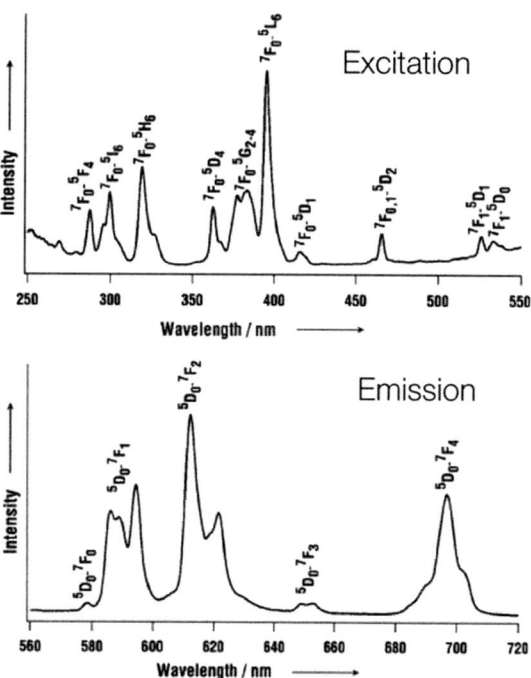

Fig. 15 Excitation and emission spectra of $Eu(OH)_{2.5}Cl_{0.5}\cdot 0.8H_2O$. Adapted with permission from [23]. Copyright 2008 Wiley-VCH Verlag GmbH & Co. KGaA

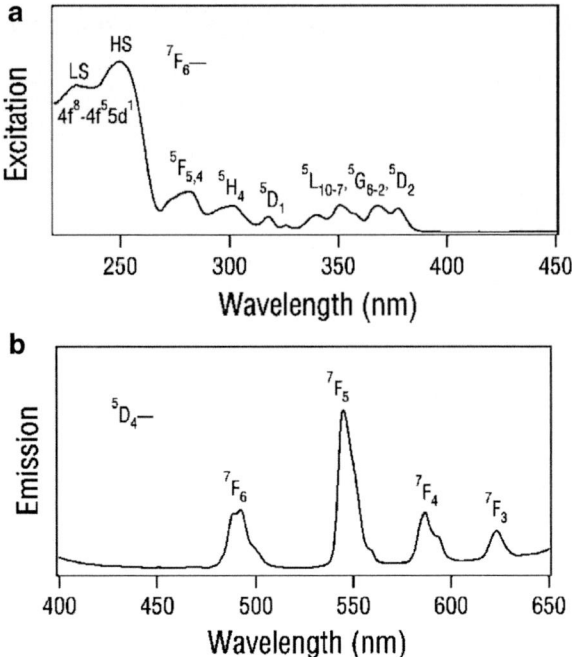

Fig. 16 The excitation and emission spectra of Tb(OH)$_{2.5}$Cl$_{0.5}$·0.8H$_2$O. Adapted with permission from [32]. Copyright 2008 American Chemical Society

splits, respectively, implying a low symmetric surrounding of the Eu^{3+} ions. This discussion is consistent with the crystallographic data, in which Eu^{3+} ions occupy sites with local symmetry groups of C_1 and C_{4v}.

In another example, green emissions of Tb^{3+} can be observed from Tb(OH)$_{2.5}$Cl$_{0.5}$·0.8H$_2$O [32]. The excitation spectrum monitored within the intra-4f^8 ^{5}D$_4$–^{7}F$_5$ lines was composed of two UV broadbands approximately at 230 and 250 nm and the lines straight from the Tb^{3+} ions. The two UV bands can be assigned to the spin-allowed (low-spin, LS) and spin-forbidden (high-spin, HS) interconfigurational Tb^{3+} f–d transitions. When the compound was excited into the UV band or directly into the intra-4f^8 levels, emissions corresponding to ^{5}D$_4$–^{7}F$_J$ ($J = 3$–6) transitions of the Tb^{3+} ion, 484, 543, 589, and 626 nm, were observed. The splitting of these emission lines may be associated with a low-symmetry coordination environment (Fig. 16).

To optimize the photoluminescence performance, Gd(OH)$_{2.5}$Cl$_{0.5}$·0.9H$_2$O was selected as a host lattice to accommodate the red emitter of Eu^{3+} in a varied concentration [35–37]. XRD data indicated the formation of the solid solution of Eu$_x$Gd$_{1-x}$(OH)$_{2.5}$Cl$_{0.5}$·0.9H$_2$O in the full composition range. The excitation spectrum showed both the intra 4f^6 transitions of Eu^{3+} and the transitions from the ground level ^{8}S$_{7/2}$ to the excited levels ^{6}D$_J$, ^{6}I$_J$, and ^{6}P$_J$ of Gd^{3+}. Upon excitation by the ^{8}S$_{7/2}$–^{6}I$_J$ transition of Gd^{3+} at 273 nm, all samples exhibited the typical ^{5}D$_0$–^{7}F$_J$

($J = 0$–4) radiative relaxational transitions of Eu^{3+} at 579, 595, 615, 651, and 701 nm. This feature indicates the effective energy transfer from Gd^{3+} to Eu^{3+}. Upon UV excitation, the energy is absorbed by the sample, inducing an electronic transition of Gd^{3+} from the ground state to the excited levels. Some of the electrons in the excited levels of Gd^{3+} are transferred to the Eu^{3+} ions followed by the $^5D_0$–^7F_J transition and result in the emission. The emission through energy transfer from Gd^{3+} to Eu^{3+} is concentration dependent. The intensity increased sharply as the Eu^{3+} content reached the optimal fraction of 0.05 and then steadily decreased upon further increasing the dopant. This phenomenon may be explained by the decrease in the effective absorption cross section of Gd^{3+} with higher Eu^{3+} content and also the concentration quenching effect. In contrast, the emission through direct excitation of Eu^{3+} at 393 nm gradually increased with the Eu^{3+} content, preferring the relaxation process over the nonradiative relaxation process through the concentration quenching effect. The intensity was nearly equal at a Eu^{3+} content of 0.3 for the two excitation pathways.

In another study, the photoluminescence properties of $(Y_{1-x}Ln_x)(OH)_{2.5}(NO_3)_{0.5} \cdot nH_2O$ (Ln = Tb or Eu) samples containing binary or ternary RE elements were studied in relation to their varied composition [44]. In addition to the excitation and emission spectra, the lifetime of photoluminescence emission was evaluated from decay profiles. The lifetimes of the Tb-containing samples were observed in the range of 0.073–0.0904 ms and for the Eu-containing samples were between 0.064 and 0.241 ms. For both systems, the lifetimes decreased with increasing dopant concentration (Table 4) (Fig. 17).

It has been demonstrated that the excitation and emission of Tb $(OH)_{2.5}Cl_{0.5} \cdot 0.8H_2O$ and $Tb(OH)_{2.5}(NO_3)_{0.5} \cdot nH_2O$ show nearly identical features [33]. The LREH-I-type compounds show anion-exchange properties. These inorganic anions are simply accommodated between the host layers by electrostatic interactions as well as hydrogen bonds. The absence of a direct bond to the phosphor centers leads to little effects on the photoluminescence properties.

$Eu_2(OH)_4SO_4 \cdot 2H_2O$ and $Tb_2(OH)_4SO_4 \cdot 2H_2O$ [39] showed characteristic red and green emissions, respectively. The excitation spectrum of the Eu sample monitored within the $^5D_0$–$^7F_2$ line showed two strong, broad peaks located in the deep UV band and a series of lines to intra-$4f^6$ transitions from the typical Eu^{3+} $4f^6$ electronic

Table 4 Photoluminescence decay data of the $(Y_{1-x}Ln_x)(OH)_{2.5}(NO_3)_{0.5} \cdot nH_2O$ samples

$(Y_{1-x}Tb_x)(OH)_{2.5}(NO_3)_{0.5} \cdot nH_2O$		$(Y_{1-x}Eu_x)(OH)_{2.5}(NO_3)_{0.5} \cdot nH_2O$	
x	τ (ms)	x	τ (ms)
0.1	0.904	0.1	0.092
0.3	0.891	0.4	0.086
0.5	0.823	0.5	0.049
0.6	0.265	0.6	0.023
0.8	0.109	0.8	0.018
1.0	0.073	1.0	0.024

Adapted from [44] with the permission of The Royal Society of Chemistry

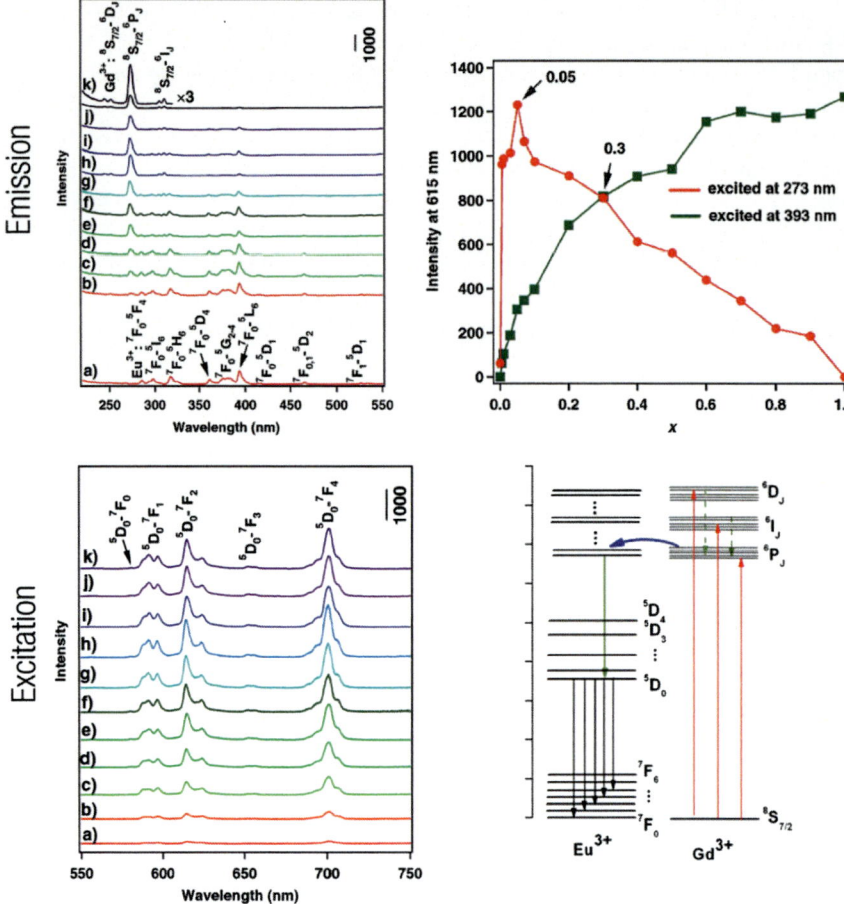

Fig. 17 The excitation and emission spectra of the solid solution of $Eu_xGd_{1-x}(OH)_{2.5}Cl_{0.5}\cdot 0.9H_2O$. The emission intensity of the 615 line versus the Eu^{3+} content and the energy levels of the Gd^{3+}–Eu^{3+} system are shown. Adapted with permission from [38]. Copyright 2010 American Chemical Society

configuration. Similar broad UV peaks were observed in a Eu^{3+}-doped YVO_3 or Gd_2O_3 matrix, which was attributed to the charge transfer from O^{2-} to Eu^{3+} [39]. Very likely, the sulfate groups may contribute to the charge transfer process. Upon excitation, typical $^5D_0$–^7F_J ($J=0$–4) transitions at 581, 589, 618, 652, and 698 nm were observed. Compared with $Eu(OH)_{2.5}Cl_{0.5}\cdot 0.8H_2O$, the line of $^5D_0$–$^7F_2$ showed a red shift by approximately 6 nm. The strong interaction between the sulfate and the Eu^{3+} ions may be responsible for this line shift. In the excitation spectrum of the Tb sample, strong spin-allowed (low-spin, LS) and spin-forbidden (high-spin, HS) interconfigurational Tb^{3+} 4f–5d transition bands were observed. The LS transition band was much stronger than the HS band, different from that in

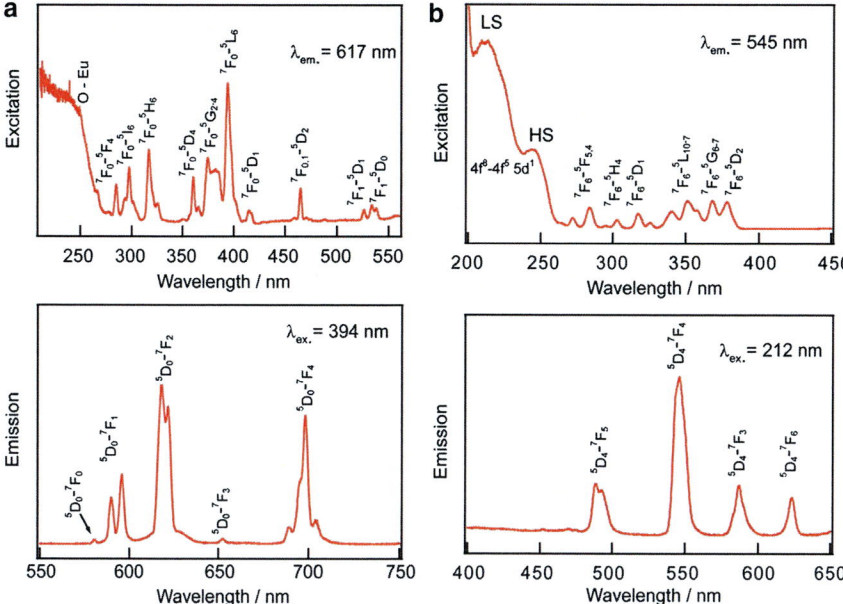

Fig. 18 The excitation and emission spectra of $Eu_2(OH)_4SO_4 \cdot 2H_2O$ (**a**) and $Tb_2(OH)_4SO_4 \cdot 2H_2O$ (**b**). Adapted with permission from [39]. Copyright 2010 American Chemical Society

$Tb(OH)_{2.5}Cl_{0.5} \cdot 0.8H_2O$ and $Tb(OH)_{2.5}(NO_3)_{0.5} \cdot 0.8H_2O$. A sharp green emission of 545 nm for Tb^{3+} was observed under UV excitation at 212 nm (Fig. 18).

$La_2(OH)_4[O_3S(CH_2)_nSO_3] \cdot 2H_2O$ was selected as a host lattice to accommodate the luminescent center of Eu^{3+} [40]. This sample showed strong O–Eu charge transfer and red emissions similar to $Eu_2(OH)_4SO_4 \cdot 2H_2O$. The $^5D_0$–$^7F_2$ line was narrow, in contrast to the splitting lines for $Eu_2(OH)_4SO_4 \cdot 2H_2O$, suggesting that the luminescent behaviors can be tuned with anions in this series. In addition, it was found that Ce^{3+} ions could be stabilized in the $La_2(OH)_4[O_3S(CH_2)_nSO_3] \cdot 2H_2O$ host. The sample showed intense emission in the near UV region, a typical feature of Ce^{3+} ions.

7.2 Oxide Phosphors Derived from LREH Compounds

Rectangular microcrystals of $(Gd_{0.95}Eu_{0.05})(OH)_{2.5}Cl_{0.5} \cdot 0.9H_2O$ were synthesized through the homogeneous alkalization reaction. These crystals could be transferred onto a substrate after trapping them at a hexane/water interface. This unique process produced a monolayer film composed of densely packed platelet crystals with their faces parallel to the substrate [36]. Upon heating at 200–1,000 °C, the hydroxide film underwent a quasi-topotactic transformation into a film of RE_2O_3. The resulting oxide film was highly oriented along the (111) axis as a consequence of

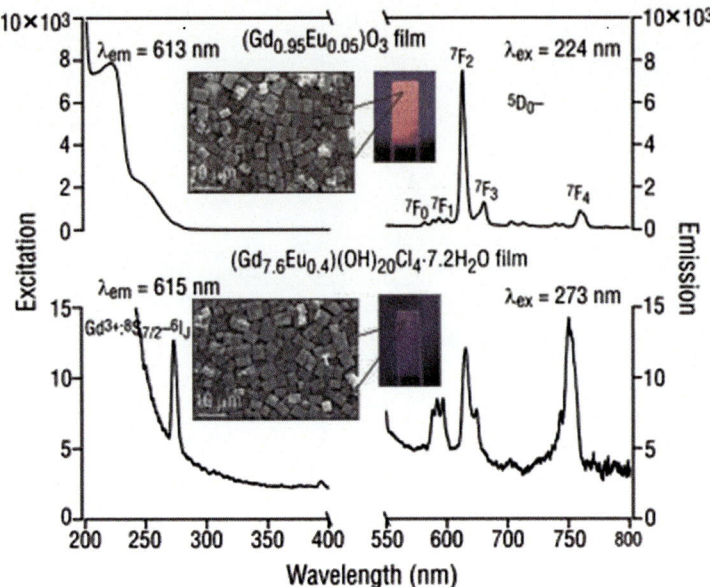

Fig. 19 The excitation and emission spectra for a film of $(Gd_{0.95}Eu_{0.05})(OH)_{2.5}Cl_{0.5} \cdot 0.9H_2O$ (*bottom*) and $(Gd_{0.95}Eu_{0.05})_2O_3$ (*top*). Insets are the corresponding SEM images and emission photographs under ultraviolet irradiation. Adapted with permission from [36]. Copyright 2009 Wiley-VCH Verlag GmbH & Co. KGaA

this unique process. The red emission was greatly enhanced by a factor ~500 compared with that of the precursor (Fig. 19).

$(La_{0.95}Eu_{0.05})_2O_2SO_4$ was derived from $(La_{0.95}Eu_{0.05})_2(OH)_4SO_4 \cdot 2H_2O$ through low-temperature dehydration and dehydroxylation reactions [39, 45]:

$$Ln_2(OH)_4SO_4 \cdot 2H_2O \rightarrow Ln_2(OH)_4SO_4 + 2H_2O \ (230 - 315 \,°C)$$
$$Ln_2(OH)_4SO_4 \rightarrow Ln_2O_2SO_4 + H_2O \ (315 - 400 \,°C)$$

$Ln_2O_2SO_4$ is usually synthesized by heating $Ln_2(SO_4)_3 \cdot 6H_2O$ or oxidizing Ln_2S_3 at high temperatures [41]. The high processing temperature (over 800 °C) and release of SO_2 through these routes are neither economically nor environmentally desirable. Apparently, the route derived from the $Ln_2(OH)_4SO_4 \cdot 2H_2O$ precursor is environment-friendly, proceeding at a much lower temperature to produce the target phase. FT-IR spectra revealed that both the precursor and the target phase have strong RE^{3+}–SO_4^{2-} chemical bonding, which should be the key reason for the low-temperature transformation. The product showed strong emission of 620 nm excited by UV light of 280 nm, the typical $^5D_0$–$^7F_2$ transition of Eu^{3+} ions (Fig. 20).

Thermal treatment of $La_2(OH)_4[O_3S(CH_2)_3SO_3] \cdot 2.2H_2O$ [40] can lead to hybrid oxides of $Ln_2O_2[O_3S(CH_2)_nSO_3]$. The sample crystallizes in an orthorhombic structure with cell parameters of $a = 0.8858(1)$, $b = 0.4250(1)$, and $c = 2.4136(4)$ nm. Parameters a and b have values close to those in $La_2O_2SO_4$, a layered

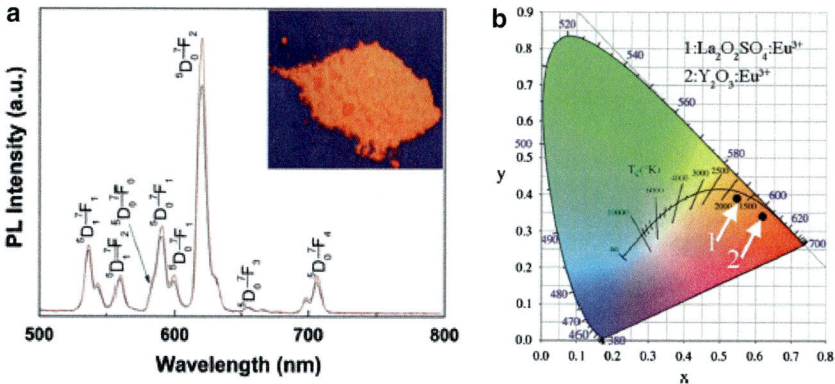

Fig. 20 The emission spectrum of $(La_{0.95}Eu_{0.05})_2O_2SO_4$ derived from corresponding $(La_{0.95}Eu_{0.05})_2(OH)_4SO_4 \cdot 2H_2O$ precursor and corresponding CIE chromaticity diagram. Adapted with permission from [45]

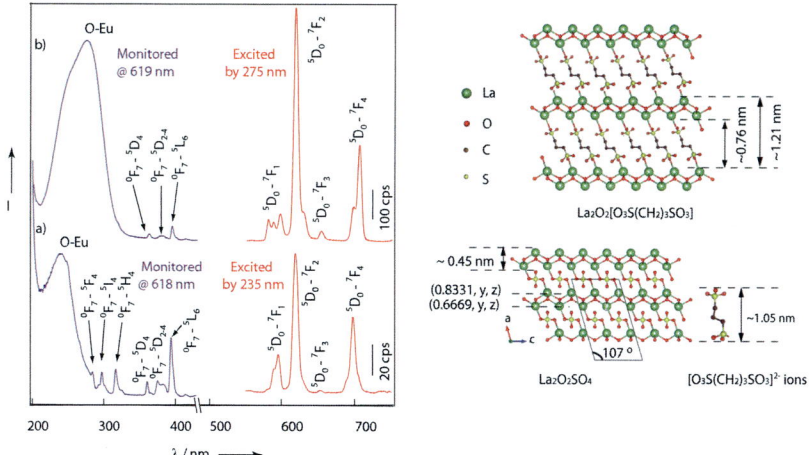

Fig. 21 Crystal structure and room temperature excitation and emission spectra for samples with doping of 5 % Eu^{3+} (**a**) $La_2(OH)_4[O_3S(CH_2)_3SO_3] \cdot 2.2H_2O$ and (**b**) $La_2O_2[O_3S(CH_2)_3SO_3]$. Adapted with permission from [40]. Copyright 2013 American Chemical Society

oxide containing infinite $\{[La_2O_2]^{2+}\}_\infty$ layers [46] [cell parameters: $a = 1.4342(1)$, $b = 0.42827(3)$, $c = 0.83853(7)$ nm, and $\beta = 107.0(1)°$]. Based on the unit cell parameter, similar $\{[La_2O_2]^{2+}\}_\infty$ layers are believed to be present in the $La_2O_2[O_3S(CH_2)_3SO_3]$ compound. The Eu^{3+}-doped hybrid oxide showed a strong emission of 619 nm when excited by UV light of 275 nm (Fig. 21).

7.3 Phosphors Based on Exfoliated Nanosheet Crystallites

LREH-I compounds can undergo swelling and exfoliation via intercalation of DS^- ions followed by treating with formamide [37]. Vigorous agitation or ultrasonic treatment helps drive the exfoliation to a completion. Atomic force microscopy (AFM) images detected extremely thin 2D crystallites with a very flat terrace after drying a diluted sample on a substrate. The observed thickness of 1.6 nm provides clear evidence for the formation of unilamellar nanosheets. In-plane XRD data showed sharp diffraction peaks compatible with the 2D lattice from the original layered compounds, confirming the intactness of the 2D layer architecture. The nanosheet suspension based on Eu^{3+} ions showed characteristic PL features similar to the precursor powder (Fig. 22).

The unilamellar nanosheets are positively charged and can be deposited onto a quartz glass by the layer-*by*-layer sequential adsorption process [29, 37]. A thin film of Eu^{3+}-based hydroxide nanosheets prepared by this method showed the typical red emission similar to that from the suspension or the pristine powder, although very weak due to its tiny quantity. A hetero-assembled film of titanium oxide and rare earth hydroxide nanosheets was found to show a drastic red emission under UV light excitation (260 nm). The titanium oxide nanosheet is believed to act like an antenna, collecting the UV light and promoting the Eu^{3+} emission [29] (Fig. 23).

A sample of $[Gd_2(OH)_5(H_2O)_x]Cl$ can be dispersed in water, forming a translucent suspension after intensive ultrasonic treatments [47–52]. The colloidal particles thus obtained can be combined with polyoxometalate anions of paradodecatungstate or polyoxomolybdate to form hierarchical microparticles. The anions were found to enhance the red emission of Eu^{3+} and the green emission of Tb^{3+} in the microparticles. Colloidal particles of $[Eu_2(OH)_5(H_2O)_x]Cl$ and $[Tb_2(OH)_5(H_2O)_x]Cl$ [47] can be transferred onto substrates by the layer-*by*-layer adsorption method, forming multilayer or hetero-assembled films. After annealing, transparent films with tunable PL features were readily obtained (Fig. 24).

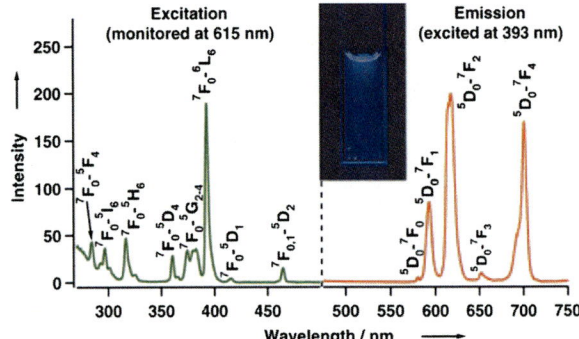

Fig. 22 The excitation and emission spectra of the suspension of nanosheets exfoliated from a $Eu_2(OH)_5DS \cdot nH_2O$ sample. Adapted with permission from [37]. Copyright 2010 Wiley-VCH Verlag GmbH & Co. KGaA

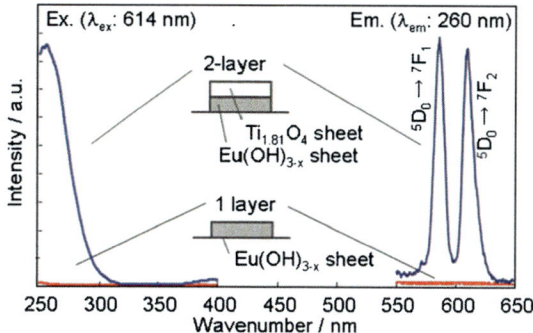

Fig. 23 The excitation and emission spectra of hetero-assembled film of titanium oxide and rare earth hydroxide nanosheets. Adapted from [29] with the permission of The Royal Society of Chemistry

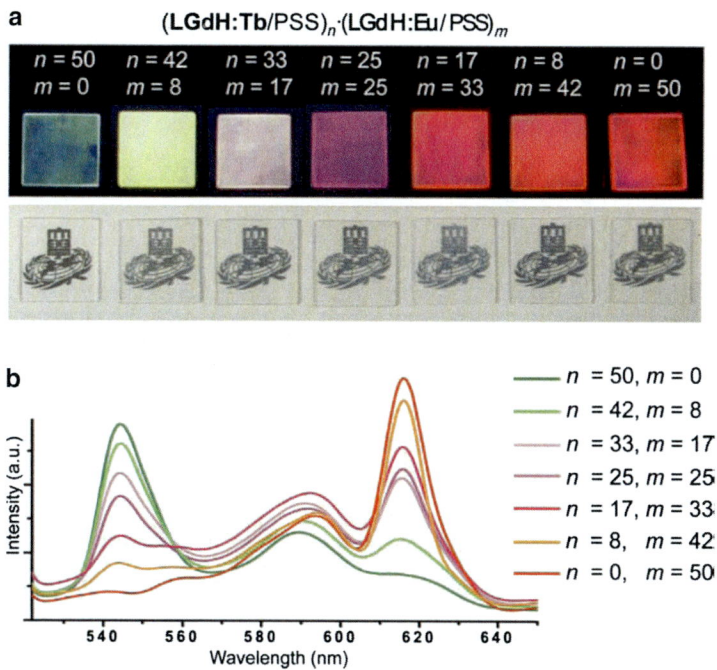

Fig. 24 The emission feature and photograph of transparent films prepared from colloidal particles of $[Eu_2(OH)_5(H_2O)_x]Cl$ and $[Tb_2(OH)_5(H_2O)_x]Cl$. Adapted with permission from [47]. Copyright 2010 Wiley-VCH Verlag GmbH & Co. KGaA

8 Summary and Outlook

This chapter overviews the background and recent advances in the development of LREH compounds, including the synthetic strategy, structural characterizations, and photoluminescence properties. LREH compounds have been recognized as a new type of cationic layered host besides the well-known LDH compounds. In the new materials, the intercalation chemistry is coupled with the unique physicochemical properties of RE elements, opening a new avenue for the design of novel multifunctional phosphors by rationally selecting RE emitting centers and anionic counter species. This is just the beginnings of the topic, and we can expect many of opportunities in this emerging field.

The structural knowledge of LREH compounds is of critical importance to understanding their fundamental properties. Though the basic lamellar topological features of LREH compounds have been demonstrated, some important aspects, for example, the hydrogen bonding networks, which will have a high impact on the intercalation activity and host–guest interaction, have not yet been thoroughly understood. Such knowledge relies on the synthesis of highly crystalline single-crystal samples for structural characterization. In this regard, improvement and development of synthetic technologies are desired to produce LREH samples with high crystallinity, particularly, single-crystal samples to satisfy structural characterizations. On the other hand, it is difficult at present to produce LREH compounds with a desired composition, for example, the synthesis of compounds across the whole RE series from La to Lu. New synthetic strategies are needed to yield such samples.

We anticipate rich opportunities when the properties of RE elements are combined with the intercalation properties of LREH compounds. Novel optical materials besides the red and green phosphors, for example, yellow, blue, or even white emission phosphors, up- or down-conversion materials, and optically active chiral crystals may be expected by incorporating novel anionic species into the rare earth hydroxide layers. Synergistic effects are expected from the rare earth hydroxide host lattice and the guest anions, which may offer great potential for the design of novel optical sensors or devices. We highlight the potential of the exfoliated nanosheet crystallites, an unusual type of 2D macromolecule-like crystals featuring positive charges, as versatile building blocks to design novel optical nanocomposites and artificial hetero-assembled structures.

References

1. Geng F, Ma R, Sasaki T (2010) Acc Chem Res 43:1177
2. Liang J, Ma R, Sasaki T (2014) Dalton Trans 43:10355
3. Gándara F, Perles J, Snejko N, Iglesias M, Gómez-Lor B, Gutiérrez-Puebla E, Monge MÁ (2006) Angew Chem Int Ed 45:7998
4. Aksel'rud NV, Ermolenko VI (1957) Zh Neorgan Khim 2:2709

5. Aksel'rud NV, Spivakovskii VB (1958) Zh Neorgan Khim 4:56
6. Aksel'rud NV (1963) Russ Chem Rev 32:353
7. Klevtsov PV, Klevtsova RF, Sheina LP (1965) Zh Strukt Khim 6:469
8. Klevtsova RF, Klevtsov PV (1966) Zh Strukt Khim 7:556
9. Klevtsova RF, Gllnskaya LA (1969) Zh Strukt Khim 10:494
10. Klevtsov PV (1969) Zh Strukt Khim 10:498
11. Carter FL, Levinson S (1969) Inorg Chem 8:2788
12. Tarkhova TN, Mironov NN, Grishin IA (1970) Zh Strukt Khim 11:556
13. Mullica DF, Sappenfield EL, Grossie DA (1986) J Solid State Chem 63:231
14. Haschke J, Eyring L (1971) Inorg Chem 10:2267
15. Haschke J (1974) Inorg Chem 13:1812
16. Haschke J (1975) J Solid State Chem 12:115
17. Haschke JM (1975) J Solid State Chem 14:238
18. Lance ET, Haschke JM (1976) J Solid State Chem 17:55
19. Lance-Gómez ET, Haschke JM (1978) J Solid State Chem 23:275
20. Lance-Gomez ET, Haschke JM (1980) J Solid State Chem 35:357
21. Haschke JM (1988) J Solid State Chem 73:71
22. McIntyre LJ, Jackson LK, Fogg AM (2008) Chem Mater 20:335
23. Geng F, Xin H, Matsushita Y, Ma R, Tanaka M, Izumi F, Iyi N, Sasaki T (2008) Chem Eur J 14:9255
24. Poudret L, Prior TJ, McIntyre LJ, Fogg AM (2008) Chem Mater 20:7447
25. McIntyre LJ, Jackson LK, Fogg AM (2008) J Phys Chem Solids 69:1070
26. Hindocha SA, McIntyre LJ, Fogg AM (2009) J Solid State Chem 182:1070
27. McIntyre LJ, Prior TJ, Fogg AM (2010) Chem Mater 22:2635
28. Southworth FY, Wilson C, Coles SJ, Fogg AM (2014) Dalton Trans 43:10451
29. Ida S, Sonoda Y, Ikeue K, Matsumoto Y (2010) Chem Commun 46:877
30. Gándara F, Puebla EG, Iglesias M, Proserpio DM, Snejko N, Monge MÁ (2009) Chem Mater 21:655
31. Louër D, Louër M (1987) J Solid State Chem 68:292
32. Geng F, Matsushita Y, Ma R, Xin H, Tanaka M, Izumi F, Iyi N, Sasaki T (2008) J Am Chem Soc 130:16344
33. Geng F, Matsushita Y, Ma R, Xin H, Tanaka M, Iyi N, Sasaki T (2009) Inorg Chem 48:6724
34. Geng F, Ma R, Matsushita Y, Liang J, Michiue Y, Sasaki T (2011) Inorg Chem 50:6667
35. Hu L, Ma R, Ozawa TC, Geng F, Iyi N, Sasaki T (2008) Chem Commun 4897
36. Hu L, Ma R, Ozawa TC, Sasaki T (2009) Angew Chem Int Ed 48:3846
37. Hu L, Ma R, Ozawa TC, Sasaki T (2010) Chem Asian J 5:248
38. Hu L, Ma R, Ozawa TC, Sasaki T (2010) Inorg Chem 49:2960
39. Liang J, Ma R, Geng F, Ebina Y, Sasaki T (2010) Chem Mater 22:6001
40. Liang J, Ma R, Ebina Y, Geng F, Sasaki T (2013) Inorg Chem 52:1755
41. Machida M, Kawamura K, Kawano T, Zhang D, Ikeue K (2006) J Mater Chem 16:3084
42. Lee K-H, Byeon S-H (2009) Eur J Inorg Chem 2009:4727
43. Nilges T (2006) Z Naturforsch 61b:117
44. Wang L, Yan D, Qin S, Li S, Lu J, Evans D, Duan X (2011) Dalton Trans 40:11781
45. Wang X, Li J-G, Zhu Q, Li X, Sun X, Sakka Y (2014) J Alloys Compd 603:28
46. Zhukov S, Yatsenko A, Chernyshev V, Trunov V, Tserkovnaya E, Antson O, Holsa J, Baules P, Schenk H (1997) Mater Res Bull 32:43
47. Yoon Y-S, Byeon S-H, Lee I (2010) Adv Mater 22:3272
48. Lee B-I, Byeon S-H (2011) Chem Commun 47:4093
49. Lee B-I, Lee S-Y, Byeon S-H (2011) J Mater Chem 21:2916
50. Jeong H, Lee B-I, Byeon S-H (2012) Dalton Trans 41:14055
51. Lee B-I, Bae J, Lee E, Byeon S-H (2012) Bull Korean Chem Soc 33:601
52. Lee B-I, Jeong H, Byeon S-H (2014) Eur J Inorg Chem 2014:3298

Layered Double Hydroxide Materials in Photocatalysis

Mingfei Shao, Min Wei, David G. Evans, and Xue Duan

Contents

1 Introduction 106
2 The Preparation of LDH-Based Photocatalysts 107
   2.1 The Construction of Photocatalytic Host Layers 107
   2.2 The Sensitization with Interlayer Guest Molecules 112
   2.3 LDH-Based Nanocomposites for Photocatalysis 114
3 The Applications of LDH-Based Photocatalysts 121
   3.1 Photocatalytic Degradation of Pollutants 122
   3.2 Photocatalytic Water Splitting 126
   3.3 Photocatalytic CO_2 Reduction 131
4 Summary and Outlook 133
References 134

Abstract Layered double hydroxide (LDH)-based photocatalysts have attracted great attention in the fields of environment and energy (e.g., degradation of pollutants, water splitting for solar fuel production), owing to their unique intercalation structure with highly dispersed metal cations and exchangeable anions, large specific surface areas, and remarkable adsorption capacities. This chapter aims to review and summarize the recent advances in the synthesis and photocatalytic applications of LDH-based materials. Typically, several important strategies have been developed for the fabrication of LDH-based photocatalysts by tuning the composition of the host layers, intercalating guest sensitizers, and constructing nanocomposites. The obtained photocatalysts exhibit excellent performances in the fields of pollutant degradation, water splitting, and reduction of CO_2 into carbon sources. The fabrication and application of LDH-based photocatalysts represent a promising direction in the development of LDH-based multifunctional materials, which will contribute to the progress of chemistry and material science.

M. Shao • M. Wei (✉) • D.G. Evans • X. Duan
State Key Laboratory of Chemical Resource Engineering, Beijing University of Chemical Technology, Beijing 100029, China
e-mail: weimin@mail.buct.edu.cn

Keywords Energy conversion • Environmental remediation • Layered double hydroxide • Photocatalysis

1 Introduction

The development of non-pollution and low-energy-consumption technologies for environmental remediation and new energy resources is an urgent task for the sustainable development of human society. With wide applications in solar energy conversion and environmental remediation (water purification and decontamination, deodorization of air, etc.), photocatalysis using semiconductors such as TiO_2 [1], ZnO [2], SnO_2 [3], Bi_2WO_6 [4], and CdS [5] have been extensively investigated. However, the easy charge (electron/hole) recombination and limited light absorption because of the relatively large band gap for most semiconductors greatly decrease their photocatalytic efficiency. In the past decade, many efforts have been developed to improve the performance of photocatalysts by nonmetal or metal doping [6, 7], morphological control [8], as well as constructing semiconductor-based composites [9–11]. Despite all this progress, the design and development of visible-light-driven photocatalysts with tunable band gap and largely enhanced photoconversion efficiency still present a significant challenge to scientists and engineers.

Layered double hydroxides (LDHs) are a large class of typical inorganic layered host materials which can be described by the general formula $[M^{II}_{1-x}M^{III}_{x}(OH)_2]^{z+}(A^{n-})_{z/n} \cdot yH_2O$ (M^{II} and M^{III} are divalent (e.g., Mg^{2+}, Zn^{2+}, Fe^{2+}, Co^{2+}, Cu^{2+}, or Ni^{2+}) and trivalent metals (e.g., Al^{3+}, Fe^{3+}, Cr^{3+}, Ga^{3+}, or In^{3+}), respectively; A^{n-} is the interlayer inorganic or organic anion compensating for the positive charge of the brucite-like layers) [12]. In addition to divalent and trivalent cations, a wide range of cations in monovalence or higher valence (e.g., Li^+, Ti^{4+}, Sn^{4+}, Zr^{4+}, etc.) may also be accommodated in the octahedral sites of host layers. The wide tunability of metal ions without altering the structure as well as anion-exchange properties of LDH materials makes them attractive candidates in catalysis [13], adsorption/separation [14], biology [15], energy storage, and conversion [16]. Recently, LDHs have been intensively investigated as promising heterogeneous photocatalysts because of their excellent intrinsic photo-response characteristics, low cost, as well as facile preparation and modification [17]. LDHs as photocatalysts are expected to possess large energy-conversion efficiency as a result of the high dispersion of active species in a layered matrix, which facilitates the charge separation. Many LDH-based photocatalysts have been developed via appropriate incorporation of photoactive components into LDH host layers, sensitizing by interlayer functional molecules and construction of nanocomposites. In addition to the development in the preparation of LDH-based photocatalysts, the applications have also been enriched from the elimination of organic pollutants and water splitting to the photocatalytic reduction of CO_2 to organic species. In this chapter, the synthesis, modification, and photocatalytic applications of the LDH-based photocatalysts have been reviewed in detail. A deep insight into understanding the role of LDH materials for the photocatalysis has also been discussed.

2 The Preparation of LDH-Based Photocatalysts

2.1 The Construction of Photocatalytic Host Layers

An advantage of LDHs is that the type of metal ions and atomic ratio between the di- and tri- or tetravalent cations in the host layer can be varied in a wide range without altering the structure of material. In recent years, researchers have found that the Ti-containing LDHs (Fig. 1) exhibit good performance in the photocatalysis [18]. Consequently, various di- and trivalent metal ions have been introduced into the Ti-containing LDHs for maximizing their photocatalytic performances. He et al. [19] reported the synthesis of NiTi–LDH for the first time by homogeneous precipitation method utilizing urea hydrolysis. The results indicate that Ti^{4+} cations are incorporated in the host layers with cyanate as the interlayer anion (Fig. 2), but the crystallinity decreases upon increasing Ti^{4+} dosage in the synthesis mixture. Ti-containing semiconductor nanocomposite can be further obtained from calcination of NiTi–LDH precursor, which shows promising photocatalytic properties [20]. It is reported that the size, shape, and morphology of semiconductors are key factors in determining their efficiency of charge separation and the photoconversion capability [4, 10, 21]. The small size of NiTi–LDH nanosheets with lateral dimensions in the range 30–60 nm has been prepared using a reverse microemulsion method, which exhibit excellent photocatalytic activity for oxygen evolution from water using visible light [22]. ZnTi–LDHs with different Zn/Ti ratio have been synthesized by coprecipitation of zinc and titanium salts from homogeneous solution, which were demonstrated as efficient visible-light photocatalysts [23]. Recently, several other Ti-containing LDH photocatalysts were also prepared, including MgTi–LDH [24], MgAlTi–LDH [25], and ZnAlTi–LDH [26], which displays promising photocatalytic activity for the water splitting.

Except for Ti-containing LDHs, Garcia et al. [27] firstly reported the high photocatalytic activity of ZnCr–LDHs, which further evoked research interest in LDH materials as photocatalysts. Compared with Ti-based LDHs, Cr-containing LDHs possess more efficiency in using the visible light. As shown in Fig. 3, ZnTi–

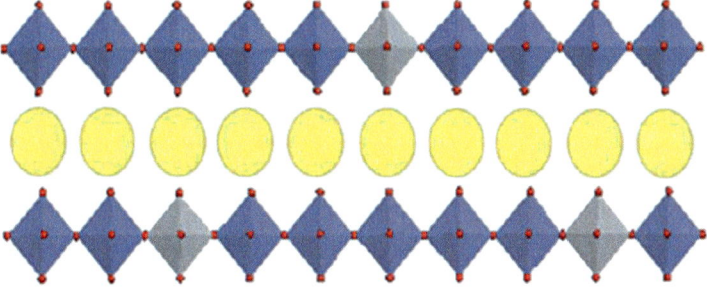

Fig. 1 Scheme of (M^{II}/Ti) LDH structure. *Blue octahedron*, M^{II}; *silver octahedron*, Ti^{4+}; *red sphere*, O; *yellow circle*, interlayer anion. Reprinted with permission from [18]. Copyright RSC

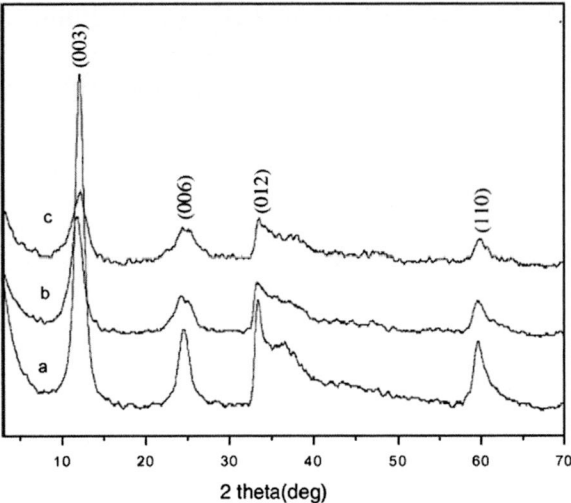

Fig. 2 Powder XRD patterns for NiTi–LDH with different molar ratios of Ni^{2+}/Ti^{4+}: (a) 5:1, (b) 5:2, (c) 3:2. Reprinted with permission from [19]. Copyright Elsevier

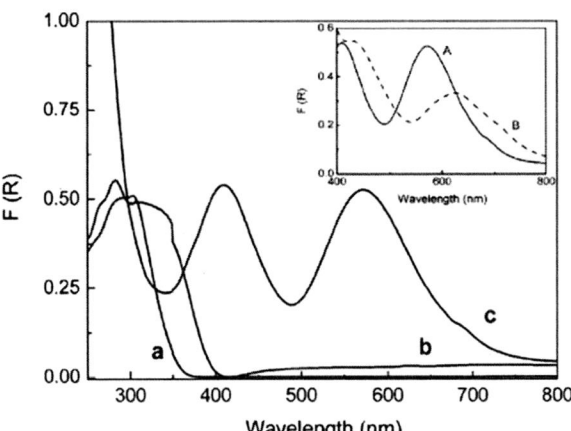

Fig. 3 Diffuse reflectance UV–vis spectra (plotted as the Kubelka–Munk function of the reflectance, R) of (a) ZnTi–LDH, (b) ZnCe–LDH, and (c) ZnCr–LDH. Inset: Diffuse reflectance UV–vis spectra of ZnCr–LDH before (A) and after (B) calcination at 450 °C. Reprinted with permission from [27]. Copyright American Chemical Society

LDH shows the most intense band occurring at 304 nm. In contrast, ZnCr–LDH exhibits two absorption maxima in the visible region at 410 and 570 nm, respectively. Various Zn/Cr atomic ratios (from 4:2 to 4:0.25) were tested for the visible-light photocatalytic oxygen generation, and the results showed that the efficiency of photocatalysts enhances asymptotically with the increase of Cr content [28]. Wei et al. [29] further reported the synthesis of visible-light-responsive MCr–X–LDHs (M = Cu, Ni, Zn; X = NO_3^-, CO_3^{2-}) by a simple and scale-up coprecipitation method. It was found that the MCr–NO_3–LDHs (M = Cu, Ni) exhibit 20 times higher photocatalytic activity than P25, owing to the high dispersion of the CrO_6 unit in the LDH matrix. Likewise, Mohapatra et al. [30] reported a series of ZnBi–LDH photocatalysts with different molar ratios of Zn/Bi. This study provided the

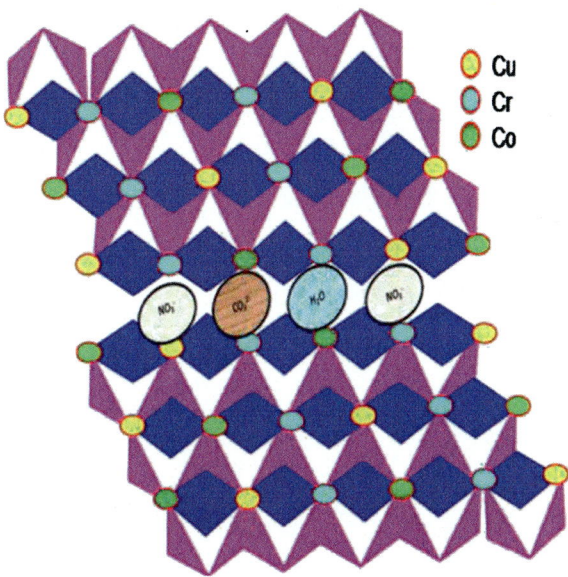

Fig. 4 A schematic diagram of CuCoCr–LDH. Reprinted with permission from [31]. Copyright American Chemical Society

evidence that the Zn(II)–O–Bi(III) units result in the generation of superoxide radicals which facilitate the photocatalytic degradation of organic pollutants.

LDHs with a cation doping capacity at the octahedral sites of brucite layers reflect the properties of a doped semiconductor toward photocatalysis. Recently, photofunctional LDHs with ternary cations were demonstrated to show synergetic results. For instance, the doping of Co^{2+} in the framework of CuCr–LDH precursor gives an enhanced photocatalytic performance. Parida et al. [31] fabricated CuCoCr–LDH with varied atomic ratio of binary cations (Cu^{2+}/Co^{2+}; the atomic ratio of (Cu + Co)/Cr is constant) and investigated the influences of Co^{2+} on structural, electronic, optical, and photocatalytic properties of CuCr–LDHs (Fig. 4). The results show that the moderate content of Co^{2+} doping definitely enhances the photocatalytic degradation of malachite green (MG). This can be attributed to the cooperative effect of binary cations and the more electron-transfer capability of Co. In addition, the amount of hydroxyl radicals can be detected by the luminescence technique using terephthalic acid (TA) as probe molecules, and the Co-doped LDH also shows more surface hydroxyl radicals. Up to now, various metal ion dopants have been introduced into the LDH host layers to tune the textural properties and photocatalytic activity, such as Fe–MgAl–LDH [32], Ni–ZnCr–LDH [33], Ga–ZnAl–LDH [34], Ti–ZnAl–LDH [26], and In–MgZn–LDH [35].

The topotactic transformation from LDHs to mixed metal oxides (MMOs) provides another effective approach for the fabrication of metal oxide nanoparticles which are grafted within an amorphous matrix, giving rise to a high dispersion of immobilized semiconductor nanoparticles since agglomeration is effectively inhibited. The large specific surface area, possible synergistic effects between the

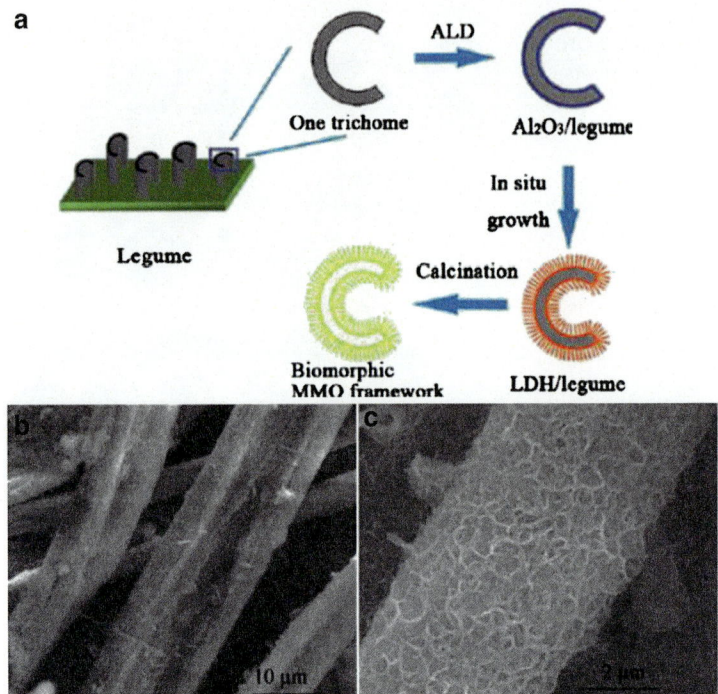

Fig. 5 (**a**) A schematic representation for the fabrication of biotemplated LDH film and MMO framework from the legume. SEM images of the in situ growth LDH/legume film at (**b**) low magnification and (**c**) high magnification. Reprinted with permission from [36]. Copyright American Chemical Society

components of MMOs further facilitates the improvement of their photocatalytic activity. Wei et al. [36] offered a mean to the fabrication of hierarchical MMO materials on a biological substrate via an atomic layer deposition (ALD) followed by in situ growth method (Fig. 5). A uniform Al_2O_3 coating was first deposited on the surface of a biotemplate with the ALD process, and the film of ZnAl–LDH was further prepared by an in situ growth technique. Subsequently, ZnAl–MMO framework was obtained by calcination of the LDH precursor, which was demonstrated as an effective and recyclable photocatalyst for the decomposition of dyes in water, owing to its rather high specific surface area and hierarchical distribution of pore size.

Wei et al. [37] further reported the preparation of ZnO nanoplatelets with a high degree of exposure of (0001) facets embedded in a hierarchical flower-like matrix using the topotactic transformation of LDH precursor (Fig. 6). It was demonstrated that ZnO nanoparticles with a high percentage exposure of (0001) facets exhibit significantly higher visible-light photocatalytic performance than other ZnO nanomaterials with fewer exposed (0001) facets. This work provides a facile approach for the fabrication of immobilized metal or metal oxide nanocatalysts

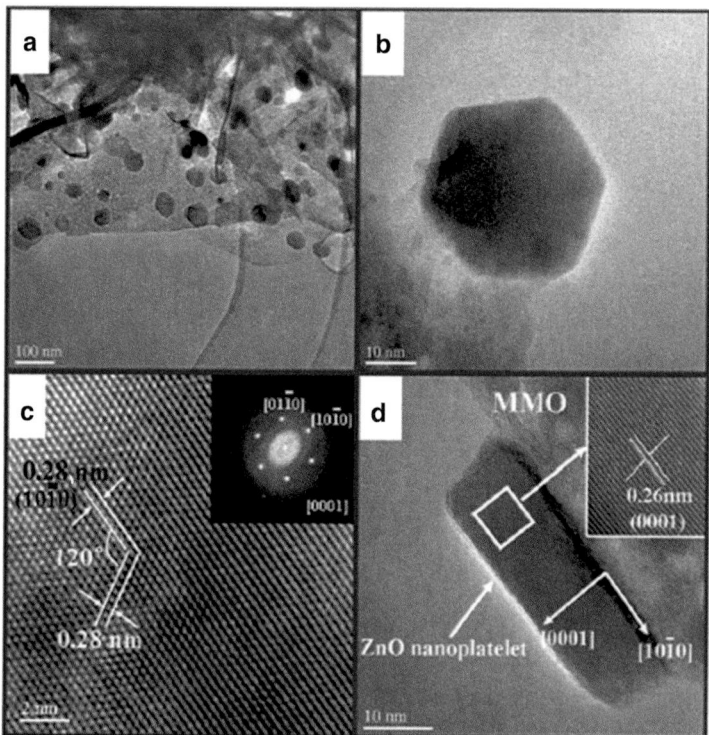

Fig. 6 HRTEM images of the MMO sample: (**a**) a MMO nanoflake with ZnO nanoplatelets on its surface; (**b**) a single-crystalline ZnO nanoplatelet; (**c**) HRTEM image of (**b**) (inset: the corresponding SAED pattern along the [0001] zone axis); (**d**) the cross section of a ZnO nanoplatelet with exposed (0001) facet embedded on the F-MMO substrate. Reprinted with permission from [37]. Copyright RSC

with preferential exposure of highly active facets. Fan et al. [38] reported the synthesis of visible-light-induced heterostructure ZnAlIn–MMO nanocomposite photocatalyst from ZnAlIn–LDH precursors. Well-dispersed amorphous In_2O_3 domains intimately attached to ZnO nanocrystallites were observed in the ZnAlIn–MMO nanocomposites, and thus a heterostructure of MMOs was formed. The ZnAlIn–MMO nanocomposites with the band gaps in the range 2.50–2.60 eV display enhanced visible-light-driven photocatalytic activity for the degradation of MB compared with single-phased In_2O_3. Seftel et al. [39] also reported a similar MMO heterostructure system consisting of nano-sized coupled ZnO/SnO_2 composite synthesized by the calcination of LDH precursors. This ZnO/SnO_2 MMO composite exhibits enhanced photocatalytic performance than the physic mixture of the two semiconductor oxides for the degradation of methyl orange (MO) dye. The enhanced photocatalytic activity was attributed to the efficient separation and transportation of photogenerated charge carriers, as a result of the novel heterostructure as well as the large specific surface area of the nanocomposites.

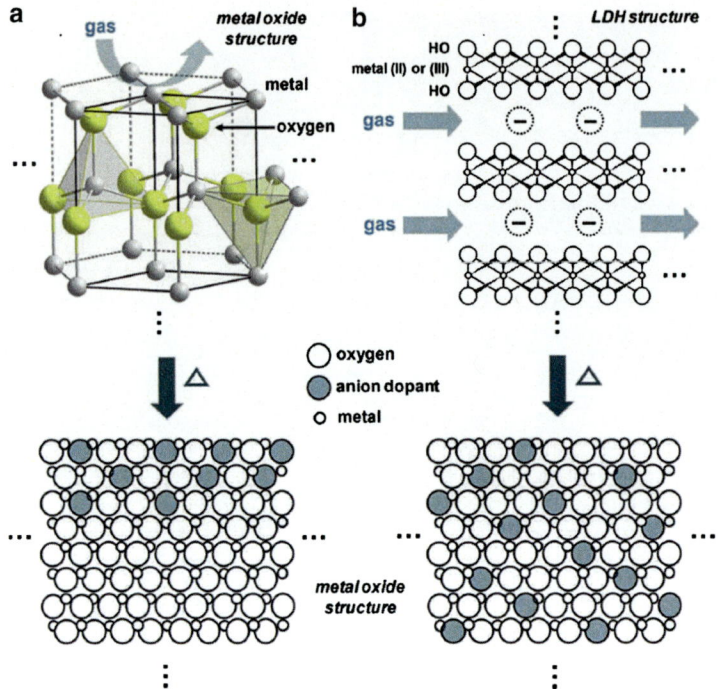

Fig. 7 A schematic diagram showing access of a nitriding gas to the precursors in the cases of (**a**) bulk ZnO structure and (**b**) ZnAl–LDH structure. Reprinted with permission from [40]. Copyright Wiley

In addition, LDH materials demonstrate the potential in the fabrication of anion-doped metal oxide nanostructures. Lee et al. [40] prepared the carbon and nitrogen co-doped MMO nanostructures by annealing a terephthalate-intercalated ZnAl–LDH under ammonia gas flow. Moreover, the interlayer gallery of LDH allows effective access of dopants to its crystal lattice for uniform doping (Fig. 7). Such co-doped MMO exhibits a significant redshift in absorption spectrum to visible-light region relative to pure MMO, resulting in band gap narrowing and improved incident-photon-to-current-conversion efficiency.

2.2 The Sensitization with Interlayer Guest Molecules

The exchangeable property of interlayer anions in LDHs evokes great interests on the construction of host–guest supramolecular structures with desirable properties and functionalities [41–44]. By an ion-exchange process, active species can be intercalated into the interlayer galleries of LDHs, and thus the improved photocatalytic performance can be achieved from the host–guest interaction or the synergistic effect between individual components. It has been demonstrated that a number of organic dyes and inorganic quantum dots can be used as efficient photosensitizers to broaden the light absorption range or to enhance the photocatalytic H_2/O_2 evolving [45, 46]. Another feasible way of creating photofunctional LDHs is the intercalation of sensitizer molecules into the interlayer space of the host. Embedding of an active species within the LDH gallery not only provides an easy-to-apply photonic material but may also minimize eventual leakage and retard degradation of sensitizer molecules. For instance, LDHs pillared by polyoxometalate (POM) ions such as $[W_7O_{24}]^{6-}$, $[PW_{12}O_{40}]^{3-}$, $[SiW_{12}O_{40}]^{4-}$, and $[W_7O_{24}]^{6-}$ have attracted considerable attention as a potentially important class of photocatalysts [47, 48]. Embedding POMs into the LDH gallery effectively overcomes some intrinsic limitations of POMs themselves as photocatalysts, such as the weak stability and hard separation from the reaction systems. Furthermore, decavanadate-intercalated ZnBi–LDHs were synthesized by the ion-exchange method (Fig. 8), which showed a high photochemical stability and photocatalytic activity toward the degradation of various organic pollutants for practical applications under solar light irradiation [49]. The study reveals that the presence of distorted VO_6 octahedron of decavanadate is essential for the photocatalysis due to the large dipole moment in the materials, which could facilitate the electron–hole separation.

Metal phthalocyanine sulfonates (MPcS) are a class of macrocyclic complex and have been widely investigated as photosensitizer for singlet oxygen generation, which possess high oxidative activity. However, their efficiency is greatly limited

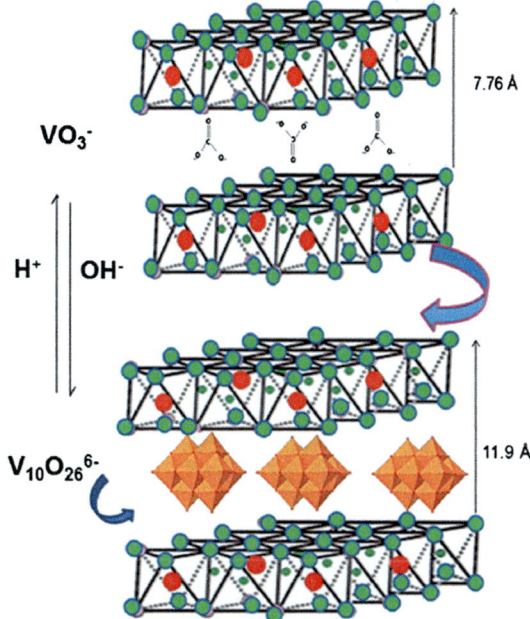

Fig. 8 A schematic representation for the nitrate/carbonate exchange with decavanadate in LDH. Reprinted with permission from [49]. Copyright RSC

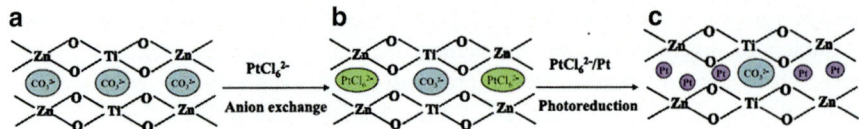

Fig. 9 A schematic illustration for the synthesis of Pt nanoparticle-intercalated ZnTi–LDH composite: (**a**) ZnTi–LDHs; (**b**) subsequent exchange of CO_3^{2-} by $PtCl_6^{2-}$; (**c**) followed by photoreduction to obtain the final Pt/ZnTi–LDHs. Reprinted with permission from [51]. Copyright Elsevier

by aggregation, resulting in enhanced self-quenching of the excited complex. Therefore, the incorporation of MPcS species into a restricted space like LDH would inhibit the dye aggregation and therefore accelerate the photosensitized reaction. Xu et al. [50] prepared PdPcS-intercalated MgAl–LDH via direct coprecipitation, and anionic dodecyl sulfonate (SDS) co-intercalation provides a chance for better dispersion of PdPcS in LDH interlayer. The as-obtained products were used as photocatalysts for the photosensitized degradation of 2,4,6-trichlorophenol (TCP) in water under visible-light irradiation. The PdPcS/SDS co-intercalated LDH displays the highest photocatalytic activity compared with pure LDH and the sample of PdPcS absorbed on the LDH surface. In addition, this photocatalyst also shows excellent stability and recyclability, without significant loss in activity after eight consecutive tests.

Nanoparticles also can be intercalated into LDH gallery for prohibiting the recombination of charge carriers. Chen et al. [51] synthesized the Pt nanoparticle-intercalated ZnTi–LDHs via ionic exchange followed by photochemical reduction in H_2PtCl_6 solution (Fig. 9). The combination of ZnTi–LDHs and Pt nanoparticles introduces some properties of Pt into photocatalysis such as excellent conductivity and controllability, which guarantees the rapid transport of charge carriers and hence improves the photocatalytic efficiency. Although some progress has been made in sensitizing LDHs using the intercalation method, the fine-tuning of guest species and microenvironments for the achievement of host–guest synergetic photocatalysis performance remains a big challenge.

2.3 *LDH-Based Nanocomposites for Photocatalysis*

It is recognized that the loading of photocatalysts with a suitable cocatalyst, such as noble metals (e.g., Pt, Ag) [52–54] and metal oxides (e.g., NiO_x, RuO_2) [55–57], is an effective method to obtain a high activity and reasonable reaction rate. Utilizing the LDH nanoplatelets as the host platform to couple with other nanoparticles for developing highly active photocatalysts has been proved as a promising approach. Recently, it is found that LDH materials can be modified by transition metal oxides under adequate reaction conditions, in which the newly introduced metal oxides are well dispersed and doped within the layered structure of LDHs. The resulting nanocomposite materials have higher specific surface area, suitable pore size distribution, as well as dramatically improved photocatalysis ability. Nano-sized TiO_2 with high photocatalytic activity is one of the most studied and used photocatalysts [58, 59]; however, the easy aggregation and health hazards by the inhalation toxicity have largely limited its practical applications. Meanwhile, it is difficult for hydrophobic organic molecules to reach the surface of TiO_2 for further photocatalytic conversion because of its intrinsic hydrophilic properties. To solve these problems, the immobilization of TiO_2 nanoparticles onto appropriate photocatalyst supports has been regarded as an effective method. LDHs are a desirable candidate for the immobilization of TiO_2 nanoparticles with the intention to provide adequate porous structure and prevent aggregation of TiO_2, thanks to the tunable composition, morphology, and surface charge density. The reconstruction process of LDHs in a dilute aqueous solution of $TiOSO_4$ at room temperature induces the formation of a bicomponent nanocomposite consisting of LDH-supported TiO_2 nanoparticles [60]. The high-magnification TEM image reveals that the small TiO_2 nanoparticles are well dispersed on the surface of platelike LDH particles (Fig. 10a, b). The corresponding selected area electron diffraction (SAED) pattern shows two faint rings for pristine LDH while powder ring diffraction patterns of nanocrystalline anatase for TiO_2/LDH sample. Guo et al. [61] further reported the synthesis of TiO_2/ZnAl–LDH composites by a one-pot in situ approach, giving contribution to the large-scale preparation of TiO_2/LDH photocatalysts.

The loading amount of TiO_2 on the photocatalytic properties of TiO_2/ZnAl–LDHs nanocomposites was further studied [62], which showed a great influence on the interactions between ZnAl–LDH and TiO_2. A proper TiO_2 loading can ensure the stability of photocatalytic system, constrain the undesired LDH memory effect,

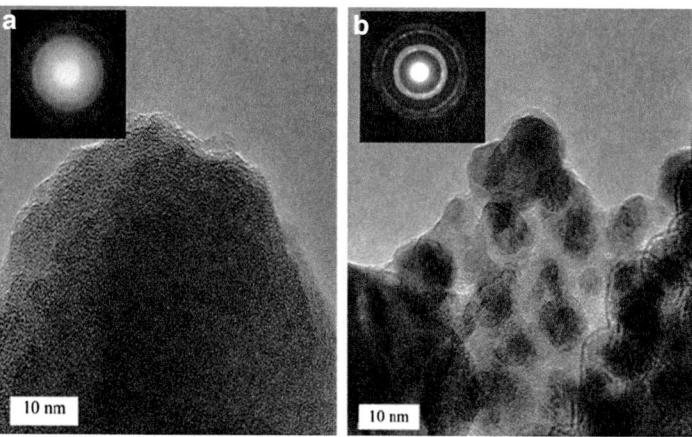

Fig. 10 Representative TEM images of (**a**) ZnAl–LDH and (**b**) TiO_2/ZnAl–LDH. Reprinted with permission from [60]. Copyright American Chemical Society

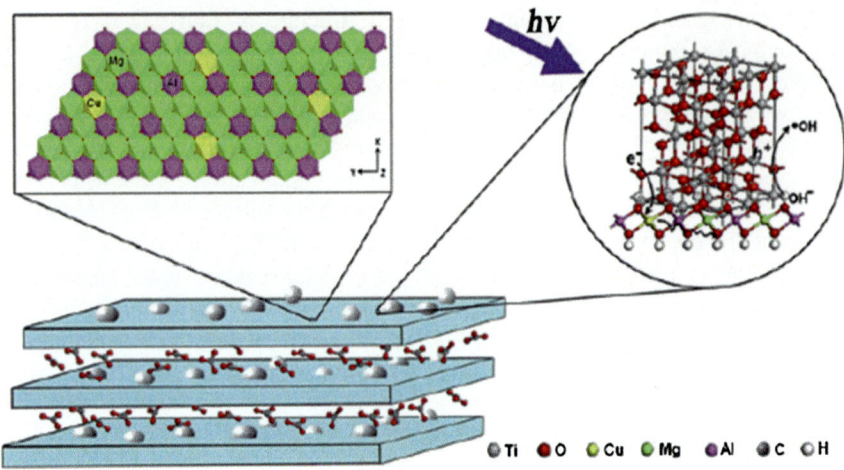

Fig. 11 A schematic illustration for the resulting TiO$_2$/CuMgAl–LDH sample and the proposed photocatalytic mechanism. Reprinted with permission from [63]. Copyright Elsevier

and enhance the photocatalytic performances upon solar light. TiO$_2$/ZnAl–LDH can be used as functional coatings on the surface of porous building mineral substrates (e.g., historical, handmade, industrial brick, and granite stone), imposing influences on the surface characteristics of the building materials (for instance, photoinduced hydrophilicity, photocatalytic activity, and self-cleaning effect). Zhang et al. [63] fabricated the LDH-supported TiO$_2$ nanoparticles by selective reconstruction of CuMgAlTi–LDH through calcination and rehydration process. CuMgAlTi–MMO can be obtained via calcination of the LDH precursor in air at 400 °C. By subsequent rehydration in deionized water at room temperature, LDH structure with Cu^{2+}, Mg^{2+}, and Al^{3+} in host layers (CuMgAl–LDH) was selectively reconstructed, accompanying with the formation of TiO$_2$ nanoparticles supported on the LDH surface (Fig. 11). Substituting partial magnesium with copper ions in the framework during the synthesis of LDH makes it possible to improve the photocatalytic efficiency.

The layered titanate coupling with LDH materials also shows enhanced photocatalytic performance. The common 2D structures of both LDH and layered titanate enable an effective physical contact and a strong electronic coupling between them. Hwang et al. [64] reported the synthesis of mesoporous heterolayered ZnCr–LDH and layered titanate nanohybrids by self-assembly between oppositely charged 2D nanosheets of ZnCr–LDH and layered titanium oxide and demonstrated their excellent photocatalytic activity toward visible-light-induced O$_2$ generation (Fig. 12). The obtained heterolayered nanohybrids possess a visible-light-harvesting ability and a highly porous structure, which are attributable to the electronic coupling between the component nanosheets. The ZnCr–LDH and layered titanate nanohybrids show highly efficient photocatalytic activity for O$_2$ production under visible illumination, which is superior to the activities of the

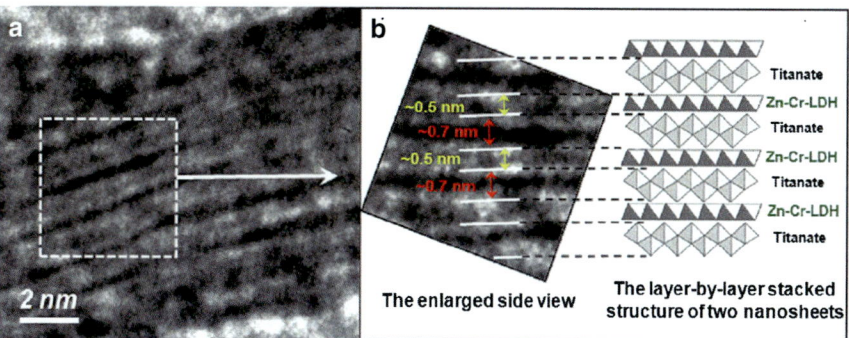

Fig. 12 (a) Cross-sectional HRTEM image of the ZnCr–LDH and layered titanate nanohybrid; (b) its enlarged view and structural model. Reprinted with permission from [64]. Copyright American Chemical Society

pristine ZnCr–LDH and layered titanate. Moreover, the chemical stability of the ZnCr–LDH is significantly improved because of the protection effect by layered titanate. In addition to the enhancement of the photocatalytic activity, the separation of catalysts from aqueous solution is also an urgent problem considering the industrial applications, especially in the treatment of river, lake, and other natural water. In this regard, magnetic separation is a very convenient approach for removing and recycling magnetic particles/composites. Recently, the combination of magnetic nanoparticles with LDHs has been developed to enhance separation and re-dispersion performance of LDHs in aqueous solution [65–67]. For instance, magnetic Fe_3O_4/ZnCr–LDH composite with a high saturation magnetization shows improved adsorption and photocatalytic activity [68]. Therefore, the construction of nanocomposites by dispersing metal oxides on LDH nanoplatelets provides an effective method to tune the physical and chemical properties of the photocatalysts in activity, separation, as well as cycling stability.

Up to now, a major limitation of achieving high photocatalytic efficiency in semiconductor nanostructures is the fast recombination of charge carriers. Metal nanoparticles exhibit unusual redox activity by readily accepting electrons from a suitable donor; such electron transfer would greatly enhance the separation of photogenerated electrons and holes and hence improve the photocatalytic efficiency [69]. Moreover, the metal cocatalysts provide reaction sites and decrease the activation energy for gas evolution in photocatalytic water splitting and CO_2 reduction [70]. Among all metal cocatalysts, platinum (Pt) has been widely used owing to its small work function and low overpotential for solar fuel production. Xu et al. [71] demonstrated the construction of photocatalytic hydrogen evolution system by using LDH-supported photosensitizer [rose bengal (RB)] and photocatalyst (Pt). Such a system offers the advantages in photocatalytic water splitting, including suppressed self-quenching by immobilizing the dye photosensitizer, efficient electron transfer between photosensitizer and catalyst nanoparticles, and good dispersion of cocatalyst nanoparticles on the LDH surface.

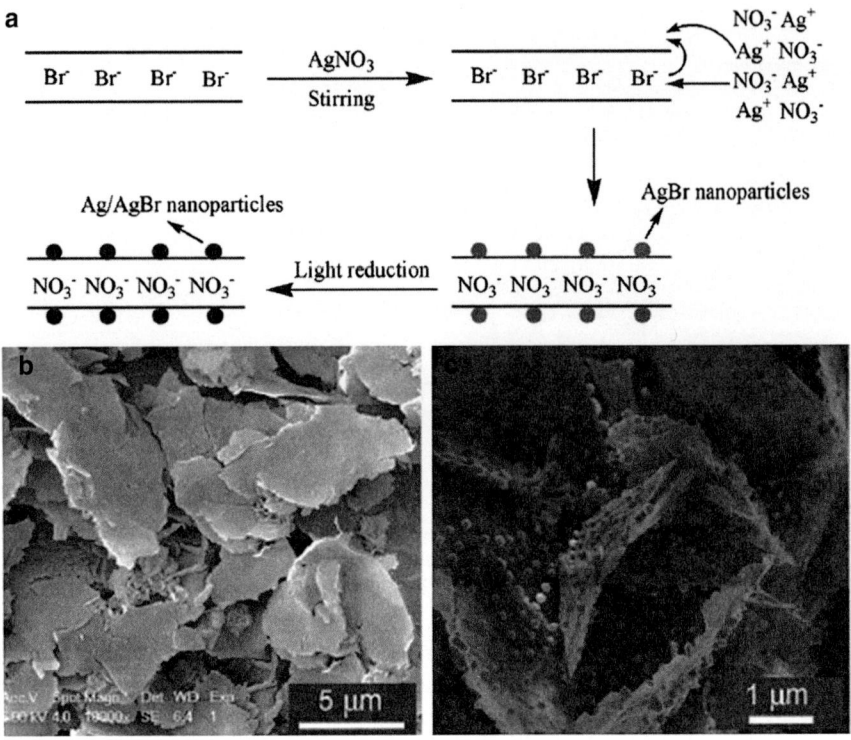

Fig. 13 (**a**) Illustration of the concept for formation of Ag/AgBr/CoNi–LDH nanocomposite by the anion-exchange precipitation method; SEM images of (**b**) CoNi–LDH sheets and (**c**) Ag/AgBr/CoNi–LDH nanocomposite. Reprinted with permission from [74]. Copyright Wiley

Recently, silver/silver halide-based (Ag/AgX, X = Br, Cl) nanomaterials have been developed as photocatalysts which display excellent plasmonic photocatalytic performances in the degradation of pollutants under visible-light irradiation [72, 73]. Ag/AgX supported on LDH substrates has shown considerable visible-light activity in the decomposition of organic pollutants with improved photocatalytic stability. Ai et al. [74] synthesized Ag/AgBr/CoNi–LDH nanocomposites by adding AgNO$_3$ solution to a suspension of CoNi–LDH (Fig. 13). CoNi–LDH was used as the nucleating agent of AgBr nanoparticles with a diameter of 50–150 nm. After partial reduction of AgBr to metallic Ag by using light illumination, the Ag/AgBr/CoNi–LDH nanocomposite was formed with negligible aggregation. The LDH surface not only acts as a substrate to anchor and grow Ag/AgBr nanoparticles but plays an important role in the stabilization of nanoparticles. Consequently, the photocatalytic activities of the obtained nanocomposite for the removal of dyes and phenol are higher than those of CoNi–LDH and Ag/AgBr. Similarly, Ag/AgCl/ZnCr–LDH composite also shows enhanced visible-light photocatalytic abilities toward degradation of organic pollutants than Ag/AgCl and ZnCr–LDHs [75]. In addition, Ag/AgCl nanoparticles

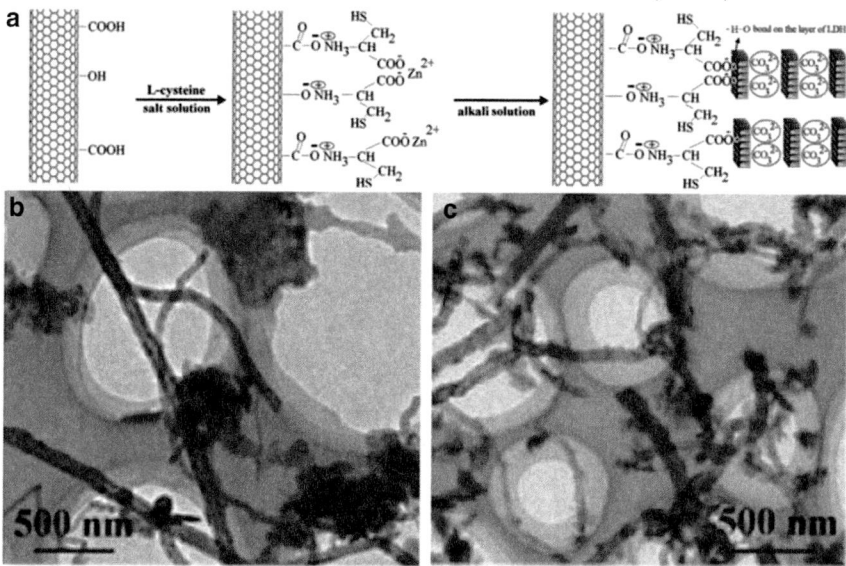

Fig. 14 (a) A schematic illustration of the synthesis pathway for in situ growth of ZnAl–LDH onto the modified CNTs in the presence of L-cysteine; TEM images of (b) LDH–CNTs and (c) LDH–cysteine–CNTs. Reprinted with permission from [77]. Copyright Wiley

grown on the ZnAl–LDH surface by using $AgNO_3$ as the silver source were also synthesized [76], which showed good antimicrobial activity against Gram-positive and Gram-negative bacteria and fungi.

Superior charge mobility of photocatalysts is beneficial to prevent the recombination of electron–hole and thus improve the catalytic efficiency. As a result, several attempts have been explored to enhance the electric conductivity of LDH-based photocatalysts by incorporation of carbon materials or conductive polymers. Li et al. [77] firstly reported the assembly of hybrid ZnAl–LDH/carbon nanotube (CNT) nanocomposites through noncovalent bonds in the presence of L-cysteine molecules (Fig. 14a). L-cysteine as bridging linker plays a double role: (1) enhancing both adhesion and dispersion of LDH nanocrystallites onto the surface of CNT matrix through the interfacial interaction and (2) inhibiting the in situ growth rate of LDH crystallites so as to obtain a reduced particle size (Fig. 14b, c). The as-assembled hybrid LDH/CNT nanocomposites exhibit excellent performance toward photodegradation of methyl orange molecules under UV irradiation.

Graphene with honeycomb-like crystal lattice has many unique properties such as high electron mobility, a large surface area, and environmental stability, which is highly desirable as a two-dimensional catalyst support [78–80]. Recently, some attempts have been made to combine graphene with LDH photocatalysts to enhance the performance due to the excellent electron-transfer capability of graphene and its visible-light-responsive activity. Graphene oxide (GO) containing negative charges can be easily combined with exfoliated LDH nanosheets by electrostatically driven

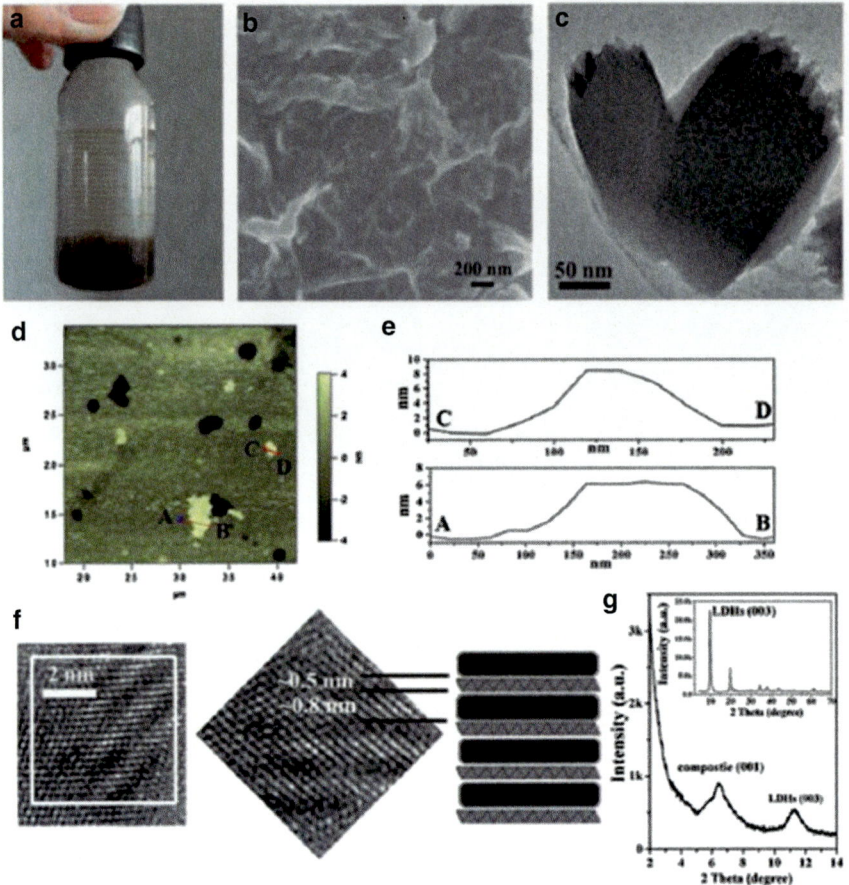

Fig. 15 CG/LDH nanohybrid prepared through layer-by-layer assembly: (**a**) photograph of nanohybrid precipitation; (**b**) SEM image of CG/LDH nanohybrid; (**c**) HRTEM image of nanohybrid; (**d**) AFM image of nanohybrid; (**e**) height profile of the AFM image; (**f**) cross-sectional HRTEM image of nanohybrid and (**g**) low-angle XRD pattern of nanohybrid powder (inset: XRD pattern of LDH). Reprinted with permission from [81]. Copyright RSC

self-assembly, resulting in a layer-by-layer ordered nanohybrid. Wu et al. [81] synthesized the ordered carboxyl graphene (CG)/ZnAl–LDH nanohybrid based on the electrostatically induced self-assembly; the porous CG/LDH nanohybrid was formed by cross-linked nanoflakes with a house-of-cards-type stacking structure (Fig. 15). The photocatalytic performance of its calcined product was significantly enhanced with improved chemical stability, which was mainly attributed to the effective electronic coupling between graphene and calcined ZnAl–LDH. Wei et al. [82] demonstrated a visible-light-driven NiTi–LDH/RGO photocatalyst fabricated by anchoring NiTi–LDH nanosheets onto the surface of RGO via an in situ growth method. It is found that NiTi–LDH nanosheets (diameter: 100–200 nm) are

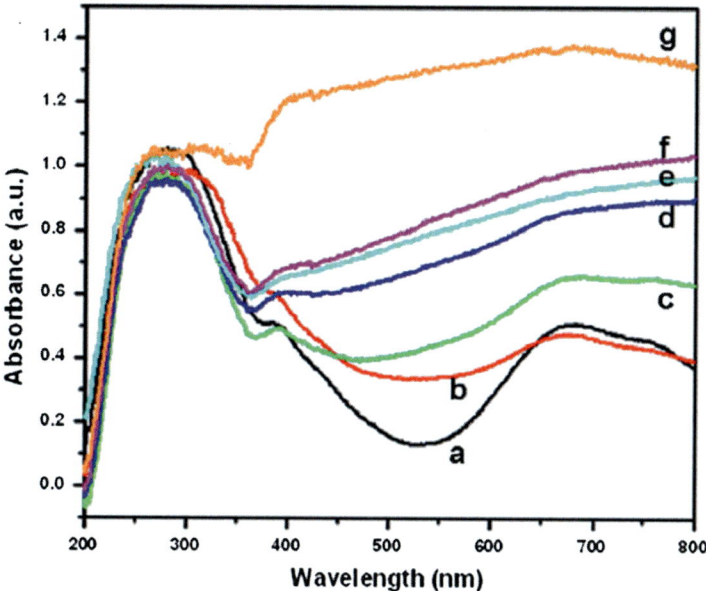

Fig. 16 UV–vis diffusion reflectance spectra of various NiTi–LDH/xRGO composites: from a to g: $x = 0$ %, 0.5 %, 1.0 %, 1.5 %, 2.0 %, 2.5 %, and 3.0 %, respectively. Reprinted with permission from [82]. Copyright American Chemical Society

highly dispersed on the surface of RGO sheets and the absorbance of the NiTi–LDH/xRGO composites in visible-light region enhances markedly with the increase of the RGO content (Fig. 16). Moreover, NiTi–LDH/RGO composites present more efficient separation of electron–hole pairs compared with the pristine NiTi–LDH. By using electron spin resonance (ESR) and Raman scattering strategies, it is shown that the electron transfer occurs from NiTi–LDH nanosheets to RGO, resulting in the largely enhanced carrier mobility and photocatalytic activity. Therefore, the combination of carbon-based materials with LDHs enhances the carrier mobility and improves the photogenerated electron–hole separation efficiency of LDHs. In addition, the large specific surface area of carbon-based materials allows sufficient exposure of LDH active sites, which would lead to superior photocatalytic performances.

3 The Applications of LDH-Based Photocatalysts

With the ever-increasing fossil fuel consumption and the consequently serious environmental pollution, the research on the utilization and conversion of solar energy has received considerable interests [83–85]. Numerous efforts have been devoted to the investigation on semiconductor-based photocatalysts with the ability of pollutant elimination as well as water splitting by using solar energy, due to their crucial role in the solar conversion process. Because of the unique properties of LDHs such as the versatility in the matrix elements and interlayer anions and the atomic-level homogeneous distribution of metal cations within the host layer, LDH materials and their derived composites (e.g., MMOs, intercalated LDHs, and LDH-based nanocomposites) have been widely studied as photocatalyst for the applications in pollutant degradation, water splitting for H_2 and/or O_2 generation, and photocatalytic synthesis.

3.1 Photocatalytic Degradation of Pollutants

Nowadays, increasing attention has been focused on the environmental pollution caused by the organic pollutants from chemical and pharmaceutical industries. The commonly used methods for the removal of pollutants (e.g., chemical oxidation, reduction degradation, biological decomposition) suffer from low reaction rate, high cost, as well as secondary pollution. The absorption process with high efficiency only simply transfers the pollutant from one medium to another, still leaving the pollutants in the environment. The photocatalytic degradation of pollutants by oxidizing organic pollutants into nontoxic inorganic micromolecules using solar energy is an environmental-friendly and cost-effective approach for the removal of pollutants [86, 87]. Recently, LDH-based photocatalysts have been widely applied to this photocatalytic degradation process.

LDHs with matrix elements containing Zn, Cr, Fe, or Ti exhibit excellent photocatalytic activity toward pollutant degradation. ZnTi–LDH with hierarchical nanostructure and large specific surface area (91.96–107.9 $m^2\ g^{-1}$) shows much higher photocatalytic activity than TiO_2, ZnO, and P25 for the degradation of methylene blue (MB) under visible-light irradiation [23]. As shown in Fig. 17, MB decomposes ~40 % under light exposure without catalyst, indicating that photolysis reaction also contributes to the degradation of dye. However, the decomposition of MB by the ZnTi–LDH reaches ~100 % under the same conditions, which is even higher than that of commercial P25. The photodegradation of MB follows the pseudo-first-order rate law; the rate constant was calculated to be 0.0407 min^{-1} ($t_{1/2} = 50$ min) with ZnTi–LDH as the photocatalyst, much larger than that of TiO_2, ZnO, as well as P25.

Cr-containing LDHs display much remarkable visible-light-induced photocatalytic activity for the degradation of sulforhodamine-B (SRB), Congo red, 2,4,6-trichlorophenol (2,4,6-TCP), and salicylic acid sodium. Wei et al. [29] proposed that the highly efficient photocatalytic performance of Cr-containing LDHs is attributed to the visible-light response originated from the d–d transition of the CrO_6 octahedron in the LDH layer. Moreover, the abundant OH groups on the surface of LDHs facilitate the generation of highly reactive hydroxyl radicals by reacting with photogenerated holes, which provides enough oxidation active sites for the degradation of pollutants. However, Parida et al. [33] put forward that the

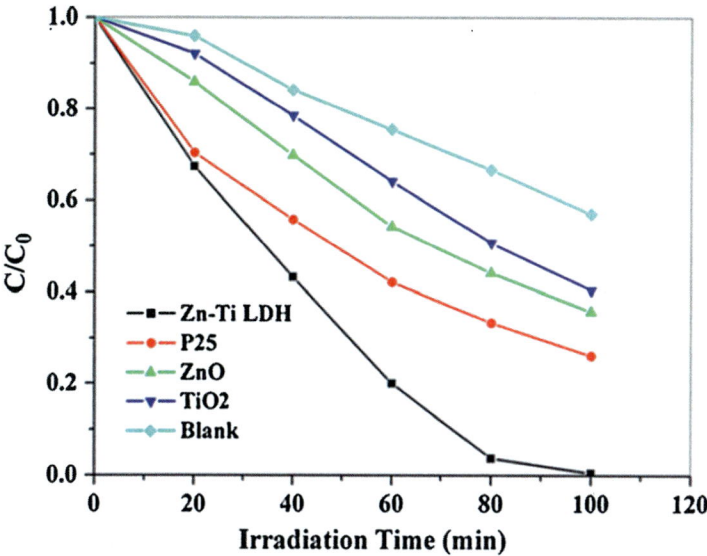

Fig. 17 Photodegradation of MB monitored as the normalized concentration vs. irradiation time under visible-light irradiation ($\lambda > 420$ nm) with the presence of different photocatalysts. Reprinted with permission from [23]. Copyright Elsevier

metal-to-metal charge transfer (MMCT) excitation of oxo-bridged bimetallic linkages of M^{II}–O–Cr^{III} in neighboring MO_6 and CrO_6 octahedra in LDHs matrix is responsible for the photocatalytic activity of Cr-containing LDHs under visible-light irradiation (Fig. 18). The presence of MMCT was confirmed by optical difference spectra (ODS); electron paramagnetic resonance (EPR) revealed that the MMCT results in the generation of superoxide anion radical, which is an effective oxidative species for pollutants. In addition, MMCT was also presented in Zn/Y and Zn/Bi LDH in the photocatalytic degradation of pollutants such as malachite green (MG), rhodamine 6G (RhG), and 4-chloro-2-nitrophenol (CNP) [30].

The photocatalytic efficiency of single-component semiconductor is seriously impeded by the rapid recombination of photogenerated electrons and holes, while coupling semiconductors with matched energy levels can allow the migration of photogenerated charge carriers from one species to the other one and therefore suppress the rapid recombination of electrons and holes. α-Ag_2WO_4/ZnCr–LDH composite with a large scale of α-Ag_2WO_4 nanoparticles dispersed on the surface of ZnCr–LDHs exhibits a higher visible-light-driven photocatalytic activity toward the degradation of rhodamine B (RhB), compared with a-Ag_2WO_4 crystals, ZnCr–LDHs, and their physical mixture [88]. In addition to the larger specific surface area of α-Ag_2WO_4/ZnCr–LDH composite, the migration of photogenerated charge carrier between the two components is responsible for the enhanced photocatalytic performance. As shown in Fig. 19a, the photogenerated electrons and holes migrate respectively to the CB of α-Ag_2WO_4 and the VB of ZnCr–LDH due to the energy

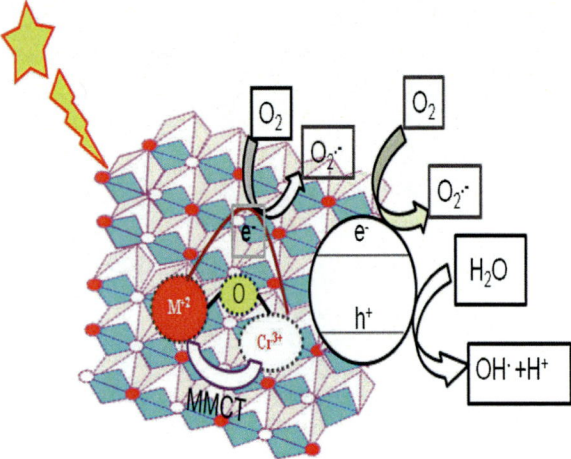

Fig. 18 A schematic representation of the proposed mechanism for the photocatalytic degradation of MO over $M^{II}Cr$–LDHs under solar light irradiation. Reprinted with permission from [33]. Copyright American Chemical Society

level difference, leading to efficient separation of photogenerated carriers. Similarly, Bi_2MoO_6/ZnAl–LDH with hierarchical heterostructure displays a higher photocatalytic activity than their single components [89]; the enhanced photocatalytic performance is ascribed to the charge transfer in the heterostructure (Fig. 19b). ZnAl–LDH inactive to visible light can accept photogenerated electrons from Bi_2MoO_6 because of its more positive CB than Bi_2MoO_6, which facilitates the effective separation of photogenerated electrons and holes. ZnAl–LDH coupled with layered titanate $H_2Ti_3O_7$ also shows a photogenerated charge carrier migration between the two components, as shown in Fig. 19c [90]. This nanohybrid exhibits a superior photocatalytic activity in the degradation of MB. Zhang et al. [91] designed novel ZnO/$ZnAl_2O_4$ nanocomposites derived from ZnAl–LDHs precursors, in which the band structure coupling of ZnO and $ZnAl_2O_4$ ensures the efficient separation of photogenerated charge carriers (Fig. 19d), accounting for the superior photocatalytic performances toward the degradation of methylene orange (MO).

For the purpose of obtaining immobilized and recyclable photocatalysts toward aqueous pollutant degradation, Wei et al. [92] fabricated oriented CuCr–LDH film on copper substrate by electrophoretic deposition method. Due to its large specific surface area and the rich macro-/mesoporous structure, the LDH film with thickness of 16.5 μm exhibits excellent visible-light-driven photocatalytic activity for the degradation of 2,4,6-TCP, SRB, and Congo red, superior to that of corresponding LDH-powered sample. The photocatalytic degradation kinetics of 2,4,6-TCP follows the pseudo-first-order kinetics. This film catalyst also displays excellent photocatalytic stability and recyclability as well as easy manipulation, facilitating the repeatable and cyclic usage of this film over a long period.

Parida et al. [93] investigated the contribution of reactive oxidative species (ROS) including hydroxyl radical (OH•), superoxide radical ($O_2^{•-}$), and singlet oxygen ($^1O_2$) in the visible-light-driven photocatalytic degradation of organic chemicals such as RhB, rhodamine 6G (R6G), and 4-CNP using LDH as

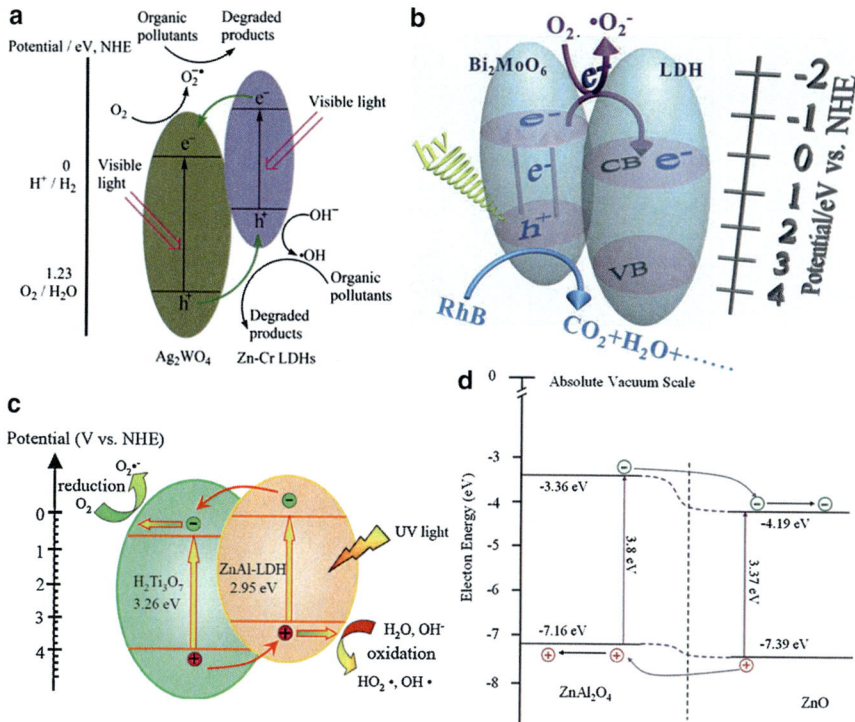

Fig. 19 Schematic photocatalytic mechanisms for (**a**) α-Ag$_2$WO$_4$/ZnCr–LDH composite under visible-light illustration; (**b**) RhB degradation over a Bi$_2$MoO$_6$/ZnAl–LDH composite under visible-light irradiation; (**c**) photoexcitation in ZnAl–Ti$_3$O$_7$ semiconductor–semiconductor interstratified heterolayered nanohybrid; (**d**) the band energy levels of ZnO and ZnAl$_2$O$_4$ as well as the transfer procedure of photogenerated e$^-$ and h$^+$ pairs. Reprinted with permission from [88–91]. Copyright Elsevier, RSC, Elsevier and Wiley, respectively

photocatalyst. The contribution of ROS was investigated by using different scavengers (isopropanol, benzoquinone (BQ), and NaN$_3$, respectively). As shown in Fig. 20, the hydroxyl radical is the predominant ROS although peroxyl radical is responsible for the degradation of all organic pollutants, while singlet oxygen (^{1}O$_2$) only plays a very minor role in the degradation process.

According to the previously reported results, the possible mechanism for efficient degradation of pollutants over the LDH-based photocatalysts was discussed [31]. The MO$_6$ (e.g., M = Ti, Cr) octahedral units in the LDH layer act as semiconductors. Photoexcitation of electrons (e$^-$) and holes (h$^+$) occurs when the energy of incident photon matches or exceeds the band gap of the MO$_6$ semiconductor. The e$^-$ and h$^+$ can react with water and dissolved oxygen on the surface of the LDH photocatalyst to form OH• and superoxide radicals O$_2$•$^-$, respectively, which are highly oxidizing species. The possible reaction process for photocatalytic degradation of pollutants by LDH is summarized by the following Eqs. (1)–(8):

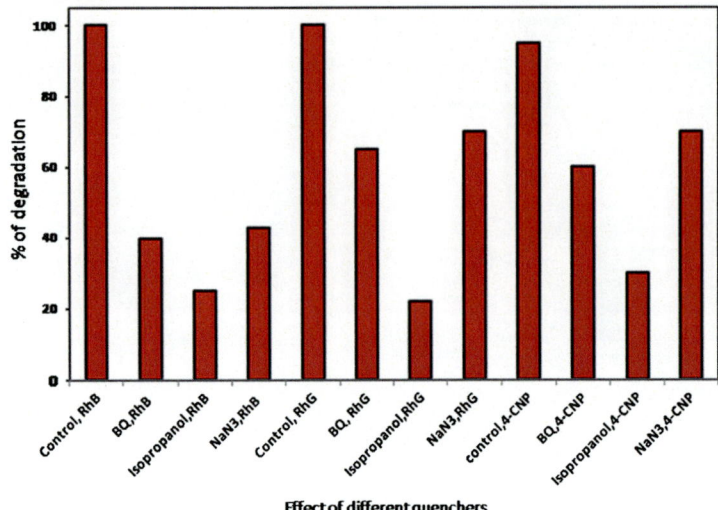

Fig. 20 Photocatalytic degradation of RhB, RhG, and 4-CNP over the ZnCr–LDH sample under visible irradiation in the presence of different quenchers. Reprinted with permission from [93]. Copyright Elsevier

$$LDH + h\nu \rightarrow h^+ + e^- \quad (1)$$

$$h^+ + H_2O \rightarrow H^+ + {}^{\bullet}OH \quad (2)$$

$$e^- + O_2 \rightarrow {}^{\bullet}O_2^- \quad (3)$$

$${}^{\bullet}O_2^- + H^+ \rightarrow HO_2^{\bullet} \quad (4)$$

$$HO_2^{\bullet} + HO_2^{\bullet} \rightarrow H_2O_2 + O_2 \quad (5)$$

$$H_2O_2 \rightarrow 2{}^{\bullet}OH \quad (6)$$

$${}^{\bullet}O_2^- + \text{pollutants} \rightarrow \text{degradation products} \quad (7)$$

$${}^{\bullet}OH + \text{pollutants} \rightarrow \text{degradation products} \quad (8)$$

3.2 Photocatalytic Water Splitting

It is very difficult for the simultaneous generation of O_2 and H_2 from a photocatalytic water splitting process due to the low efficiency caused by the easy recombination of the photogenerated electrons and holes. Therefore, great efforts have been devoted to investigating the half-reactions of photocatalytic water splitting (photocatalytic O_2 or H_2 generation) with the use of sacrificial agent, which reacts with photogenerated electrons or holes so as to suppress their recombination. In general, sacrificial agent for electrons such as $AgNO_3$ is required for photocatalytic O_2 generation, while sacrificial agent for holes such as methanol and lactic acid is necessary for photocatalytic H_2 generation. Silva et al. [27] used LDH materials in photocatalytic oxygen generation under visible light for the first time.

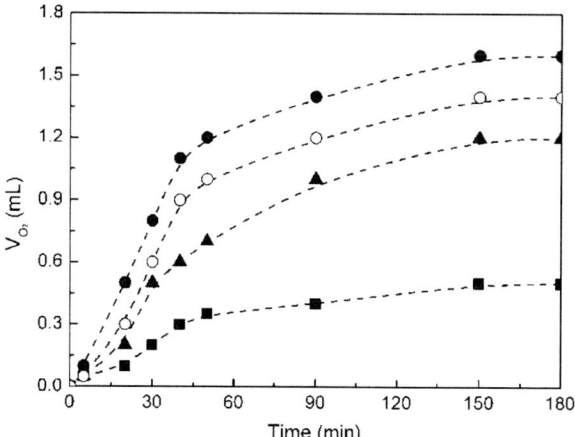

Fig. 21 Temporal profile of the volume of oxygen evolved during the irradiation of aqueous suspensions of ZnCr–LDH (*closed circle*), calcined ZnCr–LDH (*open circle*), ZnCe–LDH (*closed triangle*), and ZnTi–LDH (*closed square*) using 10^{-2} M $AgNO_3$ as the sacrificial acceptor. Reprinted with permission from [27]. Copyright American Chemical Society

Among several kinds of Zn-containing LDHs (ZnTi, ZnCe, and ZnCr–LDHs), ZnCr–LDH displays the highest photocatalytic activity (Fig. 21) with an extremely high apparent quantum yield of 60.9 % for O_2 generation. As Cr content is varied by introducing photocatalytically inactive Al species, the activity of ZnCrAl–LDH enhances gradually along with the increase of Cr^{3+}/Al^{3+} ratio in the LDH layer, and the one with the maximum possible Cr loading shows the best performance.

Titanium-embedded LDHs such as NiTi–LDH and CuTi–LDH also serve as water oxidation photocatalysts under visible light [18]. With two absorption bands in the red and blue region, both NiTi–LDH and CuTi–LDH exhibit high photocatalytic activity and produce 49 and 31 μmol of O_2 in water oxidation by using 200 mg of photocatalyst and 1 mmol of $AgNO_3$ as sacrificial agent in 5 h. The surface Ti^{3+} defects can serve as trapping sites for photogenerated electrons and suppress the electron–hole recombination. In order to increase the density of surface Ti^{3+} defects in Ti-containing LDHs, Wei et al. [22] synthesized NiTi–LDH nanosheets with lateral dimension in the range 30–60 nm. The NiTi–LDH with lateral size of ~30 nm exhibits an extraordinarily high photocatalytic activity toward O_2 generation (~2,148 μmol g^{-1} h^{-1}) under visible light with a quantum yield of ~65 % at 400 nm. CoFe–LDH has also been reported as efficient photocatalyst for O_2 generation [94]. Among the three CoFe–LDHs with different Co/Fe molar ratios, the sample containing the most iron displays the highest photocatalytic activity (~45 μmol of O_2 using 50 mg of photocatalyst in 3 h), due to the abundant Co–O–Fe bridge which suppresses the fast recombination of holes and electrons.

Recently, LDH-based nanocomposites have been reported as efficient water oxidation photocatalysts with improved performances than pristine LDH materials. Layer-by-layer assembled nanosheets of ZnCr–LDH and layered titanium oxide were employed in photocatalytic oxygen generation from water under visible light, which showed largely enhanced photocatalytic activity with a rate of ~1.18 mmol h^{-1} g^{-1} compared with pristine ZnCr–LDH (~0.67 mmol h^{-1} g^{-1}) [64]. In addition, the

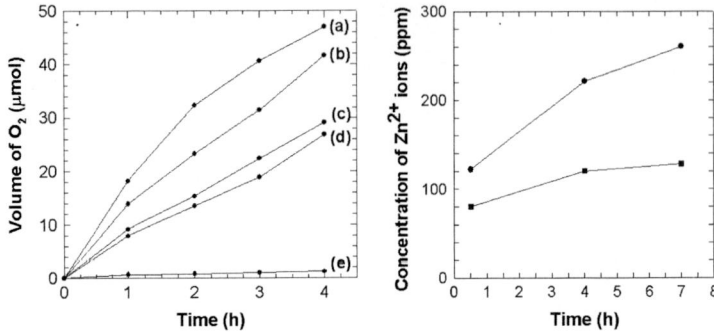

Fig. 22 (*Left*) Time-dependent photoproduction of oxygen gas under visible-light illumination ($\lambda > 420$ nm) by (a) ZCT-1, (b) ZCT-2, and (c) ZCT-3, in comparison with the data by (d) pristine ZnCr–LDH and (e) layered cesium titanate. 10 mmol of $AgNO_3$ is used as an electron scavenger. (*Right*) Time-dependent variation of zinc concentration dissolved from the ZCT-1 nanohybrid (*closed square*) and the pristine ZnCr–LDH (*closed circle*). Reprinted with permission from [64]. Copyright American Chemical Society

chemical stability of ZnCr–LDH is significantly improved via hybridization with the highly stable titanate layer (Fig. 22). Gunjakar et al. [95] reported that by the self-assembly of ZnCr–LDH nanoplatelets with graphene nanosheets, the as-obtained ZnCr–LDH/RGO nanohybrids exhibited an extremely high photocatalytic activity toward visible-light-driven O_2 generation with a rate of ~1.20 mmol h^{-1} g^{-1}, much larger than that of the pristine ZnCr–LDH (~0.67 mmol h^{-1} g^{-1}). In addition, the NiTi–LDH/RGO nanocomposite also displays an excellent activity for photocatalytic O_2 generation with a rate of 1.968 mmol h^{-1} g^{-1} and a quantum efficiency of 61.2 % at 500 nm.

LDH nanoparticles modified with photosensitizer (rose bengal, RhB) and Pt were used in photocatalytic H_2 generation and exhibited higher photocatalytic activity than the free system (without LDH). LDHs can also be directly used as photocatalyst for H_2 generation. Parida et al. [28] employed carbonate-intercalated ZnCr–LDH as photocatalyst for H_2 evolution. The ZnCr–LDH (CO_3) exhibits enhanced photocatalytic activity than other anion-intercalated ZnCr–LDHs because carbonate can be oxidized by holes to form carbonate radicals, thus suppressing the rapid recombination of photogenerated electrons and holes. Furthermore, Ni was incorporated into the host layer of ZnCr–LDH to achieve enhanced photocatalytic hydrogen evolution [33]. With a certain Ni:Zn ratio (3:1), the NiZnCr–CO_3^{2-}–LDH exhibits the best visible-light-driven photocatalytic activity, as shown in Fig. 23. It is proposed that the d–d transition of Cr^{3+} in CrO_6 is forbidden by Laporte rules due to the centrosymmetric nature of CrO_6. The visible-light absorption is attributed to the MMCT excitation of Zn^{II}/Ni^{II}–O–Cr^{III} to Zn^{I}/Ni^{I}–O–Cr^{IV}; the oxo-bridged bimetallic linkage acts as photocatalytic center for hydrogen evolution. The substitution of Ni by Zn increases the surface area and narrows the band gap energy, which increases the photocatalytic activity.

Fig. 23 Temporal changes of the amount of hydrogen evolved (μmol) with different types of LDH samples during irradiation. Reprinted with permission from [33]. Copyright RSC

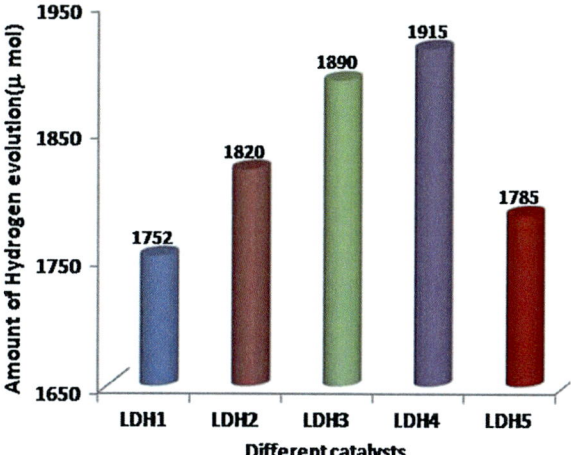

Fig. 24 H_2 evolution from aqueous solution of lactic acid as sacrificial electron donor. Conditions: 100 mg of catalyst, 300 W Xe-lamp irradiation, 150 mL of degassed lactic acid solution (0.133 vol%). Reprinted with permission from [24]. Copyright Wiley

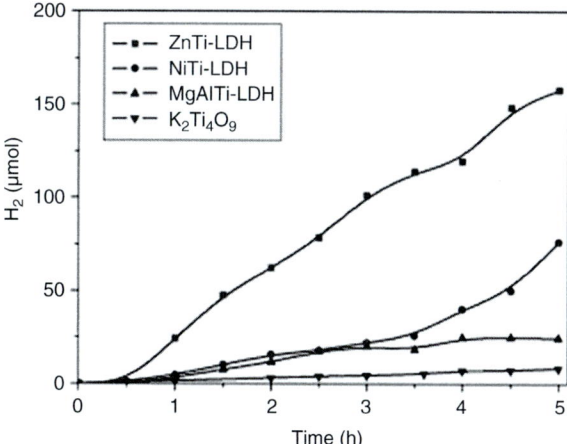

Ti- or Fe-containing LDHs have been used as photocatalyst for water splitting into hydrogen. Zhao et al. [24] reported the photocatalytic properties of MTi–LDHs (M = Ni, Zn, Mg) for H_2 generation with an H_2 production rate of 31.4 μmol h^{-1} (Fig. 24), superior to the photocatalytic activity of $K_2Ti_4O_9$. In addition, MgAlFe–LDH (CO_3) was used as efficient photocatalyst for H_2 generation from water under visible light. The LDH sample containing the highest amount of iron in the host layer was found to be the most active photocatalyst with a hydrogen evolution rate of 301 μmol g^{-1} h^{-1}. The incorporation of Fe^{3+} into Mg–Al LDH structure leads to the visible-light-driven photocatalytic activity and prevents the photogenerated electron–hole recombination.

Photoelectrochemical (PEC) water splitting which integrates photocatalytic solar energy conversion with water electrolysis can realize the overall water splitting into H_2 and O_2 without the addition of sacrificial agent. LDHs can be

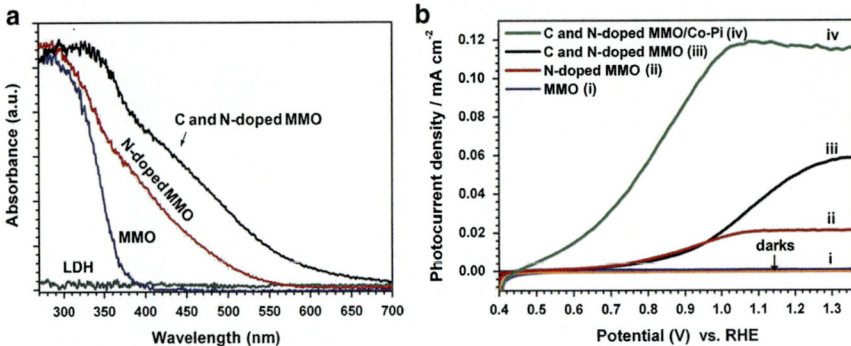

Fig. 25 (a) UV–vis absorption spectra of the ZnAl–LDH, MMO, N-doped MMO, C and N co-doped MMO material; (b) photocurrent–potential curves for ZnAl–LDH, MMO, N-doped MMO, C and N co-doped MMO, and the amorphous cobalt–phosphate (Co–Pi)-deposited C and N co-doped MMO material measured under a Xe-arc lamp (300 W) attached with a UV cut-off filter ($\lambda \geq 420$ nm) in a potassium phosphate buffer solution (0.1 M, pH 7). Reprinted with permission from [40]. Copyright Wiley

used as photocatalyst in this PEC water splitting system by the immobilization of LDH onto conductive substrates. Uniform carbon and nitrogen co-doped $ZnO/ZnAl_2O_4$ MMO was prepared by annealing terephthalate-intercalated ZnAl LDH in ammonia gas flow, followed by electrophoretic deposition on FTO glass as photocatalyst toward PEC water splitting. As shown in Fig. 25, the co-doped MMO with smaller band gap displays significantly enhanced visible-light absorption and higher photocurrent density than pure MMO. Furthermore, Cho et al. [40] reported that the ZnCr–MMO photoanode obtained by in situ synthesis method exhibits a stable and an order-of-magnitude higher PEC water splitting activity compared with electrophoretically deposited ZnCr–MMO photoanode.

The integration of promising water oxidation catalyst with a photon-absorbing substrate can provide a substantial reduction in the external power needed to drive water splitting. Transition metal-containing LDHs demonstrate potential as effective cocatalyst to increase the reaction kinetics and depress the charge recombination rate. Wei et al. [96] reported hierarchical nanoarrays containing ZnO core and CoNi–LDH nanoplatelets shell which exhibited promising behavior in PEC water splitting, giving rise to a largely enhanced photocurrent density (1.5 mA cm^{-2} at +0.5 V) as well as stability, much superior to those of ZnO-based photoelectrodes. The enhancement is attributed to the effective photogenerated electron–hole separation and fast water oxidizing kinetics originating from the intimate contact between photocatalysis-active ZnO core and electrocatalysis-active LDH shell. Moreover, the hierarchical structure enables convenient charge transfer and ion diffusion on the electrode/electrolyte interface where the water oxidation occurs.

3.3 Photocatalytic CO_2 Reduction

Conversion of the greenhouse gas CO_2 into carbon and useful carbon sources such as CH_4, CO, HCOOH, and CH_3OH is a potential strategy of solving both environmental and energy problems [97, 98]. Photocatalytic conversion of CO_2 is attractive from the viewpoint of sustainable energy production and greenhouse gas reduction. Recently, ZnCuM(III) (M = Al, Ga)–LDHs have been reported as photocatalysts to convert gaseous CO_2 to CH_3OH or CO using hydrogen as reductant under UV–visible light [99]. The results showed that ZnAl–LDH was the most active toward CO_2 photoreduction (0.16 % conversion) with a high CO selectivity (94 mol%) at a formation rate of 620 nmol h^{-1} g_{cat}^{-1}, while ZnCuGa–LDH was the most selective for CH_3OH production (68 mol%) at 0.03 % conversion and a formation rate of 170 nmol h^{-1} g_{cat}^{-1}. A mechanism for the photoreduction of CO_2 on the surface of LDHs is proposed (Fig. 26): the improved methanol selectivity by the introduction of Cu was attributed to the binding of CO_2 at the Cu–O–M (M = Zn and Ga) sites to form hydrogen carbonate species. Furthermore, the photoreduction rate and methanol selectivity of ZnCuGa–LDH were improved by the intercalation of $[Cu(OH)_4]^{2-}$ (490 $nmol_{Methanol}$ h^{-1} g_{cat}^{-1} and 88 mol%, respectively) [100]. Morikawa et al. [101] designed a reverse photofuel cell for photocatalytic reduction of CO_2 into methanol accompanied with generation of O_2. $[Cu(OH)_4]^{2-}$-intercalated ZnCuGa LDH and WO_3 were used as photoreduction catalyst of CO_2 and photooxidation catalyst of water, respectively.

Teramura et al. [102] employed a series of M^{2+}/M^{3+}–LDHs (M^{2+} = Mg^{2+}, Zn^{2+}, Ni^{2+}; M^{3+} = Al^{3+}, Ga^{3+}, In^{3+}) in the photocatalytic conversion of CO_2 into CO and O_2 in an aqueous system. As shown in Fig. 27, the LDHs exhibit superior activity: 3.21 and 17.0 µmol of CO and O_2 were generated in the presence of MgIn–LDH after 10 h of photoirradiation, while $Mg(OH)_2$ and $In(OH)_3$ exhibited much lower photocatalytic activity. The oxidation and reduction processes are listed as follows:

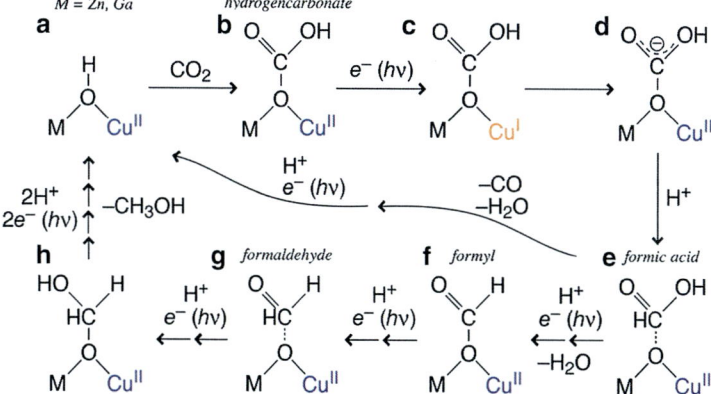

Fig. 26 Proposed photocatalytic catalytic cycle of CO_2 reduction to methanol or CO over ZnCuGa–LDH catalysts. Reprinted with permission from [100]. Copyright ScienceDirect

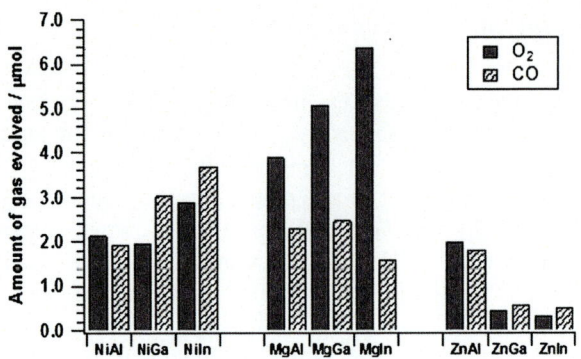

Fig. 27 Yields of O_2 and CO for the photocatalytic conversion of CO_2 in water, over various LDHs ($M^{2+}/M^{3+}=3$) after 10 h of photoirradiation (light intensity is adjusted constantly). Reprinted with permission from [102]. Copyright Wiley

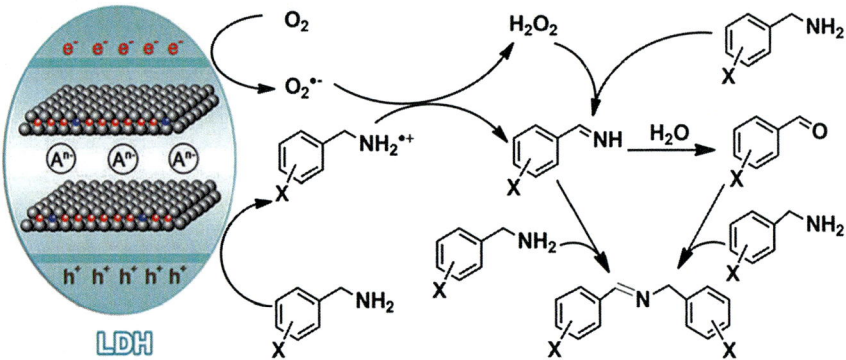

Fig. 28 Possible benzylamine aerobic oxidation reaction pathways. Reprinted with permission from [104]. Copyright RSC

$$CO_2 + 2e^- + 2H^+ \rightarrow CO + H_2O \qquad (9)$$

$$2H_2O \rightarrow O_2 + 4H^+ + 4e^- \qquad (10)$$

Recently, noble metal (Pt, Pd, Au)-loaded ZnCr–LDHs were employed in the photocatalytic reduction of CO_2 into CO under UV irradiation [103]. The 0.1 wt% Pt-loaded LDH exhibits the highest CO evolution (0.13 μmol m^{-2}). In addition to the photoreduction of CO_2, LDHs can also be used in other photocatalytic transformation processes. Yang et al. [104] employed ZnTi–LDH as a photocatalyst for the conversion of primary aromatic to their corresponding imines in aerobic atmosphere. The photocatalytic activity of ZnTi–LDH is much better than TiO_2 (P25) and ZnO; the enhancement was ascribed to the high dispersion of active TiO_6 unit in LDHs. After 5 h of photocatalytic reaction, a broad range of primary aromatic amines are converted into their corresponding imines with the conversion efficiency and selectivity closed to 100 %. The substituents are unaffected in these photocatalytic oxidation processes. A possible mechanism for the photocatalytic aerobic oxidation reaction is proposed (Fig. 28), in which $O_2^{\cdot-}$ generated from oxygen is responsible for the oxidation of aromatic amines to imines.

4 Summary and Outlook

LDHs represent one of the most promising photocatalyst materials due to their low cost, relative ease preparation/modification, and versatility in host matrix and interlayer anions. LDH-based photocatalytic materials have been rationally designed and synthesized by well-developed methodologies: (1) taking advantage of the flexible tunability of metal cations in LDH host layer, various photoactive species (e.g., Zn, Cu, Ti, Cr, Fe, In) have been introduced into LDHs, resulting in improvements in electronic microstructure, light absorption property, and surface characteristics of LDH-derived photocatalysts; (2) the ion-exchange ability of interlayer anions allows further sensitization with organic dyes and inorganic nanoparticles, which expands the light response range and facilitates the photocatalytic reactions; (3) the layered structure and positively charged surface of LDHs make them as candidates for supporting other functional materials so as to achieve a synergetic effect in photocatalysis. In addition, the assembly or integration of LDHs with hierarchical nanomaterials affords multifunctional nanocomposites with improved photocatalytic performance. The high adsorption capabilities of LDH materials enrich the reactant concentration on the surface of LDH-based photocatalysts, accelerating the photocatalytic reaction rate. The obtained LDH-based photocatalysts have been widely used for the photocatalytic degradation of pollutants, water splitting for H_2 and O_2 generation, as well as reduction of CO_2. Therefore, LDH-based photocatalysts can be potentially used as effective and efficient photocatalysts for large-scale environmental purification as well as solar to chemical energy conversion.

Although new and exciting applications of LDH-based photocatalysts are springing up at an increasing rate, great challenges still remain in the design, preparation, and application of LDH-based photocatalysts: (1) the intrinsic semiconductor properties of LDHs and the synergetic effects among different metal ions in the host layer are not well understood; (2) the methods to achieve the precise tailoring of electronic structure and architecture of LDH-based photocatalysts on multi-scale levels need to be further developed; (3) the detailed insights into the photocatalytic processes of LDH-based photocatalysts as well as further enhancements in photocatalytic efficiency deserve an intensive study. Therefore, in the future, it is worthy to prepare high-quality and cost-effective LDH-based photocatalysts with careful control over composition, structure, and morphology. Moreover, the understanding on active sites, reaction kinetics, and the synergy between two or more active components of LDHs for light absorption and carrier transportation are very significant and challenging. In addition, the application extension in the field of energy conversion (e.g., water splitting and photocatalytic synthesis) would be welcome, in view of the intrinsic catalytic advantages of LDH-based materials and a great demand for the new energy resources and synthetic technologies.

Acknowledgments This work was supported by the 973 Program (Grant No. 2011CBA00504), the National Natural Science Foundation of China (NSFC), the Beijing Natural Science Foundation (2132043), and the Scientific Fund from Beijing Municipal Commission of Education. M. Wei particularly appreciates the financial aid from the China National Funds for Distinguished Young Scientists of the NSFC.

References

1. Linsebigler AL, Lu G, Yates JT Jr (1995) Chem Rev 95:735
2. Zheng YH, Zheng LR, Zhan YY, Lin XY, Zheng Q, Wei KM (2007) Inorg Chem 46:6980
3. Niu MT, Huang F, Cui LF, Huang P, Yu YL, Wang YS (2010) ACS Nano 4:681
4. Fu HB, Pan CS, Yao WQ, Zhu YF (2005) J Phys Chem B 109:22432
5. Tada H, Mitsui T, Kiyonaga T, Akita T, Tanaka K (2006) Nat Mater 5:782
6. Asahi R, Morikawa T, Ohwaki T, Aoki K, Taga Y (2001) Science 293:269
7. Wang C, Xie ZG, deKrafft KE, Lin WL (2011) J Am Chem Soc 133:13445
8. Kamat PV (2007) J Phys Chem C 111:2834
9. Williams G, Seger B, Kamat PV (2008) ACS Nano 2:1487
10. Subramanian V, Wolf EE, Kamat PV (2004) J Am Chem Soc 126:4943
11. Yao Y, Li G, Ciston S, Lueptow RM, Gray KA (2008) Environ Sci Technol 42:4952
12. Khan AI, O'Hare D (2002) J Mater Chem 12:3191
13. Sels B, De Vos D, Buntinx M, Pierard F, Mesmaeker KD, Jacobs P (1999) Nature 400:855
14. Shao MF, Ning FY, Zhao JW, Wei M, Evans DG, Duan X (2012) J Am Chem Soc 134:1071
15. Gu Z, Thomas AC, Xu ZP, Campbell JH, Lu GQ (2008) Chem Mater 20:3715
16. Shao MF, Ning FY, Zhao YF, Zhao JW, Wei M, Evans DG, Duan X (2012) Chem Mater 24:1192
17. Liu J, Zhang G (2014) Phys Chem Chem Phys 16:8178
18. Lee Y, Choi JH, Jeon HJ, Choi KM, Lee JW, Kang JK (2011) Energy Environ Sci 4:914
19. Shu X, Zhang W, He J, Gao F, Zhu Y (2006) Solid State Sci 8:634
20. Shu X, He J, Chen D (2008) J Phys Chem C 112:4151
21. Tian J, Zhao Z, Kumar A, Boughton R, Liu H (2014) Chem Soc Rev 43:6920
22. Zhao Y, Li B, Wang Q, Gao W, Wang CJ, Wei M, Evans DG, Duan X, O'Hare D (2014) Chem Sci 5:951
23. Shao M, Han J, Wei M, Evans DG, Duan X (2011) Chem Eng J 168:519
24. Zhao Y, Chen P, Zhang B, Su DS, Zhang S, Tian L, Lu J, Li Z, Cao X, Wang B, Wei M, Evans DG, Duan X (2012) Chem Eur J 18:11949
25. Mendoza-Damián G, Tzompantzi F, Mantilla A, Barrera A, Lartundo-Rojas L (2013) J Hazard Mater 263P:67
26. Wang X, Wu P, Huang Z, Zhu N, Wu J, Li P, Dang Z (2014) Appl Clay Sci 95:95
27. Silva CG, Bouizi Y, Fornes V, Garcıa H (2009) J Am Chem Soc 131:13833
28. Parida K, Mohapatra L (2012) Dalton Trans 41:1173
29. Zhao Y, Zhang S, Li B, Yan H, He S, Tian L, Shi W, Ma J, Wei M, Evans DG, Duan X (2011) Chem Eur J 17:13175
30. Mohapatra L, Parida KM (2014) Phys Chem Chem Phys 16:16985
31. Parida K, Mohapatra L, Baliarsingh N (2012) J Phys Chem C 116:22417
32. Parida K, Satpathy M, Mohapatra L (2012) J Mater Chem 22:7350
33. Baliarsingh N, Mohapatra L, Parida K (2013) J Mater Chem A 1:4236

34. Prince J, Tzompantzi F, Mendoza-Damián G, Hernández-Beltrán F, Valente JS (2015) Appl Catal B 163:352
35. Huang L, Chua S, Wang J, Kong F, Luo L, Wang Y, Zou Z (2013) Catal Today 212:81
36. Zhao Y, Wei M, Lu J, Wang ZL, Duan X (2009) ACS Nano 3:4009
37. He S, Zhang S, Lu J, Zhao Y, Ma J, Wei M, Evans DG, Duan X (2011) Chem Commun 47:10797
38. Lan M, Fan GL, Sun W, Li F (2013) Appl Surf Sci 282:937
39. Seftel EM, Puscasu MC, Mertens M, Cool P, Carja G (2014) Appl Catal B 150:157
40. Cho S, Jang JW, Kong KJ, Kim ES, Lee KH, Lee JS (2013) Adv Funct Mater 23:2348
41. Yan D, Lu J, Ma J, Wei M, Evans DG, Duan X (2011) Angew Chem Int Ed 50:720
42. Wang Q, O'Hare D (2012) Chem Rev 112:4124
43. He S, An Z, Wei M, Evans DG, Duan X (2013) Chem Commun 49:5912
44. Zhang H, Pan D, Duan X (2009) J Phys Chem C 113:12140
45. Takanabe K, Kamata K, Wang XC, Antonietti M, Kubota J, Domen K (2010) Phys Chem Chem Phys 12:13020
46. Parayil SK, Baltrusaitis J, Wu CM, Koodali RT (2013) Int J Hydrogen Energy 38:2656
47. Guo Y, Li D, Hu C, Wang Y, Wang E (2001) Int J Inorg Mater 3:347
48. Guo Y, Li D, Hu C, Wang E, Zou Y, Ding H, Feng S (2002) Microporous Mesoporous Mater 56:153
49. Mohapatraab L, Parida KM (2014) Phys Chem Chem Phys 16:16985
50. Xiong Z, Xu Y (2007) Chem Mater 19:1452
51. Chen G, Qian S, Tu X, Wei X, Zou J, Leng L, Luo S (2013) Appl Surf Sci 293:345
52. Wang C, Yin L, Zhang L, Liu N, Lun N, Qi Y (2010) ACS Appl Mater Interfaces 2:3373
53. Height MJ, Pratsinis SE, Mekasuwandumrong O, Praserthdam P (2006) Appl Catal B Environ 63:305
54. Li H, Bian Z, Zhu J, Huo Y, Li H, Lu Y (2007) J Am Chem Soc 129:4538
55. Corma A, Atienzar P, Garcia H, Chane-Ching JY (2004) Nat Mater 3:394
56. Kato H, Kudo A (2003) Catal Today 78:561
57. Sato J, Kobayashi H, Ikarashi K, Saito N, Nishiyama H, Inoue Y (2004) J Phys Chem B 108:4369
58. Yu JC, Yu JG, Ho WK, Zhang LZ (2001) Chem Commun 19:1942
59. Jing LQ, Sun XJ, Shang J, Cai WM, Xu ZL, Du YG, Fu HG (2003) Sol Energ Mater Sol Cells 79:133
60. Carja G, Nakajima A, Dranca S, Dranca C, Okada K (2010) J Phys Chem C 114:14722
61. Shao L, Yao Y, Quan S, Wei H, Wang R, Guo Z (2014) Mater Lett 114:111
62. Huang Z, Wu P, Lu Y, Wang X, Zhu N, Dang Z (2013) J Hazard Mater 246–247:70
63. Lu R, Xu X, Chang J, Zhu Y, Xu S, Zhang F (2012) Appl Catal B 111–112:389
64. Gunjakar JL, Kim TW, Kim HN, Kim IY, Hwang SJ (2011) J Am Chem Soc 133:14998
65. Zhang H, Pan D, Zou K, He J, Duan X (2009) J Mater Chem 19:3069
66. Ay AN, Zumreoglu-Karan B, Temel A, Rives V (2009) Inorg Chem 48:8871
67. Mi F, Chen X, Ma Y, Yin S, Yuan F, Zhang H (2011) Chem Commun 47:12804
68. Chen D, Li Y, Zhang J, Zhou J, Guo Y, Liu H (2012) Chem Eng J 185–186:120
69. Cushing SK, Li JT, Meng FK, Senty TR, Suri S, Zhi MJ, Li M, Bristow AD, Wu NQ (2012) J Am Chem Soc 134:15033
70. Feng XJ, Sloppy JD, LaTemp TJ, Paulose M, Komarneni S, Bao NZ, Grimes CA (2011) J Mater Chem 21:13429
71. Hong J, Wang Y, Pan J, Zhong Z, Xu R (2011) Nanoscale 3:4655
72. Wang P, Huang BB, Qin XY, Zhang XY, Dai Y, Wei JY, Whangbo MH (2008) Angew Chem Int Ed 120:8049
73. Han L, Wang P, Zhu CZ, Zhai YM, Dong SJ (2011) Nanoscale 3:2931
74. Fan H, Zhu J, Sun J, Zhang S, Ai S (2013) Chem Eur J 19:2523
75. Nocchetti M, Donnadio A, Ambrogi V, Andreani P, Bastianini M, Pietrellac D, Latterini L (2013) J Mater Chem B 1:2383

76. Sun J, Zhang Y, Cheng J, Fan H, Zhu J, Wang X, Ai S (2014) J Mol Catal A Chem 382:146
77. Wang H, Xiang X, Li F (2010) AIChE J 56:768
78. Eda G, Chhowalla M (2010) Adv Mater 22:2392
79. Rao CV, Reddy ALM, Ishikawa Y, Ajayan PM (2011) ACS Appl Mater Interfaces 3:2966
80. Mathkar A, Tozier D, Cox P, Ong PJ, Galande C, Balakrishnan K, Reddy ALM, Ajayan PM (2012) J Phys Chem Lett 3:986
81. Huang Z, Wu P, Gong B, Fang Y, Zhu N (2014) J Mater Chem A 2:5534
82. Li B, Zhao Y, Zhang S, Gao W, Wei M (2013) ACS Appl Mater Interfaces 5:10233
83. Fujishima A, Honda K (1972) Nature 238:37
84. Kato H, Asakura K, Kudo A (2003) J Am Chem Soc 125:3082
85. Armaroli N, Balzani V (2007) Angew Chem Int Ed 46:52
86. Muruganandham M, Swaminathan M (2006) J Hazard Mater 135:78
87. Takeuchi M, Kimura T, Hidaka M, Rakhmawaty D, Anpo M (2007) J Catal 246:235
88. Zhu J, Fan H, Sun J, Ai S (2013) Sep Purif Technol 120:134
89. Li H, Deng Q, Liu J, Hou W, Du N, Zhang R, Tao X (2014) Catal Sci Technol 4:102
90. Lin BZ, Ping S, Zhou YI, Jiang SF, Gao BF, Chen YL (2014) J Hazard Mater 280:156
91. Zhao XF, Wang L, Xin X, Ler XD, Xu SL, Zhang FZ (2012) AIChE J 58:573
92. Tian L, Zhao Y, He S, Wei M, Duan X (2012) Chem Eng J 184:261
93. Mohapatra L, Parida KM (2012) Sep Purif Technol 91:73
94. Kim JS, Lee Y, Lee DK, Lee JW, Kang JK (2014) J Mater Chem A 2:4136
95. Gunjakar JL, Kim IY, Lee JM, Lee NS, Hwang SJ (2013) Energy Environ Sci 6:1008
96. Shao M, Ning F, Wei M, Evans DG, Duan X (2014) Adv Funct Mater 24:580
97. Matsuoka M, Kitano M, Takeuchi M, Tsujimaru K, Anpo M, Thomas JM (2007) Catal Today 122:51
98. Chen XB, Shen SH, Guo LJ, Mao SS (2010) Chem Rev 110:6503
99. Ahemed N, Shibata Y, Taniguchi T, Izumi Y (2011) J Catal 279:123
100. Ahmed N, Morikawa M, Izumi Y (2012) Catal Today 185:263
101. Morikawa M, Ogura Y, Ahmek N, Kawamura S, Mikami G, Okamoto S, Izumi Y (2014) Catal Sci Technol 4:1644
102. Teramura K, Iguchi S, Mizuno Y, Shishido T, Tanaka T (2012) Angew Chem 124:8132
103. Katsumata KI, Sakai K, Ikeda K, Carja G, Matsushita N (2013) Mater Lett 107:138
104. Yang XJ, Chen B, Li XB, Zheng LQ, Wu LZ, Tung CH (2014) Chem Commun 50:6664

Bio-Layered Double Hydroxides Nanohybrids for Theranostics Applications

Dae-Hwan Park, Goeun Choi, and Jin-Ho Choy

Contents

1 Introduction 138
2 Design of Bio-LDH Nanohybrids 141
   2.1   Intercalation Routes to Bio-LDH Nanohybrids 141
   2.2   Structural Features and Chemical Bonding Nature 143
3 Molecular Engineering for Multifunctional Bio-LDH Nanohybrids 148
   3.1   Bio-LDH Nanohybrids with Targeting Functions 150
   3.2   Bio-LDH Nanohybrids with Imaging Functions 156
4 Bio-LDH Nanohybrids for Theranostics Applications 162
   4.1   Gene-LDH Theranostics In Vitro and In Vivo 162
   4.2   Drug–LDH Theranostics In Vitro and In Vivo 166
5 Summary and Outlook 170
References 173

Abstract Bio-related science has been one of the most important fields in the viewpoint of health and welfare of human being. A biofunctionalization of inorganic nanoparticles becomes more and more important in the near future, since it allows what is called theranostics, a new concept of next-generation medicine that combines therapeutics and diagnosis. Ever since the notion of intercalative bioinorganic nanohybrids was created based on the layered double hydroxide (LDH) material, a variety of bio-LDH nanohybrids have been developed to provide new opportunities toward theranostic biomedical applications. In this chapter, we review the structural chemistry of several bio-LDH nanohybrids available for gene and drug delivery system and focus on the platform strategies for advanced theranostic system. The cellular trafficking mechanism of LDH and cancer therapeutic efficacy along with diagnostic imaging of the bio-LDH nanohybrids are also discussed in in vitro as well as in vivo studies.

D.-H. Park • G. Choi • J.-H. Choy (✉)
Department of Chemistry and Nano Science, Center for Intelligent NanoBio Materials (CINBM), Ewha Womans University, Seoul 120-750, Republic of Korea
e-mail: jhchoy@ewha.ac.kr

Keywords Bio-LDH • Biomedical • Diagnosis • Drug delivery system • Gene delivery system • Imaging • Layered double hydroxide (LDH) • Nanohybrid • Nanomedicine • Targeting • Theranostics • Therapy

1 Introduction

In material chemistry, the "intercalation" denotes a chemical process, where guest species are reversibly inserted into an interlayer space of two-dimensional (2D) inorganic hosts to form new heterostructured hybrid materials [1–4]. Atoms, ions, molecules, and clusters have been successfully intercalated into 2D lattices of diverse inorganic solids, from natural clay minerals to synthetic materials such as layered double hydroxides (LDHs), transition metal oxides and chalcogenides, and carbonaceous layered materials [5–9].

Since the intercalation chemistry for "bioinorganic nanohybrids" was proposed for the first time in 1999 [10], the pioneering works of Choy's group have led to not only a paradigm shift in the research fields of nanoengineering and nanomedicine but also the tremendous developments for new multidisciplinary convergence science and technology (NT-BT-IT-CT) and multiplex functional nanohybrid systems [11–16].

For over a decade, it has been very successful in developing 2D bioinorganic nanohybrids via intercalation reaction, based on *chimie douce*, of biofunctional guest species into LDHs for innovative nano-biomedical breakthroughs in drug delivery, therapeutics, diagnostics, and even forensics [17–21].

The chemical formula of synthetic LDHs, in general, is described as $[M^{2+}_{1-x} M^{3+}_x(OH)_2]^{x+}(A^{m-})_{x/m} \cdot nH_2O$, where M^{2+} is a divalent metal cation, M^{3+} is a trivalent metal cation, and A^{m-} is an interlayer anion [22–25]. LDHs are also known as anionic clays, since their chemistry and crystal structure are very closely related to those of natural mineral, hydrotalcite, where its chemical composition is fixed like $Mg_6Al_2(OH)_{16}CO_3 \cdot 4H_2O$ with the Mg/Al ratio of 3 and the interlayer anion of CO_3^{2-}. To describe soft chemistry of LDHs, it is required to understand, for a start, the structure of brucite, $Mg(OH)_2$, where the divalent Mg^{2+} ion is coordinated with six hydroxo (OH^-) ligands to form an octahedron, and thus formed octahedrons are bound to one another by sharing edges to form infinite sheets, those which are then stacked on the top of each other to build up lamellar structure through strong hydrogen bonding. In the LDH structure, however, divalent cations, such as Mg^{2+}, Zn^{2+}, Ca^{2+}, and Fe^{2+}, are partially substituted with trivalent ones such as Al^{3+} and Fe^{3+}, so that the positive layer charge can be generated. In order to compensate the positive layer charge in LDH lattice, exchangeable anionic species should be simultaneously stabilized in the interlayer space of LDHs, which can be then further exchanged with other anionic species from atomic ions to molecular clusters—and even supramolecular anions.

The earliest example of medical application of LDHs is thought to be a pharmaceutical active ingredient for mainly antacid or anti-pepsin agent [26–28]. Pharmacodynamic effect of a synthetic LDH in the form of hydrotalcite is described as relief for pain and disorders of indigestion, heartburn, and other symptoms related to excess stomach acid, by raising the pH of stomach contents to above 3.5. Commercially available drug product contains 500 mg of hydrotalcite in one chewable tablet, which is indicated equivalent to a neutralization capacity of 13 mval HCl [28].

The recent studies on medical applications of LDHs have aimed to demonstrate the feasibility of LDHs and their intercalative nanohybrids as multiplex bio-LDH drug delivery hybrid systems for storage, transport, and subsequent controlled release of bioactive molecules such as genes, drugs, and nutrients and cosmetic ingredients with potential therapeutic functions in deceased environments [29–45]. To explore such hybrid systems, attempts have been made on the basis of four different synthetic strategies being well known, (1) coprecipitation, (2) ion exchange, (3) calcination–reconstruction, and (4) exfoliation–reassembling, and to understand chemical bonding nature between host LDH and guest molecules, intercalation and de-intercalation mechanisms, and kinetics along with controlled release behaviors of bioactive molecules out of LDHs in a simulated body fluid. These studies have taken advantages of LDHs such as anion-exchange capacity, variable–chemical compositions, biocompatibility, crystallization–dissolution characters, and host–guest interactions.

For example, the negatively charged bioactive molecules (nucleosides, ATP, DNA, antisense, vitamers, etc.) or anticancer drugs (methotrexate, 5-fluorouracil, doxorubicin, etc.) have been electrostatically incorporated into the interlayer space of positively charged LDHs via various intercalation routes to form charge-neutralized bio-LDH nanohybrids. The LDHs can play a role as a reservoir to stably protect chemical and biological integrity of bioactive molecules from severe physical, chemical, and biological environments arising from the change in parameters like temperature, light, moisture, pH, enzyme, etc. It has been also proved that thus prepared bio-LDHs could be used as delivery–carrier systems, since they can easily be permeated inside the cells by reducing the electrostatic repulsion between anionic biomolecules in LDHs and negatively charged cell membrane giving rise to enhanced cellular transfection efficiency. Since LDHs are fairly insoluble in neutral and basic pH domains, but become soluble at low pH, one can intentionally recover drugs or biomolecules intercalated in LDHs by dissolving them in a weakly acidic medium like a cellular cytoplasm and/or induce sustained release of guest molecules out of LDHs via ion-exchange reaction in electrolyte solution.

A most pioneering work, in this regard, was the delivery of therapeutic DNA stands into cancerous cells by means of the LDH carrier. Once DNA–LDH nanohybrids penetrate into the cell, DNA molecules stabilized in LDH would be then slowly discharged in cytosolic electrolyte not only by ion-exchange reaction but also by dissolution of LDH nanoparticles in acidic intracellular compartments such as late endosome and lysosome, where the pH becomes lower than 5. The first report on antisense oligonucleotide technology, a new gene therapy, has boosted popular interest in LDH-based new delivery system. It has been, therefore, highly expected that LDHs can act as a new type of nonviral vector to effectively reserve and deliver therapeutic genes or drugs into specific targets in the human body, especially tumor cells.

According to the US NIH road map for medical research in 2002, "nanomedicine" is considered as a new pathway to discover highly specific medical intervention at the molecular scale for curing disease or repairing damaged tissues [46, 47]. Furthermore, as appeared in the MIT 10 Technology Review in 2006, two research fields related to nano-bio convergence, such as nanomedicine and nanobiomechanics, were announced as the emerging ones out of ten technologies [48]. What is nanomedicine? The nanomedicine is nothing but a compound word, which is formed from the words "nano" and "medicine" and now becomes an important field of nano convergence. Nanomedicine comprises of six research fields such as (1) drug delivery, (2) drug and therapy, (3) in vivo imaging, (4) in vitro diagnostics, (5) active implants, and (6) biomaterials. Among them, drug delivery has been considered as the most intensively studied field in terms of the number of publications and patents in the world [49]. Current research trends in nanomedicine are moving to the development of personalized medicine for most cancers based on the theranostic strategies. As coined in 2002, the term "theranostics" is defined as bio-functionalized materials at nanoscale for the combination modality of therapy and diagnostic imaging at the same time [50, 51]. Such a convergence system is considered as an emerging platform technology to realize a significant increase in therapeutic efficacy with an accurate detection of early-stage cancer through selective biodistribution and enhanced pharmacokinetics. Such integrated approaches could not only strengthen a number of advantages but also make up for drawbacks of traditional drug delivery systems (DDS), and, namely, allow to overcome biological barriers such as blood–brain barrier (BBB), vascular endothelium, or augmented osmotic pressure states around cancer diseases by primarily tracking the position of cancer tissues and eventually cancer cells and subsequently targeted delivery with controlled release manner to maintain an effective and nontoxic concentration level of therapeutic agents for a relatively long period of time [52, 53].

In addition to therapeutic applications, the bio-LDH nanohybrids have been engineered with contrast reagents to promote imaging and diagnostic functions, such as fluorescence imaging, optical imaging, magnetic resonance imaging (MRI), and X-ray computed tomography (CT) imaging. More recently, some challenging works have been made to modify or functionalize the surface of LDHs in order to induce the advanced theranostic functions; drug–LDH nanohybrids can (a) target passively tumor cells along with fluorescence imaging in vivo cancer models based on the enhanced permeability and retention (EPR) effect and (b) target actively tumor cells on the basis of receptor-mediated endocytosis through targeting moieties and (c) the metal ions in LDH can be isomorphously substituted in order to have photoluminescence or magnetic properties and/or LDH nanoparticles can be hybridized with functional metal ones for multimodal imaging and be intercalated

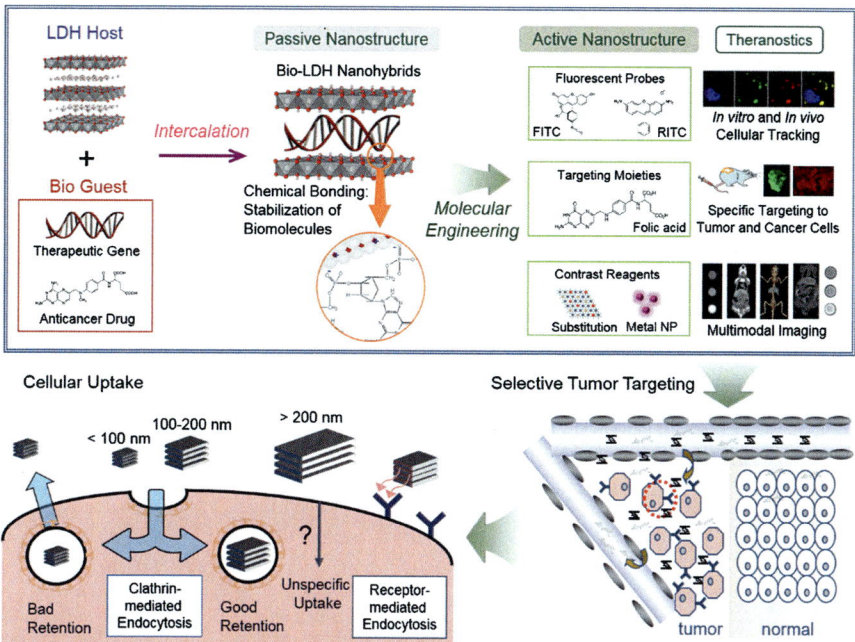

Fig. 1 Schematic illustration of the multifunctional bio-LDH system for theranostics applications and their targeted transfer mechanism into tumor and cancer cell

with drug or gene molecules to form drug/gene-LDHs for new drug delivery systems simultaneously with imaging functions.

In this review, we present an overview of multifunctional bio-LDH nanohybrids along with their synthetic methods, structural features, chemical bonding nature, and surface modifications. As shown in Fig. 1, the bio-LDH systems with passive targeting function developed for cancer treatments and emerging challenges in active theranostics strategies for the design of selective tumor targeting and imaging, anticancer activity, pharmacokinetics, and biodistributions will be highlighted with dramatic experimental findings in in vitro and in vivo studies.

2 Design of Bio-LDH Nanohybrids

2.1 Intercalation Routes to Bio-LDH Nanohybrids

Recently, great attentions have been made to the intercalative nanohybrids originated from LDHs because of their unique 2D nature in terms of crystal structure and internal and external surface properties. Especially their intercalation property allows new potential in exploring novel bioinorganic nanohybrids. LDHs consist

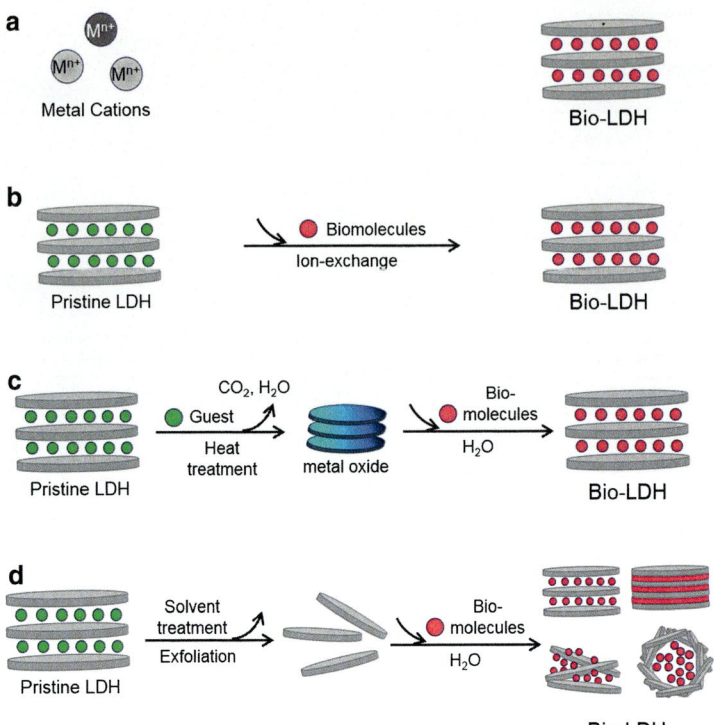

Fig. 2 Intercalation routes to bio-LDH nanohybrids. (**a**) coprecipitation, (**b**) ion-exchange, (**c**) calcination-reconstruction, and (**d**) exfoliation-reassembling (reproduced from [6])

of positively charged hydrotalcite-like layers of metal hydroxide and negatively charged molecules in the interlayer space satisfying the charge neutrality condition. Due to their high anionic-exchange capacity, various biomolecules with negative charge can readily be immobilized in the interlayer space of LDHs depending on the synthetic routes such as coprecipitation, ion exchange, calcination–reconstruction, and exfoliation–reassembling (Fig. 2) [6]. Such intercalative nanohybrids could be stabilized mainly by electrostatic, partly by coordinative bondings, and molecular interactions such as hydrogen bonding and van der Waals at the interfaces between LDH layers and interlayer biomolecules.

As described, the coprecipitation is a simple and economic way of preparing bio-LDH nanohybrids. In this process, the formation of 2D framework of LDH is made upon base titration of metal salt precursors in the presence of anionic biomolecules in a solution, and therefore, anions can simultaneously be incorporated inside the LDH lattice. The pH of a solution is generally adjusted to a coprecipitation region of corresponding metal hydroxides, whose range can also be decided after considering the pKa values of functional groups in biomolecules. The hydrothermal treatment can be subsequently made to improve the crystallinity and to control the particle size of LDHs. The ion-exchange route to LDH nanohybrids is a traditional way to replace interlayer anions with other guests like biomolecules. The anion-exchange capacity of LDHs is a critical factor for the ion-exchange reaction, and it depends on effective layer charge per unit area (charge density and charge distribution), particle size, and chemical compositions of LDHs. Some bio-LDH nanohybrids can be prepared by calcination-reconstruction method, thanks to the structural memory effect of LDHs. When calcined at around 400–500 °C, the pristine LDH particles are decomposed into mixed metal oxides with a formula [$M^{2+}_{1-x}M^{3+}_{x}O_{1+0.5x}$] due to the dehydroxylation reaction. The resulting mixed metal oxides can return topotactically to the pristine LDH upon rehydration and also turn out to be bio-LDH in the presence of biomolecules in an aqueous solution. The exfoliation–reassembling technique is also used for the encapsulation of bulky and large-sized biomolecules. Since the chemical exfoliation method to single-LDH nanosheets was recently developed, the reassembling reaction to bio-LDH hybrids would be actually quite simple and straightforward when the positively charged LDH sheets and negatively charged biomolecules were coexisting in an aqueous solution. More recently, several attempts were made to explore new morphologies with core–shell, hollow, and porous nanostructure and/or multilayered thin films under mild conditions.

2.2 Structural Features and Chemical Bonding Nature

Generally, structural features of intercalative nanohybrids based on LDHs are systematically characterized with combination of X-ray diffraction (XRD), X-ray absorption spectroscopy (XAS), nuclear resonance spectroscopy (NMR), molecular dynamics simulations, and microscopic tools. In particular, powder XRD is the main tool to analyze the crystal structure of 2D nanohybrids together with electron density mapping estimated from the well-developed (00l) reflections, which are the dominating feature of most intercalative nanohybrids corresponding to the basal spacing (d) that is increased to higher values upon intercalation. The d spacings of LDH nanohybrids can be calculated from Bragg's Law: $d = n\lambda/2\sin\theta$, where d is the distance between basal planes of LDHs, λ is the wavelength of X-ray, and θ is the diffraction angle. The d spacing reflects size, orientation, and interlayer arrangement of the charge-balancing guest molecules, whatever they are, inorganic or organic or biomolecules. Moreover, one-dimensional (1D) electron density map along the crystallographic c-axis can be obtained via Fourier transform analysis from the relative large number of (00l) reflections. From those results, one can suggest the intracrystalline structures, namely, the molecular orientations and arrangements of guest molecules in the interlayer space projected along the c-axis.

In this section, structural features of DNA–LDH nanohybrids are shown, and theoretical principles and experimental findings are also described in detail for better understanding of intercalation mechanism as well as physicochemical properties of bio-LDH nanohybrids for various medical applications.

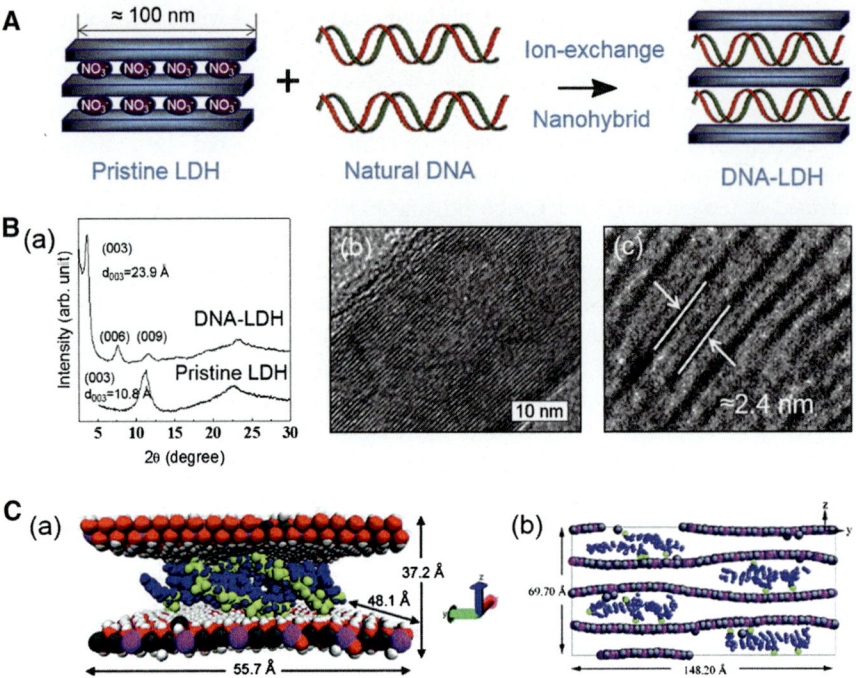

Fig. 3 (**A**) Schematic illustration of the ion-exchange route to DNA-LDH nanohybrid with layered structure (reproduced from [29]) (**B**) (a) XRD patterns and cross-section TEM images of the (b) pristine [Mg$_2$Al(OH)$_6$][NO$_3$] LDH and (c) DNA-LDH (reproduced from [13]) (**C**) Snapshots of computer simulations of the (a) intercalation of linear duplex DNA strands into [Mg$_2$Al(OH)$_6$][Cl] LDH and (b) partially ion-exchanged DNA in Daumas-Hérold configuration (reproduced from [54, 55])

As shown in Figs. 3 and 4, Choy et al. have reported two reaction routes to incorporate DNA molecules into [Mg$_2$Al(OH)$_6$] LDH (DNA–LDH nanohybrids) [10, 13]. One is the conventional way of intercalating natural DNA molecules into [Mg$_2$Al(OH)$_6$][NO$_3$] (DNA–LDH), namely, the ion-exchange reaction (Fig. 3A), and the other novel method is the encapsulation of designed DNA molecules into the core of spherical LDH nanoshell reassembled from the exfoliated [Mg$_2$Al(OH)$_6$]$^+$ LDH nanosheets (DNA@LDH) via exfoliation–reassembling reaction (Fig. 4A). For better understanding of these two DNA–LDH nanohybrids, the structural difference between earlier 2D DNA–LDH and recent core–shell DNA@LDH is visualized with XRD patterns and cross-section TEM images.

In the ion-exchange reaction, the d spacing was increased from 10.8 to 23.9 Å along the c-axis by replacing nitrate ions with DNA molecules (Fig. 3B-a). It corresponds to the van der Waals thickness of DNA molecules (≈20 Å) with double helix structure oriented parallel to the basal plane of LDH. The periodic lattice fringes caused by the (00l) serial planes were also observed in both pristine LDH and 2D DNA–LDH, exhibiting well-ordered layer structure with the d spacings of 0.8 nm for the pristine LDH and 2.4 nm for 2D DNA–LDH,

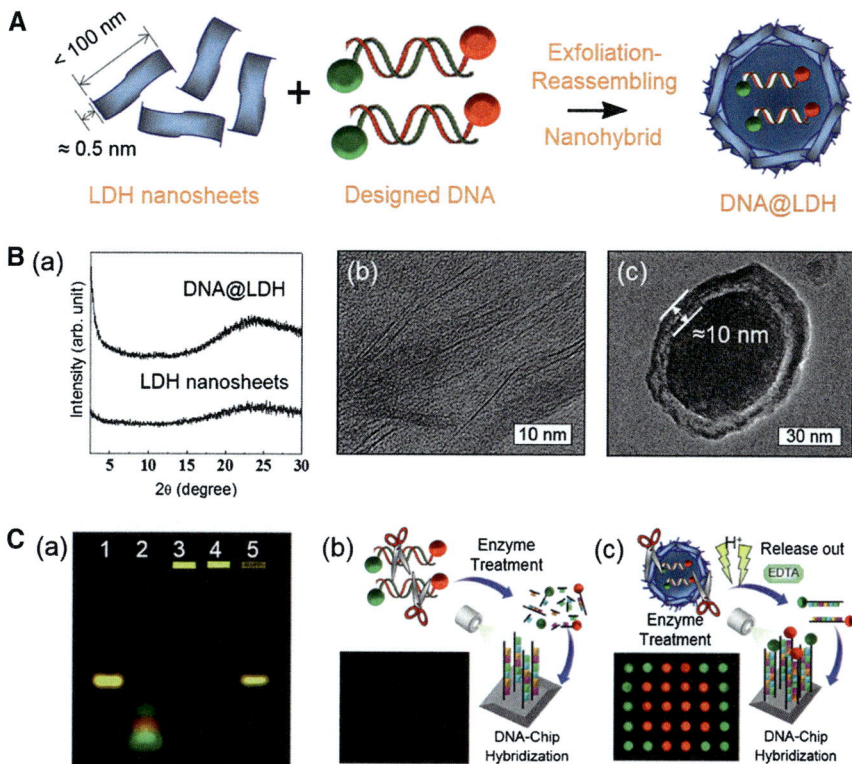

Fig. 4 (A) Schematic illustration of the exfoliation-reassembling route to DNA@LDH nanohybrid with core-shell structure (reproduced from [29]) (B) (a) XRD patterns and cross-section TEM images of the (b) [Mg$_2$Al(OH)$_6$]$^+$ LDH nanosheets and (c) DNA@LDH (reproduced from [13]) (C) (a) gel electrophoresis of DNA@LDH; lane 1: positive control of DNA; lane 2: DNA after DNase I treatment, lane 3: positive control of DNA@LDH, lane 4: DNA@LDH after DNase I treatment, lane 5: DNA released from DNA@LDH after DNase I treatment, DNA-chip assay images of the (b) DNA after DNase I treatment and (c) DNA released from DNA@LDH after DNase I treatment (reproduced from [13])

respectively (Fig. 3B-b, c). It was experimentally found that the helical DNA molecule is slightly more twisted than the intact DNA to fit the demanded area per unit charge of DNA (27.9–29.7 Å^2/e$^-$) into the equivalent area of the hexagonal LDH (25.0–26.6 Å^2/e$^-$) [56]. The spectra of circular dichroism (CD) supported the intracrystalline structure of the DNA–LDH based on the interlayer orientation of phosphate groups of DNA adjusted to aluminum site of LDH lattice. The collapse of CD band of DNA at 275 nm was observed not only for 2D DNA–LDH but also for the intact DNA in an aqueous solution with high-salt concentration, where electrostatic attraction is maximized between negative charge of DNA and positive charge of LDH to satisfy the charge neutrality condition.

Recent computer simulation studies on DNA–LDH nanohybrid have suggested the possible mechanism of DNA intercalation, which allows getting an insight on intracrystalline structure and a rationale for thermodynamic stability of DNA in

LDH (Fig. 3C) [54, 55]. As a result of molecular dynamic modeling, it was found that the LDH layers are flexible enough to deform around bulky DNA molecules. The calculated XRD patterns for all the simulated systems were in closer agreement with experimental results of Choy et al., who used ion-exchange method to intercalate DNA strands into Mg_2Al LDH [10], but were slightly different from those of Desigaux et al., who intercalated DNA strands into Mg_2Ga LDH by coprecipitation method [57]. It was theoretically demonstrated that the negatively charged phosphate groups of DNA are preferentially clustered closer to positively charged aluminum sites than magnesium sites along the LDH lattice. It was also found that the structural stability of intercalated DNA in LDH is significantly enhanced as compared to that of free DNA in bulk water because almost Watson–Crick hydrogen bonds can be retained in the Daumas–Hérold arrangement.

On the other hand, the XRD pattern of the core–shell DNA@LDH showed broad feature in the range of low Bragg's angle ($2\theta < 15°$), indicating the formation of amorphous metaphase (Fig. 4B-a). Its morphology turned out to be fairly spherical with uniform structure, and the particle size was determined to be around 120 nm. By cross-section TEM image, the DNA@LDH was found to be chemically well-defined core–shell nanostructure, which consisted of rationally designed DNA molecules in the core with a size of 100 nm and spherical LDH nanoshell with a wall thickness of 10 nm reassembled with LDH nanosheets (Fig. 4B-b, c). The spherical nature was induced by turbostratic multilayer reassembling of several tens of LDH nanosheets around DNA molecules. Such phenomena are often observed in the transformation of 2D nanosheets into 1D nanotubes by curling up gradually in a salt solution.

This DNA@LDH with spherical shape was the first example of core–shell-type bioinorganic nanohybrid, which is completely different from layer-by-layer type bioinorganic nanohybrids. It is, therefore, worthy to note that the exfoliation–reassembling route can provide facile method to encapsulate any size of biopolymers like DNA with genetic information and even supramolecular DNA architectures functionalized with nanoparticles. Structure and morphology of thus prepared core–shell nanohybrids would be a cytoskeleton and sponge-like heterostructure at nanoscale or a wire or worm appearance at the submicron scale, which could be controlled by flexibility of charge density and distribution as well as steric property of biomolecules.

For both 2D DNA–LDH and spherical DNA@LDH, the fragile and soft DNA molecules were found to be physically, chemically, and biologically stable inside LDH containers even after treatment with heat, oxidative radical stress, and endonuclease DNase I enzyme. As well known, the MgAl LDH is crystallized under basic condition but slowly dissolved in acidic environment, especially below pH 5. Thus, the intercalated and encapsulated DNA could be safely recovered from the LDHs without any alterations of their structural integrity by treating them with an acidic solution pH $\leq$ 5 in the presence of EDTA, a chelating agent, to trap Mg^{2+} and Al^{3+} ions dissolved from the LDH. The designed DNA released out from LDH nanoshell exhibited the same electrophoretic mobility as that of the free DNA (Fig. 4C-a). The unique base sequence of DNA was also protected from the enzymatic hydrolytic cleavage via a microarray chip assay, where single-stranded DNA with complementary sequence was embedded in the each spot to detect the target DNA (Fig. 4C-b, c). These results could be extended to a wide range of bio-inspired applications, such as advanced gene delivery system, biomedical diagnostics, and bioinformatics.

As shown in Fig. 5, Choy et al. came up with an idea that this research could be further explored to a new convergence system through combination of nanotechnology, biotechnology, and information and communication technology (NT-BT-

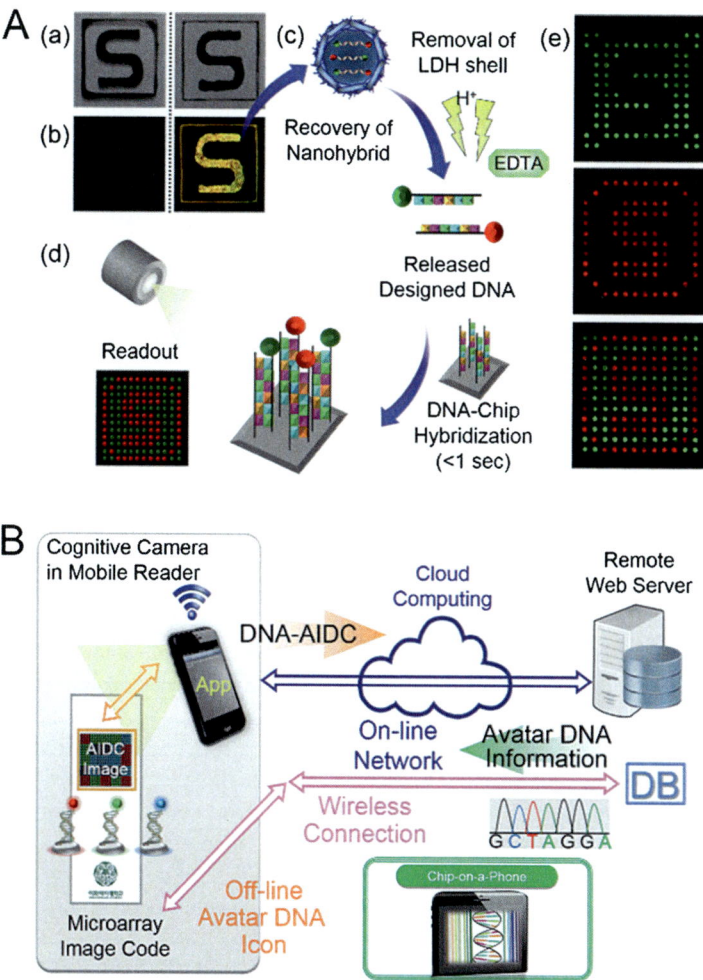

Fig. 5 (**A**) Nano-forensics with DNA@LDH. (a) Photographs of the [S] marking by pure ink (*left*) and by DNA@LDH ink (*right*), (b) fluorescence microscopy images, (c) schematic procedure for DNA detection, (d) readout images of the microarray (reproduced from [13]) (**B**) Schematic illustration of the Avatar DNA nanohybrid system (reproduced from [15])

ICT) [13, 15]. The DNA–LDH nanohybrid could be used as a genetic molecular coding system consisting of four steps: encoding, encrypting, decrypting, and decoding. This result was conceptually successful, but supplementary strategies were needed for practical application: (1) design of rational DNA with probing agents for on-site detection, (2) complete protection of bulky DNA molecules by LDH, and (3) rapid analyzing tool at extremely low concentration. For those reasons, Choy et al. suggested the nano-forensics as a newly adopted integrative concept that combines nanotechnology with forensic science. In this application, the DNA@LDH nanohybrid showed its potential as a security marking in DNA-based authentication (Fig. 5A). The integration of DNA@LDH with biosensor assay like a microarray detection was found to give solutions to the rapid and accurate DNA coding system at concentration as low as 100 pm within a time of 1 s. More recently, "Avatar DNA@LDH" nanohybrid was developed and then combined with electrical mobile smartphone device for more practical accessibility than the previous system (Fig. 5B). Information encoded in DNA sequence could be decoded through camera scan of smartphone for connecting an off-line DNA to an online web server network to provide secret message, index, or URL from database library. This NT-BT-ICT platform is continuously exploring for strategic vision detection and recording at molecular level to coordinate and integrate convergence channels.

3 Molecular Engineering for Multifunctional Bio-LDH Nanohybrids

The various bio-LDH nanohybrids have attracted much attention especially in controlled gene and drug delivery systems to maintain their therapeutic dosages during cancer treatments. Although they hold great promising aspects in biomedical applications, there are still limitations and drawbacks of using conventional bio-LDH nanohybrids in a passive type for advanced theranostics applications. The development of effective biomedical applications requires clear understanding of physical and chemical properties of LDHs that affect the interaction, toxicity, and fate in biological environments. Besides, simple bio-LDH nanohybrids should be upgraded to multiplex bio-LDH nanohybrids to meet the requirements of advanced theranostics system. Those active nanostructures involve the multifunction of an extended circulation time together with imaging tracking functions and tumor-targeted delivery for on-demand release to minimize harmful effects on normal cells.

Recently, engineering of bio-LDH nanohybrids have been explored for targeted delivery and functional imaging, as both internal and external surfaces could be further chemically functionalized with additional agents [58–60]. Figure 6 shows the evolution of chemical modification techniques at nanoscale level to improve bioavailability of bio-LDH and to impart new diagnostics and sensing functions.

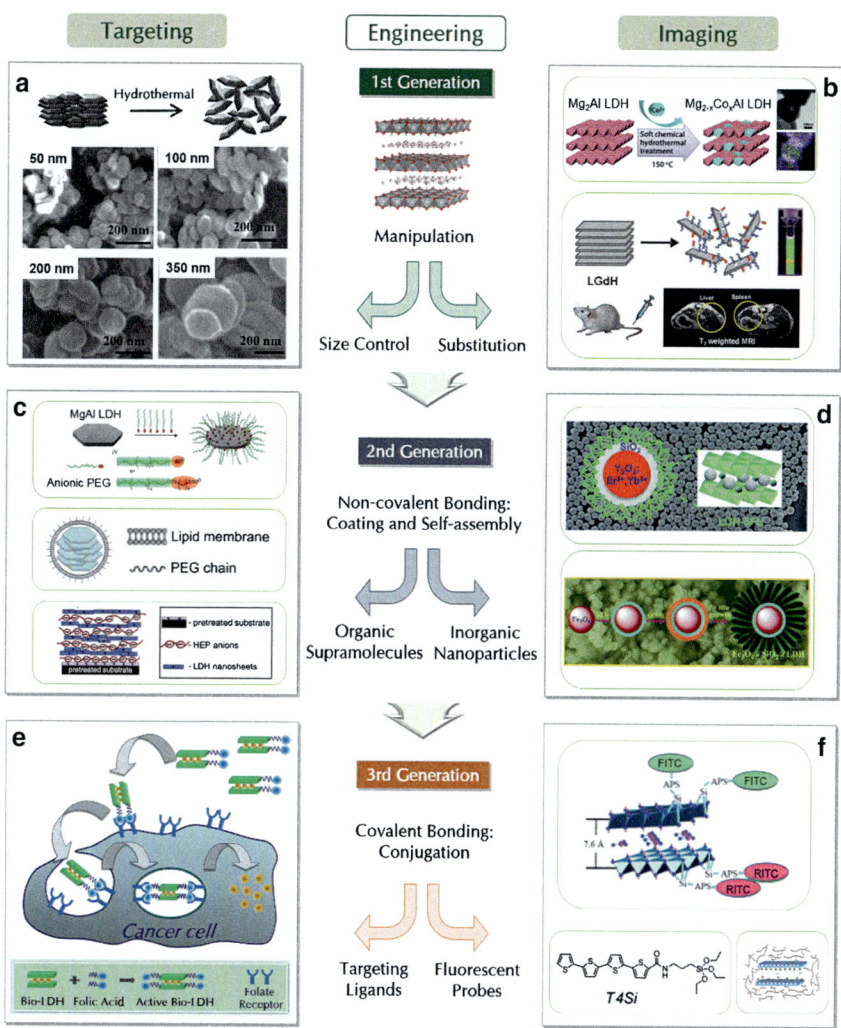

Fig. 6 Evolution of molecular engineering techniques of the bio-LDH nanohybrids for targeting and imaging functions. First generation: manipulation of bio-LDH nanohybrids. (**a**) Size control of LDH by hydrothermal treatment (reproduced from [61]) and (**b**) Intercalation of photo-functional agents or substitution of metal cations LDH layers (reproduced from [62, 63]) Second generation: non-covalent coating and self-assembly: (**c**) covering with organic supramolecules (reproduced from [64–66) and (**d**) assembly with inorganic nanoparticles (reproduced from [67, 68]) 3rd generation: covalent conjugation. (**e**) Grafting of active targeting ligands (reproduced from [37]) and (**f**) grafting of fluorescent probes (reproduced from [36, 69, 70])

First generation is the manipulation of structural properties of bio-LDH nanohybrids: (a) precise control of composition, size, and morphology of LDHs to enhance its passive targeting function and reduce its toxicity and (b) typical intercalation of photofunctional anions into LDH matrix or substitution of isomorphous metal cations into LDH lattice for fluorescent, optical, and MR imaging.

Second generation is the non-covalent coating and self-assembly: (c) covering of the bio-LDH materials with organic molecules or polymers to improve their bioavailability and biodistribution and (d) immobilization or incorporation of third functional components with photoluminescence and magnetic properties onto external surface of the bio-LDH nanohybrids for imparting novel multifunctionality. Third generation is the covalent conjugation of bio-LDH nanohybrids with (e) specific moieties for active targeting and (f) fluorescent probes for imaging and tracing both in vitro and in vivo.

3.1 Bio-LDH Nanohybrids with Targeting Functions

To obtain optimal therapeutic efficacy, therapeutic agents should be delivered to the intended sites, i.e., specific tissues or intracellular compartments, by two ways of accumulation process: passive targeting and active targeting [59, 60].

Generally, passive targeting is based on enhanced permeability and retention (EPR) effect caused by leaky vasculature around solid tumors. Thus, selective biodistribution is determined by blood circulation and extravasation after administration of specific nanohybrids through oral or injection route. Such passive targeting requires a long circulation half-life for sufficient delivery, and therefore, it is largely dependent on the particle size, surface charge, and other biophysical properties.

As investigated in many reports, the LDH nanoparticles with a size of around 80–200 nm and hydrophilic surface property were able to be accumulated in the tumor sites rather than other organs and tissues [37, 39, 40]. Most nanoparticles are synthesized with organic capping molecules to control their colloidal stability (dispersion), chemical functionality, surface modification, or biocompatibility. One of the solutions to manipulate the size of platelike LDH nanoparticles is the hydrothermal treatment at different temperature and aging time [61, 71, 72]. Oh et al. have successfully synthesized four different Mg_2Al LDHs in the range of 50–350 nm (Fig. 6a) [61]. The formation of crystal nuclei or nucleation of LDH can be generalized on the basis of crystal growth mechanism by the following equation: $L = (cv/8N)^{1/3}$, where L is the crystal size, c is concentration, v is molecular volume, and N is the number of crystal. The hydrothermal conditions for LDHs with an average particle size of 50 nm, 100 nm, 200 nm, and 350 nm were 80 °C for 6 h, 100 °C for 12 h, 200 °C for 24 h, and 200 °C for 48 h, respectively. In this way, the LDH nanoparticles with a narrow size distribution can be homogeneously dispersed in an aqueous solution enough to prepare stable colloidal suspension for injectable drug delivery system [73].

Indeed, Choy's group have intensively reported that size-controlled bio-LDH nanohybrid intercalated with anticancer drug, methotrexate (MTX)-LDH, which could target to tumor, and then strongly inhibit cancer cell growth compared to the controls. The MTX-LDH nanohybrid had a well-developed stacking structure with

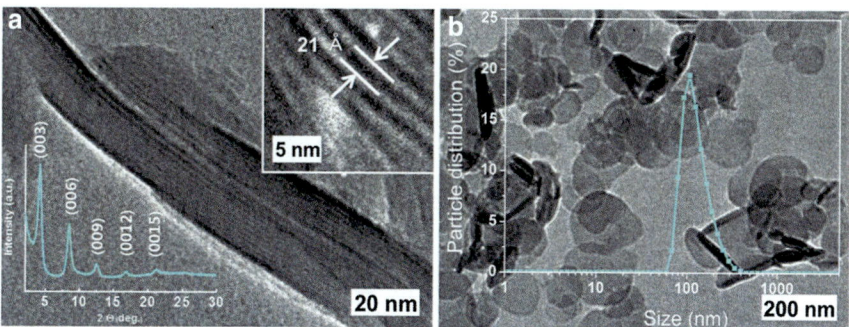

Fig. 7 (**a**) TEM image (side view) and XRD pattern (*inset image*), and (**b**) particle size distribution and HR-TEM image of MTX-LDH nanohybrid (reproduced from [41])

a interlayer distance of 21.0 Å between the periodic lattice fringes from the (00*l*) planes and a platelike morphology with an average particle size of 100 nm (Fig. 7). According to the in vitro cell culture experiments, the drugs or genes intercalated in LDHs were effectively transferred into the cell by clathrin-mediated endocytosis [32].

The anticancer efficiency and cellular uptake of MTX-LDH nanohybrids compared to MTX itself were evaluated by bioassay using *N*-methyl-*N*-nitro-*N*-nitrosoguanidine (MNNG) human osteosarcoma (HOS) cells. The IC_{50} value (at which concentration, 50 % of inhibition is measured) of MTX-LDH in MNNG/HOS cells was about 2.5 lower than that of MTX only, indicating that MTX-LDH is more effectively suppressing the proliferation of cancer cells than MTX only [74, 75].

Generally, an efficacy of new chemotherapeutic agents should be validated for their therapeutic responses in appropriate in vivo tumor models before preclinical and clinical studies for the FDA approval [76]. The xenograft mice model subcutaneously implanted with human tumor is useful to evaluate therapeutic efficacy and, in other cases, to identify targeting functions by predictive biomarkers. These mice models are technically straightforward and generate relatively successful results. On the other hand, the orthotopic models can provide an organ microenvironment in which the tumor with metastatic properties grows similar to clinical cases, so that such important information leads to more reliable translation for successful clinical trials.

Most recently, Choy's group have successfully demonstrated in vivo anticancer effect of the MTX-LDH nanohybrid in the orthotopic and xenograft tumor models [40, 41]. Especially, the 2-dimensional inorganic nanovector was used for injectable nanomedicine in real animal orthotopic breast cancer model for the first time. The pharmacokinetic properties, biodistributions, antitumor activities, and even survival rates and body weights of mice were systemically evaluated after injection of MTX-LDH into each tumor mice model (Fig. 8). The MTX-LDH hybrid system showed targeted tumor delivery and remarkable tumor inhibition effect with low toxicity in both in vivo cases based on the EPR effect. As shown in Fig. 8A-a, B-a,

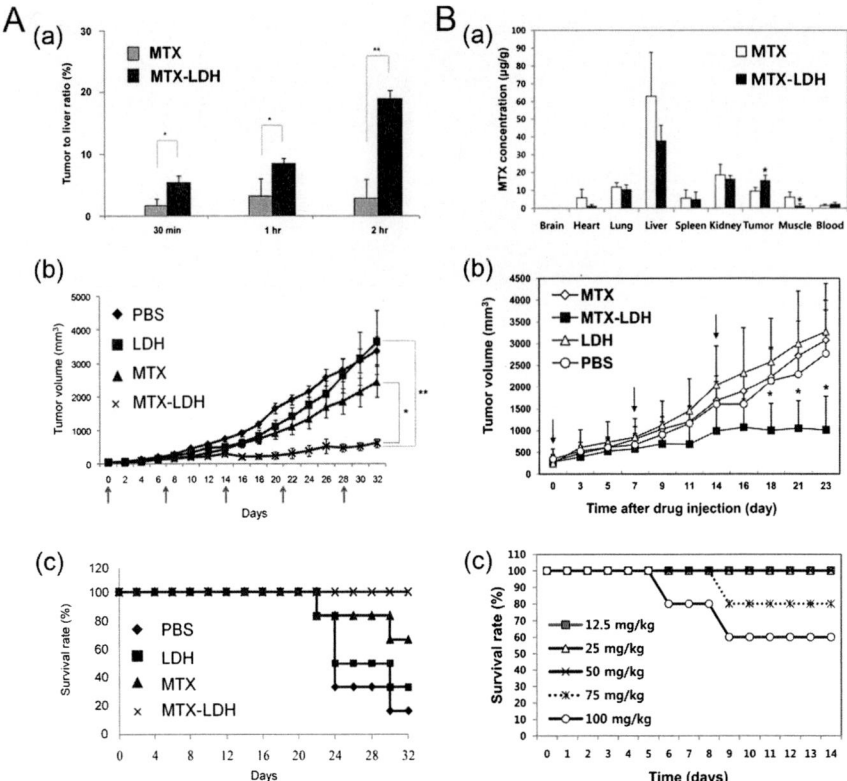

Fig. 8 (**A**) Orthotopic mice model by IP injection and (**B**) Xenograft mice model by IV injection: (a) biodistributions, (b) anti-tumor activity, and (c) survival rates of each tumor-bearing mice model (reproduced from [40, 41])

the tumor-to-liver bioaccumulation ratio for MTX-LDH compared with free MTX represents a 2.6-fold and a 2.4-fold increase at 1 h in the orthotopic and xenograft tumor models, respectively. Interestingly, the challenges to breast cancer (MCF-7/mot)-bearing orthotopic model represented a sixfold increase in the accumulation of tumor-to-liver ratio for MTX-LDH compared with free MTX at 2 h after intraperitoneal (IP) injection (Fig. 8A-a). Since the tumor-to-liver ratio is a very important indicator of both the therapeutic efficacy and the safety profile, the enhanced tumor-to-liver ratio for MTX-LDH indicates its high potential as an effective and safe systemic therapy for the treatment of cancer patients. As expected, the MTX-LDH showed enhanced therapeutic efficacy with 74.3 % reductions in tumor volume compared with free MTX (Fig. 8A-b). The groups treated with MTX-LDH or MTX exhibited survival rates of 100 % and 66.6 %, respectively (Fig. 8A-c). In the study on HOS-bearing xenograft model, tumor growth was significantly suppressed by treating with MTX-LDH compared to the control group (Fig. 8B-b). Furthermore, the MTX-LDH did not show acute toxicity up to dose of 50 mg/kg, and its LD_{50} value was determined to be higher than 100 mg/kg (Fig. 8B-c).

In the literature, zeta (ζ) potentials of MgAl LDHs are generally from +20 to +30 mV when the LDHs are monodispersed in an aqueous solution [13]. The electrokinetic potentials of the drug–LDH nanohybrids, however, were largely decreased in the simulated body fluid (SBF); thus the redispersed ones can be incompletely suspended with serious aggregates of average submicron size [39]. Such phenomena were resulted from the drastic subtraction of repulsive force between the nanohybrids. In the physiological condition, the positively charged drug–LDH is likely to tend to aggregate each other as well as to bind with nonspecific foreign substances in the bloodstream which are commonly negatively charged when clinically administrated into the human body. To increase the water solubility, bioavailability, circulation time, and accumulation in the targeted sites, those nanohybrids can be additionally coated with hydrophilic surfactants, lipid, enteric polymers, biopolymers, and biodegradable copolymers, e.g., tween [77], dextran [64, 78], phospholipid [79], polyethylene glycol (PEG) [80], PEGylated phospholipid [65], Eudragit® L100 [81], Eudragit® S100 [82], poly (lactic-co-glycolic acid) (PLGA) [83, 84], and heparin [66] (Fig. 6c).

When the LDH particles were coated with nontoxic hydrophilic polymer, anionic PEG derivatives, zeta potentials were changed from a positive value to a negative value in the range -20 to -25 mV depending on the feeding amount of the PEG [80]. Thus, obtained LDH/PEG nanocomposites showed a highly stable water stability and enhanced dispersity due to the steric hindrance of PEG and the electrostatic repulse of LDH.

Most recently, Yan et al. reported the PEGylated phospholipid-coated LDH (PEG-PLDH) with an average particle size of 130 nm for improvement of pharmacokinetic and antitumor activity in drug delivery system (Fig. 9) [65]. The LDH loaded with anticancer drug, methotrexate (MTX), was used coated with an anionic phospholipid, dioleoylphosphaidic acid (DOPA). The association of positively

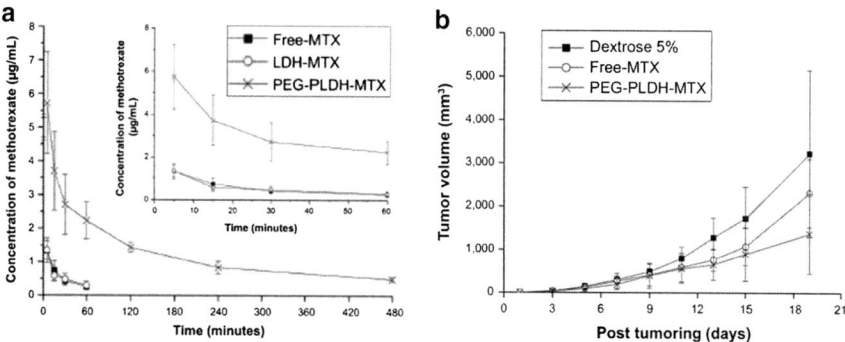

Fig. 9 In vivo improvement of pharmacokinetic and antitumor activity of the LDH by coating with PEGylated phospholipid. (a) Plasma concentration-time curves after IV administration to Sprague–Dawley rats. (b) Tumor growth rate in mice model after IV administration (reproduced from [65])

charged nanomaterials with negatively charged plasma proteins adversely influences their pharmacokinetics and pharmacodynamics. The negatively charged phosphatidic acid group of DOPA in the PEG-lipid conjugates readily formed liposomes around the LDH-MTX, and it aided materially in the shielding of positive charge of the LDH-MTX. It resulted in slower blood clearance and more effective passive target delivery. From in vivo pharmacokinetic study, a larger area under the concentration–time curve (AUC) and longer half-life were observed in the PEG-PLDH-MTX (Fig. 9a). The PEG-PLDH-MTX also showed the prolonged circulation time of 8 h after intravenous (IV) injection into the tail vein of rats. In vivo tumor model of S-180, the antitumor efficacy of PEG-PLDH-MTX, was significantly enhanced compared with 5 % dextrose, free MTX, and LDH-MTX (Fig. 9b). Those results can be facilitated by the longer blood circulation time and the lesser interactions with non-targeted biological environments such as endothelial cells, plasma proteins, and immune systems.

Some LDH–polymer hybrid films have been reported for blood compatibility and tissue regeneration. Chakraborti et al. developed LDH/PLGA films co-loaded with antibacterial and bisphosphonate for guided tissue regeneration membrane in periodontal therapy [83, 84]. The bio-LDH/PLGA hybrids were able to achieve significant increase in alkaline phosphatase activity and bone nodule formation bioavailability by providing better controlled drug-release kinetics, and therefore, they would be suitable implant matrices for supporting osteoblast proliferation and differentiation. Shu et al. developed nacre-like heparin/LDH ultrathin films via bottom-up layer-by-layer (LBL) deposition technique [66]. It simultaneously showed high-performance mechanical properties (modulus to 23 GPa) and good blood compatibility from the hemolysis and dynamic clotting time tests, which may find new medical applications.

On the other hand, the active targeting can be achieved through ligand receptor-mediated accumulation and uptake in cancer cells (Fig. 6e) [37]. Such more efficient delivery could be facilitated by modifying and engineering those nanohybrids with ligand moieties for specific interaction and recognition of target cells at molecular level. A wide variety of receptor-specific targeting moieties include small molecules, aptamers, peptides, and proteins (antibodies), and they can be covalently grafted to or grafted from the surface of nanohybrids during synthesis process.

The silylation, also known as silane grafting, is one of the most efficient and conjugation methods to introduce reactive functional group onto the surface of LDHs [69, 85–94]. The reaction of abundant hydroxyl ($-OH$) groups on the LDH surfaces with nucleophilic groups of organosilanes or aminosilane derivates to form Si-O-Si networks is a general way not only to modify the physicochemical properties of the LDH surface but also to covalently bond with further components, either directly or after activation of the graft via coupling agents.

Park et al. firstly attempted and reported surface modification of ZnCr LDH with silane, 3-aminopropyltriethoxysilane (APS), by covalent bonding [85]. A major difficulty of LDHs in surface modification is their high-surface charge density

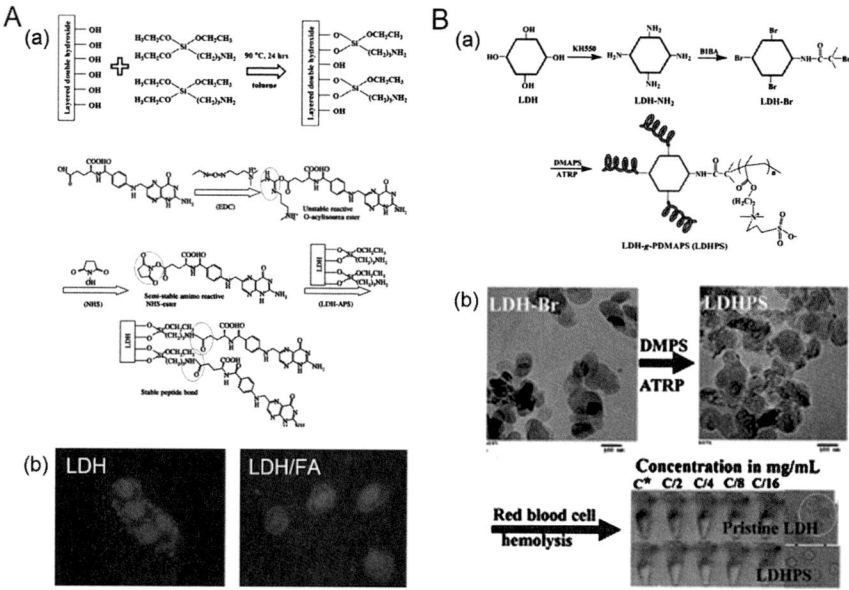

Fig. 10 (**A**) (a) LDH nanoparticles conjugated with FA and (b) fluorescence microscopy images of KB cells treated with either fluorescence-labeled LDH or LDH/FA (reproduced from [37]) (**B**) TEM images and hemolysis assay results of the LDH and LDH/PS (reproduced from [86])

(ca. one positive charge per 25 Å^2 in the molar ratio of Mg^{2+}/Al^{3+}, $x = 0.32$) [95] compared with other clays (e.g., from 0.6 to 1.4 per O_{20} unit for the unit cell charge of smectite clays) [96]. In order to overcome such a problem, a two-step approach was taken in their study: pre-intercalation of long alkyl-chain organic molecules, sodium dodecyl sulfate (SDS), to improve the hydrophobicity of LDH surface and then addition of cationic surfactant, N-cetyl-N,N,N-trimethylammonium (CTAB), in organic polar solvent to introduced silane molecules to condense with the –OH groups of LDH.

More recently, there are some studies on the LDH modification for biology and medical purposes. Oh et al. reported for the first time both LDH nanoparticles and MTX-LDH nanohybrid conjugated with active targeting moiety molecules for enhanced drug delivery system (Fig. 10A) [37]. This process involved previous step of the APS-based silylation for surface activation followed by grafting reaction of cancer cell-specific ligand, folic acid (FA). It was proven that the formation of silylated LDH products occurred mostly on the outer surface of them with a higher population of bidentate (T^2) and tridentate (T^3) bonding characters. The FA ligands were then covalently bounded onto the surface of both LDH (LDH/FA) and MTX-LDH (MTX-LDH/FA) through thiourea bond, i.e., by the reaction between the NH_2 moieties of APS with the amino-reactive NHS-ester bond of FA coupled with N-hydroxysuccinimide (NHS). The MTX-LDH/FA showed higher cellular uptake ability and anticancer drug efficacy than MTX-LDH in the folate receptor overexpressing KB cell line, which was resulted from the targeting function of ligand conjugation and subsequent ligand–receptor-mediated endocytosis. Chen group suggested a self-assembled type of FA-conjugated LDH nanoparticles. In their study, APS FA was attached to SDS-LDH nanoparticles under ultrasonication treatment, where CTAB was used to extract SDS by forming hydrophobic salts between CTAB and SDS in the interlayer of the LDH. Those nanoparticles loaded with MTX (27.4 wt%) also exhibited strong cell penetration and improved efficacy of MTX in KB cells.

As mentioned above, the blood compatibility of LDH surfaces is of crucial issue for in vivo applications. Well-defined hemocompatible LDH conjugated with poly (sulfobetaine) was developed by Mao group (Fig. 10B) [86]. Surface-initiated atom transfer radical polymerization (ATRP) of zwitterionic polymer, 3-dimethyl (methacryloyloxyethyl) ammonium propane sulfonate (DMPS), has been successfully employed to the LDH nanoparticles (LDH/PS) for anticoagulation and effectiveness in resisting thrombosis. From the several investigations for the blood compatibilities, such as coagulation tests, complement activation, platelet activation, hemolysis assay, morphological changes of red blood cells (RBCs), and cytotoxicity assay, the hemocompatibility of LDH/PS was found to be substantially enhanced without causing any hemolysis.

3.2 Bio-LDH Nanohybrids with Imaging Functions

Molecular imaging combined with employment of nanomaterials can provide new functionalities such as unique fluorescence, optical, magnetic, or radioactive properties for enhanced contrast, signal amplification, controlled quantification, and targeted biodistribution. Such imaging functions can be induced by LDH nanoparticles for disease diagnostics and tumor visualization in a biological system. In order to impart such functions to LDH, several ways have been reported so far: (1) intercalation of organic fluorophores or rare-earth complexes into LDH matrix, (2) substitution of radioactive elements or rare-earth metals into LDH lattices, (3) self-assembly with metal or metal oxides nanoparticles, and (4) conjugation of LDHs with photofunctional components.

Choy's group firstly investigated not only in vitro cellular uptake behaviors but also in vivo biodistribution using MgAl LDH nanoparticles with either green fluorescent dye, fluorescein isothiocyanate (FITC), or red fluorescent dye, rhodamine B isothiocyanate (RITC) (Fig. 6f) [11, 33, 35, 36, 38].

According to the in vitro cell culture experiments, the drugs or genes intercalated in LDHs were effectively transferred into the cell by clathrin-mediated endocytosis [33]. What would happen if inorganic nanovehicles, such as LDHs, were in contact with cellular membranes? A challenging study has been systematically made in vitro and in vivo, especially involving cellular behaviors such as intercellular uptake mechanism [33] and intracellular trafficking pathway [35] and

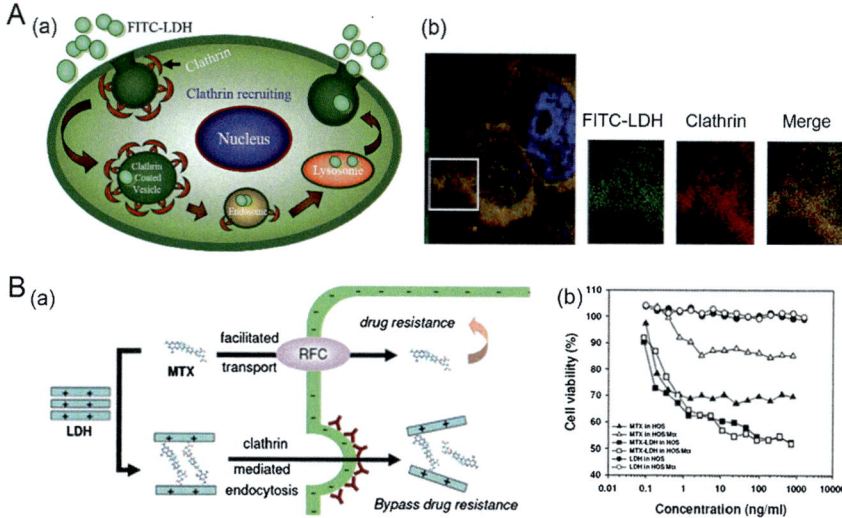

Fig. 11 (**A**) Intercellular uptake mechanism of the LDH nanoparticles: (a) schematic illustration of the proposed clathrin-mediated endocytosis and (b) fluorescence microscopic images of MNNG/HOS cells treated with FITC-LDH (*green*) (reproduced from [33]). (**B**) The mechanism of action of MTX-LDH for bypassing drug resistance and (b) comparative growth inhibition profile of LDH, free MTX, and the MTX-LDH in wild-type HOS and HOS/Mtx cell lines (reproduced from [34])

biocompatibility and safety depending on the physicochemical properties and particle size of LDH nanovehicles [42–44]. According to the immunofluorescence microscopic studies in in vitro, it was clearly proven that LDH nanoparticles were internalized inside the cells through the clathrin-mediated endocytosis (Fig. 11A). When the same experiments were made with the drug only and the drug–LDH hybrid, respectively, the drug molecules inside the cell were detected several times more for the latter than the former, thanks to the clathrin-mediated endocytosis. Though no targeting ligands or antibodies were grafted on the surface of LDHs, a passive targeting effect could be observed due to the self-assembling of clathrin-related membrane proteins to form vesicles and vacuoles effectively wrapping alien LDH nanoparticles to transfer them inside the cells. However, such a selective permeation of LDH nanovehicle in in vitro is no more valid if the particle size is beyond ~300 nm [36].

One more thing to underline is that the LDH nanovehicle could bypass the drug resistance [34]. According to the cellular permeation mechanism, MTX can be internalized in the cell through reduced folate carrier (RFC) and folate receptor sites, since it behaves as a folate antagonist. When MTX is once internalized in the cell, it conjugates to the dihydrofolate reductase, deactivates its enzymatic reaction involved in folate cycle (conversion of FH2–FH4), consequently inhibits the DNA synthesis and cell division, and finally drives the cells to death. If MTX molecules were encased in LDH by intercalation reaction, namely, dressed in camouflage mask with LDH, the MTX-LDH hybrid drug could be internalized inside the cell not through RFC and folate receptor site but through the clathrin-mediated endocytosis (Fig. 11B). Therefore, more MTX could be delivered for the MTX-LDH hybrid than for the intact MTX in the MTX-resistant cell culture lines, since MTX must suffer from drug resistance due to the repeated administrations with a high dosage. The diffusion rate and pathway of therapeutic gene and drug out of the LDHs in endosome is also controlled by intercellular/intracellular trafficking pathway of LDHs.

Choy's group suggested the intracellular fate and trafficking mechanism of LDH nanoparticles employing different-sized LDHs (50 nm and 100 nm) conjugated with FITC (Fig. 12) [35]. Intracellular uptake was investigated in HOS cells by quantitatively analyzing the amount of nanoparticles co-localized in specific intracellular subcompartments such as early endosomes, late endosomes, lysosomes, Golgi complex, and endoplasmic reticulum, respectively. These organelles are involved in transport, storage, release, and degradation. Once LDH nanoparticles are taken up by cells, they follow endocytosis and exocytosis pathways; one is a typical endocytic pathway represented as early endosome → late endosome → lysosome, and the other is an exocytic pathway represented as early endosome → Golgi apparatus → endoplasmic reticulum → Golgi apparatus. The 100 nm LDH nanoparticles were massively discovered in the exocytic Golgi apparatus, whereas the 50 nm ones were almost equally observed in lysosomes and Golgi apparatus. From the results, it is apparent that most of the 100 nm LDH nanoparticles escaped typical endolysosomal degradation. Therefore, it is concluded that the size of LDH nanoparticles plays key roles in intercellular uptake and intracellular trafficking to maximize therapeutic efficacy of drug or gene intercalated in LDHs. Especially, the 100 nm LDH nanoparticles can be more useful in gene delivery system, since they can steer clear of the lysosome containing diverse hydrolysis enzymes to digest nucleic acid.

In vivo tissue distribution of LDH nanoparticles in normal mice model was studied by Choy's group to obtain information on distribution in target-specific organs and permeation of blood–brain barrier (BBB) [38]. To study tissue distribution, they prepared 100 nm LDH nanoparticles conjugated with RITC (LDH/RITC) via silane-coupling reaction and thiourea bond formation and then intraperitoneally injected into female Balb/c mice. As shown in Fig. 13a, the LDH/RITC was largely accumulated in the kidney, liver, lung, and spleen due to reticuloendothelial system (RES) uptake, but not virtually observed in the brain and heart. According to histopathological examination, no remarkable abnormality, morphological damage, and inflammation response were observed in overall tissues in the presence of effective dose (100 mg/kg), which could be considered of practical medical use in drug delivery system (Fig. 13b).

On the other hand, radioisotopes have been directly substituted into LDH lattices. Musumeci et al. have directly doped radioactive $^{57}Co^{2+}$ and $^{67}Ga^{3+}$ into LDH for generation of radiolabeled LDHs [97]. Recently, Oh et al. were successful in preparing Co^{2+}-substituted MgAl LDH with a regular particle size under hydrothermal conditions, suggesting possible application in positron emission tomography (PET) (Fig. 6b) [62]. In addition, MR contrasting agents such as Gd(III) metal

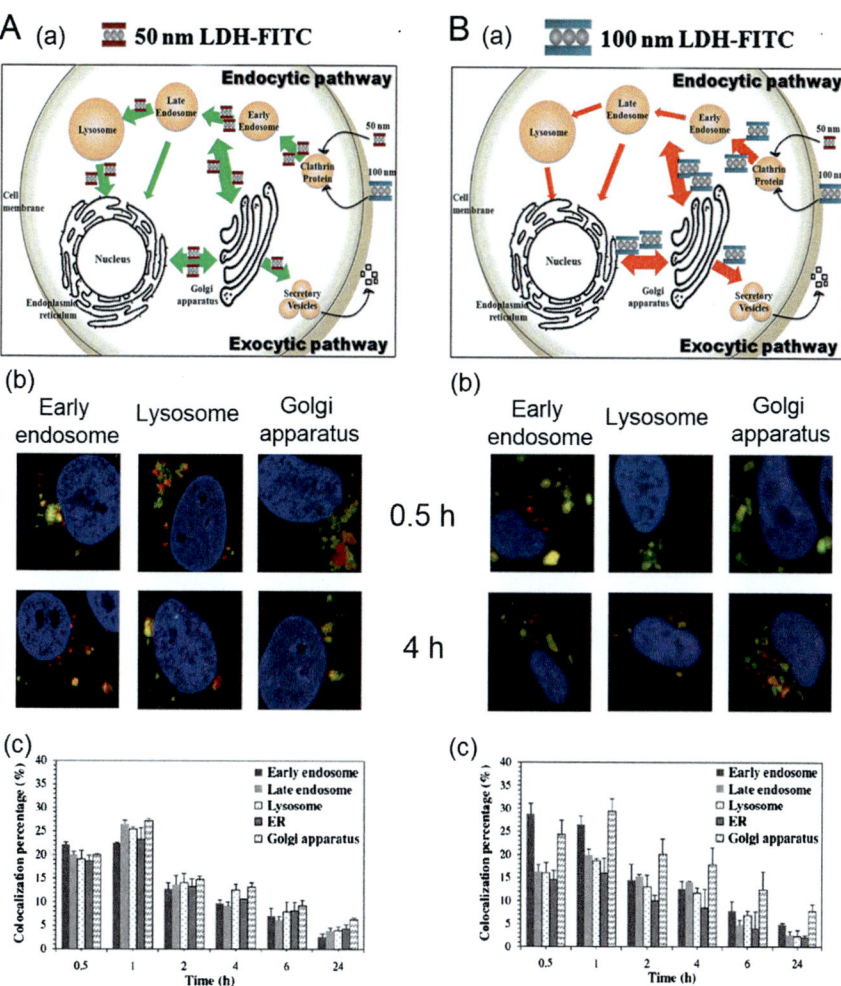

Fig. 12 Intracellular trafficking pathway of the LDH nanoparticles with a size of (**A**) 50 nm and (**B**) 100 nm: (a) schematic diagram of endocytic and exocytic pathways, (b) confocal microscopic images of the HOS cells treated with LDH-FITC (*green*) of 50 and 100 nm co-localized (*yellowish*) in specific organelles (*red*), early endosome, lysosome, and Golgi apparatus at 0.5 and 4 h, respectively, and (c) quantitative co-localization analysis of the LDH-FITC nanoparticles (reproduced from [35])

cation and its complex, Gd(III)-DTPA, have been incorporated into LDHs by substitution or intercalation [98, 99]. Yoo et al. demonstrated Gd-doped LDHs as multifunctional nanoparticles and as a novel delivery platform for microRNA therapeutics [100]. Byeon et al. have developed another type of LDH containing only Gd metal (LGdH) for bimodal imaging function (Fig. 6b) [63, 101]. From the results of cytotoxicity test and T_2-weighted MRI experiment in vivo, the stable colloid of LGdH nanosheets modified with FITC and PEG was found to be

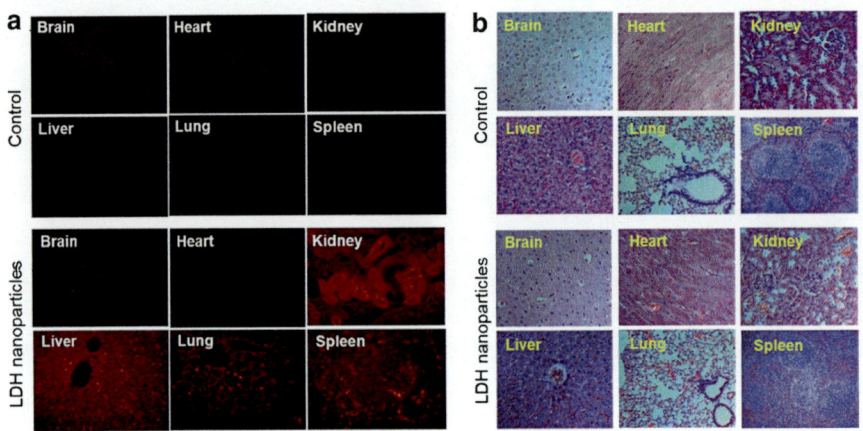

Fig. 13 (**a**) Tissue distribution and (**b**) histopathological examination of the LDH/RITC with a size of 100 nm in mice after IP injection with a dose of 600 mg/kg (reproduced from [38])

biocompatible and effective as a multimodal probe for MR and photoluminescence imaging.

Lanthanide (Ln) substitution and intercalation of lanthanide complexes have been made in order to develop new LDHs with photoluminescence property. Several Ln(III) cations, e.g., Eu^{3+}, Tb^{3+}, Yb^{3+}, Tb^{3+}, Nd^{3+}, and Sm^{3+}, have been substituted into various LDHs lattice. For example, Musumeci et al. reported Tb-doped MgAl LDH material with strong green fluorescence property [102]. Chen et al. reported photoluminescence properties of Eu-doped ZnAl LDH. Typical $^5D_0 \rightarrow {}^7F_J$ ($J = 0$–4) transitions of Eu^{3+} were clearly observed due to the structural substitution of the Ln in the LDH layers [103]. Rocha's group developed multi-wavelength luminescent LDH materials by co-doping Eu^{3+} and Nd^{3+} into hydrocalumite-type LDH [102]. The complex has been reported to emit a large vis/NIR spectral region with potential biomedical imaging applications. Recently, Posati et al. were able to prepare nano-sized LDH doped with Eu^{3+}, Yb^{3+}, Tb^{3+}, or Nd^{3+} in confined environment following microemulsion method [104]. Furthermore, there are also some reports on photoluminescent LDHs along with intercalation of organic sensitizers to enhance the degree of excitation [105–107]. Organic sensitizers with strong UV-absorption property have commonly been intercalated to convert light energy into the excited states of Ln cations. Such incorporation of the photofunctional anions into the Ln-doped LDH matrix can provide stable and robust photostability. It is proposed that good photoluminescent LDH nanoparticles with modified targeting ligands may become one of the novel optical imaging materials.

The bio-LDH nanohybrids can be extended by self-assembling with third functional inorganic components including metal or metal oxide particles for biological applications [67, 68, 108–111]. Chen et al. suggested new design of bio-LDH nanohybrid intercalated with chemotherapeutic drug, 5-fluorouracil (5FU), at the same time functionalized with luminescent properties, especially, the near-infrared

(NIR) upconversion luminescence ($Y_2O_3:Er^{3+},Yb^{3+}$@SiO_2@LDH-5FU) via in situ growth method [67]. The bio-LDH nanohybrids not only exhibited strong red fluorescence under the excitation of 980 nm laser for in vitro cell imaging but also showed superior anticancer efficacy in cancerous cells. Duan's group developed core–shell microspheres containing three functional components, Fe_3O_4@SiO_2@LDH, via in situ growth method for magnetic separation of proteins [68]. This report provides a facile approach to prepare hierarchical functionalized nanostructures, which have potential bio-imaging applications.

Posati et al. reported LDH nanoparticles conjugated with organic oligothiophene fluorescent compound (T4Si) by using direct microwave-assisted silylation for producing fluorescent and biocompatible nanocomposites [69]. Wei et al. recently developed innovative chitosan-coated LDH nanoparticles intercalated with a NIR fluorescent dye indocyanine green (ICG), contrast agent with US Food and Drug Administration (FDA) approval for clinical use by electrostatic interaction for in vivo NIR optical imaging [70]. The coating of chitosan (CS) was achieved on the external surface of LDHs-NH_2-ICG by using glutaraldehyde as a cross-linked agent (Fig. 14A). Typical huge accumulation of LDHs-NH_2-ICG-CS in the liver and lungs was observed in vivo biodistribution through noninvasively optical imaging (Fig. 14B). From NIR optical image results of dissected organs from a

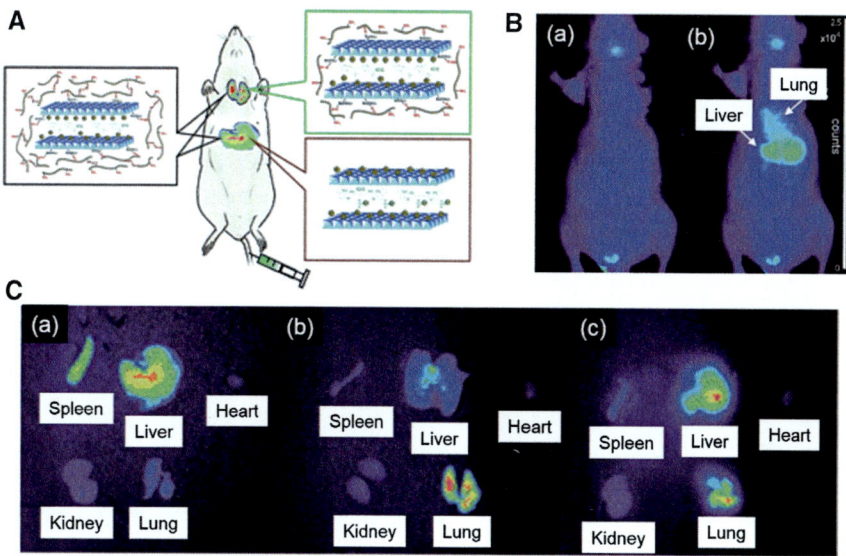

Fig. 14 (A) Schematic illustration of the LDHs-NH_2-ICG-CS and their biodistributions. (B) In vivo optical images of LDHs-NH_2-ICG-CS-2 in nude mice (a) before and (b) after (1.5 h) IV tail vein injection. (C) Ex vivo optical images after (3 h) IV injection of (a) LDHs-NH_2-ICG, (b) LDHs-NH_2-ICG-CS-1, and (c) LDHs-NH_2-ICG-CS-2 (reproduced from [70])

sacrificed mouse (Fig. 14C), it was demonstrated that the degree of coating for the LDHs-NH$_2$-ICG can offer organ-specific drug delivery systems; uncoated LDHs-NH$_2$-ICG targeted liver and spleen, mono chitosan-coated one (LDHs-NH$_2$-ICG-CS-1) targeted lungs, and double chitosan-coated one (LDHs-NH$_2$-ICG-CS-2) targeted liver and lungs. Those bio-LDH nanohybrids could function as high-potential theranostics materials for both clinical cancer diagnosis and chemotherapy.

4 Bio-LDH Nanohybrids for Theranostics Applications

4.1 Gene-LDH Theranostics In Vitro and In Vivo

As mentioned above, LDH nanoparticles are quite useful as delivery vehicle to target the cell nucleus, in which the information of therapeutic genes can be transmitted, and subsequently function of the cell can be regulated by expression of specific genes [112, 113]. Choy et al. suggested for the first time the use of As-myc-LDH nanohybrid in nonviral gene therapy [11]. They were successful in intercalating antisense oligonucleotide, As-myc with sequence of 5′-(AACGTTGAGGGGCAT)-3′, to effectively transport it to HL-60 cancer cells (Fig. 15A). The HL-60 cells treated with As-myc-LDH exhibited time-dependent inhibition on cell proliferation, indicating nearly 65 % of inhibition on the growth compared to the untreated cells. Since it has already been proven in various cell culture lines that LDH itself is neither harmful to cell growth or division nor critical to cell membrane, the suppressive effect of leukemia cell growth is purely affected by delivery ability of LDH. It was also reported that the growth inhibition effect is time and dose dependent.

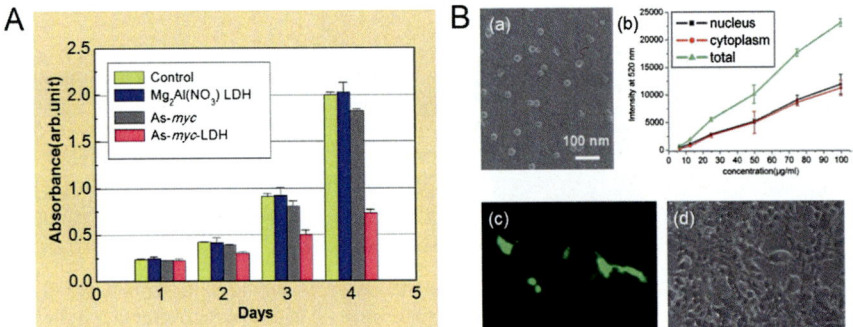

Fig. 15 (**A**) Effect of As-myc-LDH nanohybrid on the growth of HL-60 (reproduced from [11]) (**B**) (a) SEM image of the 20 nm Mg$_3$Al LDH nanoparticles, (b) cellular uptake curves for the LDH-FITC with NSC 34 cells, (c) fluorescence image and (d) bright field image of cells with pEGFP-N1 expressed after transfection (reproduced from [114])

Giannelis's group reported intercalation, delivery, and expression of the gene-encoding green fluorescence protein (GFP) by using LDH in various types of cells including 9L glioma cells, JEG3 choriocarcinoma placental cells, and cardiac myocytes [115]. All cells expressed the gene with up to 90 % transfection efficiency. Lu and O'Hare recently developed fluorescein isothiocyanate isomer I labeled Mg_3Al LDH nanoparticles with size as small as ca. 20 nm and, at the same time, demonstrated their cellular uptake and low toxicity for potential gene delivery (Fig. 15B) [114]. As concentration of the LDH/FITC nanoparticles was increased, the amount of intracellular distribution in nucleus and cytoplasm was linearly increased, indicating an enhanced nucleus translocation of the LDH/FITC nanoparticles. From TEM and confocal images, it was found that 20 nm LDHs nanoparticles were internalized into the cytoplasm with subsequent enrichment in the cellular nucleus, while LDH nanoparticles larger than 20 nm were only located in the cytoplasm. The 20 nm LDH nanoparticles loaded with plasmid DNA (pEGFP-N1) were then successfully transfected into NSC 34 cells for the expression of GFP.

Xu's group has intensively reported that small and regular LDH nanoparticles can efficiently deliver therapeutic small interfering RNA (siRNA) molecules into the cytoplasm of cancer cells for siRNA-based mRNA degradation [116, 117]. More recently, they suggested co-delivery system of Allstars Cell Death siRNA (CD-siRNA) and anticancer drug (5-FU) through LDH nanoparticles for further enhanced efficacy of cancer treatment (Fig. 16A) [118]. Fluorescence image data clearly demonstrated entry of red siRNA 456-5-FU/LDH nanohybrids into cancer cells with subsequent delivery of 5-FU drug and simultaneous transfection of siRNA gene into the cytoplasm via clathrin-mediated endocytosis. It was also clear that the proposed combination treatment of the CD-siRNA/5-FU/LDH nanohybrid achieved more enhanced efficacy than single treatments of either CD-

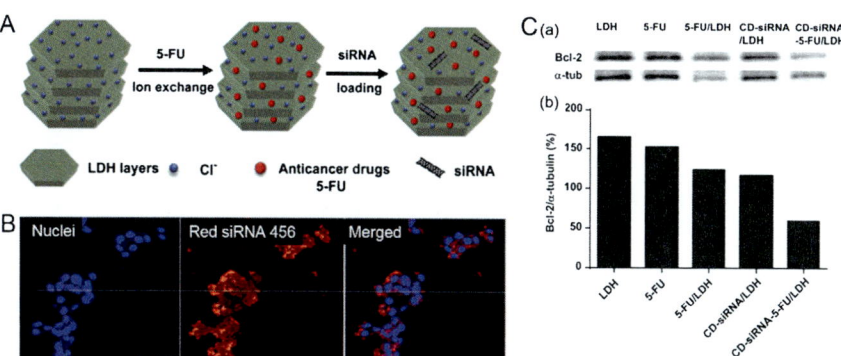

Fig. 16 (A) Schematic diagram of the LDH co-delivery system loaded with 5-FU and siRNA. (B) Confocal microscopy images of MCF-7 cells treated with red siRNA 456-5-FU/LDH nanohybrid (C) Suppression of Bcl-2 protein expression in MCF-7 cells treated with CD-siRNA/5-FU/LDH nanohybrid: (a) western blot analysis and (b) quantitative evaluation of the expression levels of Bcl-2 protein against α-tubulin (reproduced from [118])

siRNA-LDH or 5-FU/LDH in the suppression of Bcl-2 protein expression in MCF-7 cells.

Although the combination of chemotherapeutic genes and drugs showed high potential as a novel chemotherapy approach, it is hypothesized that effective theranostics would be promising as an ideal delivery system that can incorporate therapeutic agents with imaging probes. Currently, Choy's group is exploring theranostic siRNA-LDH nanohybrids system for targeted cancer therapy and imaging function (Fig. 17) [119]. Two groups of siRNA-LDH nanohybrids were developed for the in vivo delivery of siRNA with advanced cancer treatment. One is a passive type of siRNA-LDH nanohybrids for passive delivery based on EPR effect followed by clathrin-mediated endocytosis. The other is active type of siRNA-LDH/FA nanohybrids conjugated with folic acid ligand aiming at folate receptor (FR)-mediated active cancer targeting. All of the nanohybrids were made with 100 nm LDH nanoparticles conjugated with fluorescent probe, FITC (LDH/FITC and LDH/FITC/FA), via silane coupling. The negatively charged siRNA was self-assembled with positively charged LDH or LDH/FA by electrostatic interaction at weight ratio of 1:25 and 1:50, respectively. Each siRNA-HTN and siRNA-LDH/FA was uniform platelets with a size of 100 nm enough to avoid endolysome degradation (Fig. 17a). In vitro Survivin inhibition efficacy of siRNA-LDH in KB cell was observed to be 50 % in Survivin mRNA expression regardless of the existence of free FA in the media, indicating internalization of siRNA-LDH via clathrin endocytosis pathway (Fig. 17b). In contrast, siRNA-LDH/FA downregulated Survivin expression by more than 70 % under FA-free condition in KB cell line, in which the receptor binding endocytosis occurred completely between free FA and FA conjugated with siRNA-LDH/FA. Both the LDH/FITC and LDH/FITC/FA showed selective tumor targeting over other organs including lungs, liver, and spleen, which are the main sites for the accumulation of typical nanoparticles by RES uptake (Fig. 17c). Fluorescence intensity of tumor treated with LDH/FITC/FA was 1.5-folds higher than that of the one treated with LDH/FITC. It indicated that the FA ligand conjugated onto the surface of LDH nanoparticles could offer Survivin siRNA uptake by tumors in a short period of time while minimizing nonspecific uptake of siRNA by undesired tissues. Treatment of LDH/FA hybridized with Survivin siRNA showed not only high transfection in FR-overexpressing KB cells in vitro followed by effective Survivin mRNA expression but also significant downregulation of tumor volume in vivo after intraperitoneal injection into tumor xenograft nude mice bearing KB cells (Fig. 17d).

Yan et al. developed LDH nanoparticles incorporated with antigen and co-stimulatory ligand for protein-based vaccination purposes to induce both cell-mediated (Th1) and antibody-mediated (Th2) immune responses [120]. In such a system, polarized immune response was modulated with a toll-like receptor ligand, CpG, by immobilizing them onto the LDH nanoparticles, and a model antigen, ovalbumin (OVA), was used to deliver LDH nanoparticles into antigen-presenting cells for validation of adjuvant activity (LDH-CpG-OVA) (Fig. 18A). In vivo adjuvant activity of LDH nanoparticles was investigated by subcutaneously (SC) injecting each sample into C57BL/6 female mice followed by determining

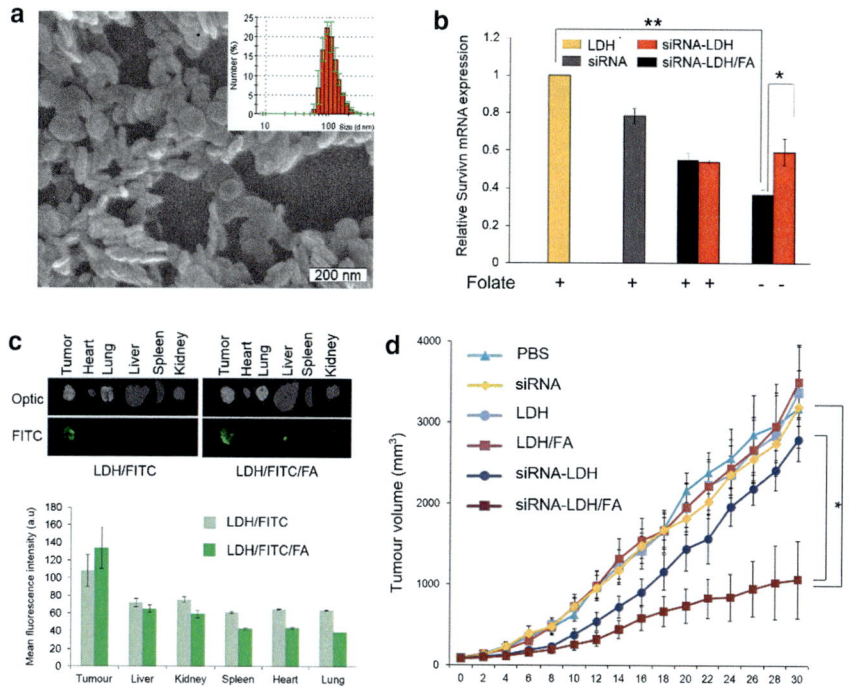

Fig. 17 (**a**) SEM image and size distribution of the siRNA-LDH/FA. (**b**) In vitro quantitative RT-PCR analysis of Survivin in KB cell after treatment with siRNA-LDH or siRNA-LDH/FA in the presence of free FA or in the absence of free FA in the media. (**c**) In vivo fluorescence images (*top*) and quantitative analysis of tumors (*bottom*) after (2 h) IP injection of the LDH/FITC (*left*) and LDH/FITC/FA (*right*). (**d**) Anti-tumour efficacy of xenograft tumor mice bearing KB cells via IP injection of the siRNA-LDH and siRNA-LDH/FA (Tumor sizes are presented as means ± SE. $*p < 0.05$ vs. siRNA-LDH treated group; $**p < 0.01$ vs. PBS-treated group.) (reproduced from [119])

the specific anti-OVA antibody titer in serum with ELISA. Co-transportation of OVA and CpG with LDH nanoparticles prompted much higher levels of both IgG1 and IgG2a antibodies (Fig. 18B). Significant induction of Th1-polarized immune response by LDH-CpG-OVA was observed; especially, LDH-CpG-OVA produced almost fourfold higher antibody than commercial adjuvant aluminum hydroxide phosphate (Alum) conjugated with CpG and OVA (Alum-CpG-OVA) at 35 days. Interestingly, adjuvant effect of LDH-CpG-OVA on antitumor response to OVA-expressing B16/F10 tumor cells positively contributed in the reduction of tumor growth more than Alum-CpG-OVA (Fig. 18C).

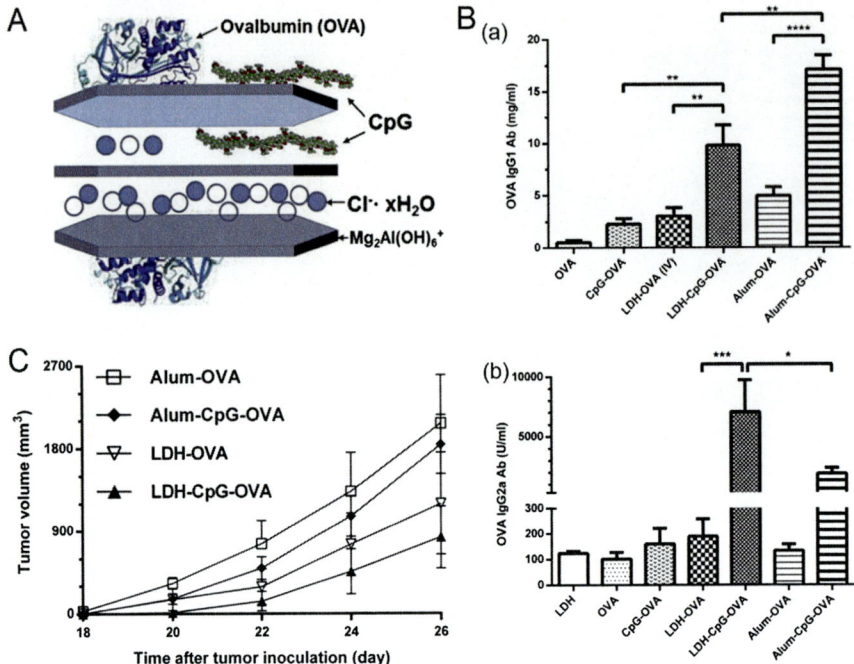

Fig. 18 (**A**) Schematic illustration of the LDH-based adjuvant-antigen hybrid system with OVA surface-adsorbed and CpG surface-adsorbed or intercalated. (**B**) Histograms of the OVA-specific serum (a) IgG1 and (b) IgG2a production in C57BL/6 female mice at 35 days following immunization with OVA coupled with LDH or Alum, with or without CpG (data are expressed as mean ± S.E.M. ***$p < 0.001$, **$p < 0.01$, *$p < 0.05$.). (**C**) Comparison of LDH and Alum adjuvant activity on OVA-specific anti-tumor protective immunity (reproduced from [120])

4.2 Drug–LDH Theranostics In Vitro and In Vivo

A variety of pharmaceutically active drugs including anticancer drugs, anti-inflammatory drugs, and cardiovascular drugs have been hybridized with LDH nanoparticles for sustained drug release, prolonged drug half-life, increased drug accumulation, and enhanced drug efficacy [121, 122].

Most recently, Zhu et al. reported novel platinum (Pt)-loaded LDH nanoparticles for efficient and cancer cell-specific delivery of a Pt-based anticancer drug (Fig. 19A) [123]. The disuccinatocisplatin (DSCP) is a member of the most effective Pt(IV) cisplatin prodrug for cancer treatments including testicular, ovarian, cervical, and non-small cell lung cancer. However, such Pt(IV) complex can lead to high toxicity, severe side effects, and drug resistance. The cytotoxicity of cisplatin prodrug in normal cells was reduced after intercalation. It was found that cancer cells were able to uptake LDH-Pt(IV) nanohybrid, which subsequently bonded to genomic DNA, stopped cell cycle, and induced apoptosis (Fig. 19B).

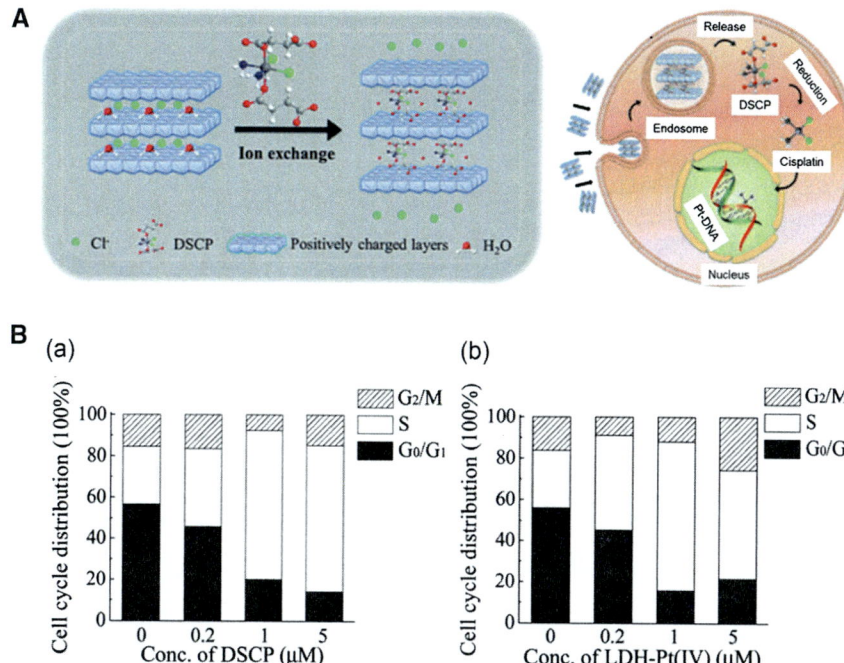

Fig. 19 (**A**) Schematic diagram of the intracellular pathway of LDH-Pt(IV). (**B**) Cell cycle arrest of A2780 cells after treatment with different concentrations of (a) DSCP and (b) LDH-Pt(IV) (reproduced from [123])

Li et al. developed hierarchically structured core–shell nanospheres with multifunction such as controlled release rate of anticancer drug (doxorubicin (DOX), fluorescence imaging by FITC, magnetic susceptibility from magnetic nanoparticles (MSPs), and folate-targeting capability (DOX@MSPs/Ni-LDH-folate) [124]. Interestingly, they found acidic cytoplasm as a key factor for the dramatic anticancer efficacy of DOX@MSPs/Ni-LDH-folate in only cancer cells (Fig. 20). In the acidic cytoplasm, the amide bond of DOX-COOH inside the interlayer of LDH could be cleaved, and thus protonated DOX drug could be quickly released out from LDH into cytoplasm. The results were apparent in cancer cells (Hela cell) but not in normal cells (HEK 293 T cell). This unique release capability of nanohybrid has a great potential in selective drug delivery system only after tumorous cell uptake.

Shi's group has successfully demonstrated diagnosis and simultaneous drug delivery monitored by noninvasive visualization in vitro and in vivo [45]. They developed platform material, Gd-doped LDH nanoparticles followed by deposition of Au nanoparticles (LDH-Gd/Au), serving both as drug vehicle and diagnostic agent (Fig. 21A). The LDH-Gd/Au was able to transport DOX into the cancer cell via endocytosis followed by intracellular release of loaded DOX in the acidic

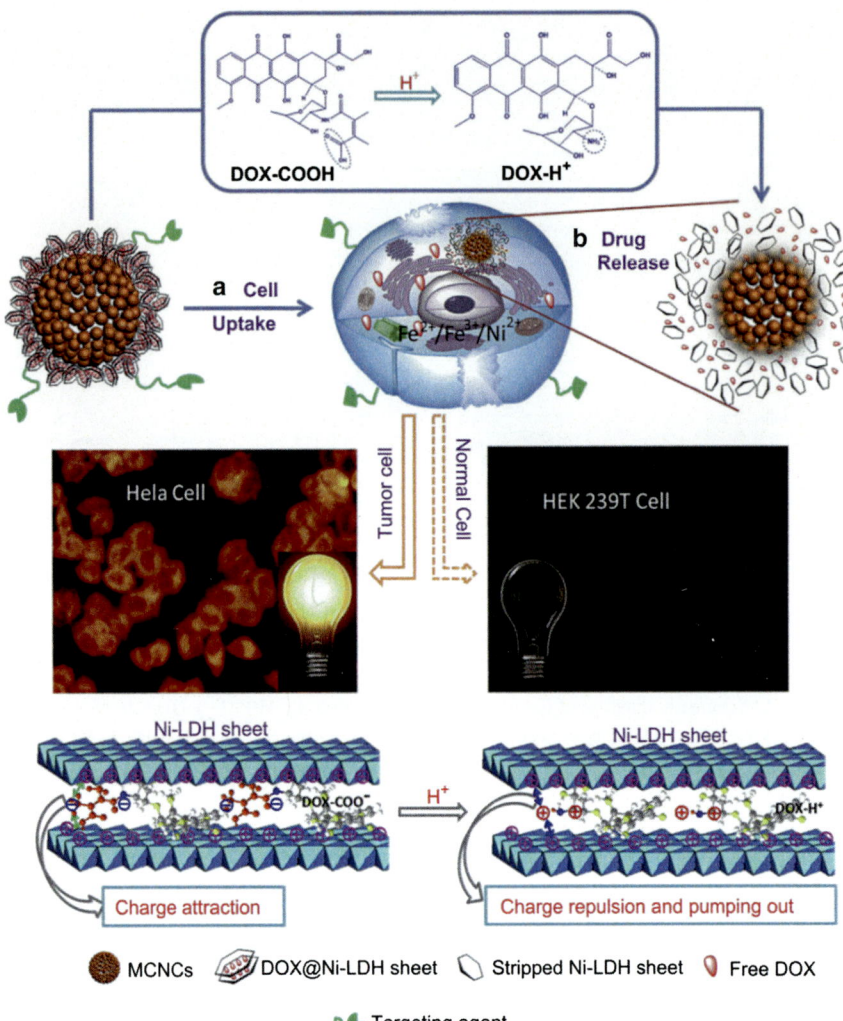

Fig. 20 Schematic illustration of the plausible DOX release mechanism of DOX@MSPs/Ni-LDH-folate in cytoplasm of tumor cells. (**a**) The different fluorescent behaviors of Hela and HEK 293 T cells incubated with the nanohybrid. (**b**) The mechanism of pumping positive-charged DOX out of the LDH nanoparticles (reproduced from [124])

cytoplasm and inducing death of cancer cells. The pharmaceutical activity of DOX loaded inside LDH-Gd/Au was much larger than that of free DOX in the HeLa cell line (Fig. 21B). In addition, LDH-Gd/Au nanohybrid exhibited not only better in vitro CT and T_1-weighted MRI performance than commercial agents, Iobitridol and Magnevist, but also feasible and effective in vivo dual imaging function. It was also demonstrated that LDH-Gd/Au nanohybrids were gradually accumulated in the

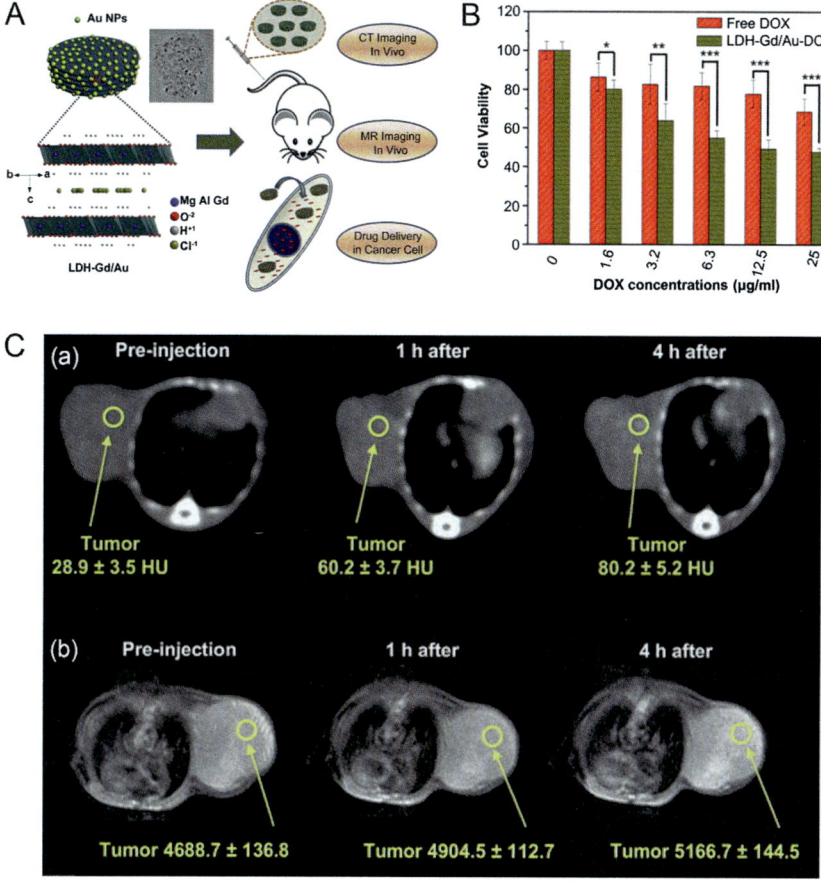

Fig. 21 (A) Schematic illustration of the LDH-Gd/Au as a theranostics platform for dual modal CT-MR imaging and anti-cancer drug delivery (inset: TEM image of the LDH-Gd/Au). (B) HeLa cell viabilities when exposed to free DOX and LDH-Gd/Au-DOX at different concentrations ($*p < 0.05$, $**p < 0.01$, $***p < 0.001$). (C) (a) CT and (b) T_1-weighted MR images of tumor after IV injection of LDH-Gd/Au-heparin in 4 T1 murine breast tumor-bearing mice for 0 h, 1 h and 4 h (reproduced from [45])

liver and spleen through nonspecific RES uptake mechanism in normal mice. After surface modification of LDH-Gd/Au nanohybrid with heparin, in vivo CT and MRI were carried out on 4 T1-murine breast tumor-bearing mice via intravenous injection. It was clearly found in dual modal CT-MR imaging that LDH-Gd/Au-heparin nanohybrid selectively targeted the tumor in vivo through EPR effect and was then uptaken by cancer cells for pH-responsive anticancer drug delivery (Fig. 21C). Consequently, the theranostics system of bio-LDH nanohybrids is no longer limited to other multifunctional nano-biomaterials with respect to structural features, controlled release, diagnostics, and targeting functions.

5 Summary and Outlook

The LDHs hold a number of strategic properties, such as unique host–guest interactions, tunable particle size, biocompatibility, high loading capacity of biomolecules, protection of genes and drugs within interlayer galleys during the delivery process, pH-controlled solubility, and sustained release. Accordingly, substantial efforts have been made by Choy's group and many scientists to develop soft-chemical intercalation methods and to demonstrate uptake of LDH nanoparticles into the nucleus of cancer cells for efficient drug and gene delivery. Dynamic computer simulations have provided information on structural features of bio-LDH nanohybrids with detailed information on molecular arrangement of the biomolecules inside interlayer space and have supported chemical interactions at the interface between biomolecules and LDH layers including charge density and distribution, energy functions, influence of interlayer water on the stability of structure, etc.

The manipulation and size controlling of LDHs are of crucial importance in the designing of new biomedical hybrid materials. The most fascinating feature is that the LDH nanoparticles with a regular size of 100 nm play key roles in clathrin-mediated intercellular uptake to overcome drug resistance and in the intracellular trafficking to escape endolysosomal degradation in cancer cells. The surface properties of LDHs are actively functionalized via molecular engineering techniques, such as covalent coupling of cancer cell-specific ligands. Therefore, such improvements in bio-LDH nanohybrids extend their biomedical applications to theranostics applications for cancer therapy and simultaneous functional diagnosis. These integrated and multifunctional approaches can significantly improve efficiency of targeted delivery and imaging system both in vitro and in vivo. The representative LDH nanoparticles for various theranostics purposes are briefly summarized in Table 1.

In this review, we highlighted rapid progress and outstanding figures for both passive and active bio-LDH nanohybrids for theranostics applications with respect to two aspects—systematic targeting of therapeutic agents and imaging of tumor and cancer cells—which thus far have been demonstrated to possess high potential for clinical trials. However, further new platforms with intelligent communicating system may still be necessary to increase patient compliance and reduce nonspecific effects on normal cells in cancer patients. The ideal LDH nanoparticles and next generation of their bionanohybrids would be challenged in order to compromise several critical issues shared among different nanoparticles and biosystems for efficient clinical use following the approval by FDA.

Table 1 A brief summary of representative theranostics applications of LDH nanoparticles

LDH host	Theranostics agents	Molecular engineering	Application	Key feature	Refs.
MgAl	–	Size control	Biosafety study	Biocompatibility	[73]
MgAl	FITC	Intercalation and size control	Intercellular imaging	Intercellular uptake mechanism study: clathrin-mediated endocytosis	[33]
MgAl	FITC	Size control and silane coupling	Intracellular imaging	Intracellular fate and trafficking mechanism study: endolysosomal escape for only 100 nm	[35]
MgAl	MTX	Size control	Anticancer drug delivery without repeated administrations with a high dosage	Bypass drug resistance	[34]
MgAl	MTX, folic acid, and FITC	Silane grafting	Cancer cell-specific drug delivery with active targeting moiety	Ligand receptor-mediated cellular uptake in cancer cells	[37]
MgAl	MTX	Size control	Anticancer drug delivery in in vivo xenograft model	EPR targeting effect: 2.4-fold tumor-to-liver accumulation	[41]
MgAl	MTX	Size control	Anticancer drug delivery in in vivo orthotopic model	EPR targeting effect: 2.6-fold tumor-to-liver accumulation	[40]
MgAl	MTX	PEGylated phospholipid coating	Anticancer drug delivery and pharmacokinetic study in vivo	Longer blood circulation time and EPR targeting effect	[65]
MgAl	DMPS	ATRP	Hemocompatibility study	Blood compatibility and anti-biofouling property	[86]
MgAl	RITC	Size control and silane coupling	Biodistribution study	Large accumulation in kidney, liver, lung, and spleen by RES uptake, but not in brain and heart	[38]
MgAl	Indocyanine green and chitosan	Intercalation and covalent coating	In vivo NIR optical imaging	Organ-specific drug delivery system with noninvasive optical imaging	[70]
MgAl	As-myc (antisense)	Intercalation	Gene-based cancer therapy	The first work on gene delivery system with high transfection efficiency	[11]

(continued)

Table 1 (continued)

LDH host	Theranostics agents	Molecular engineering	Application	Key feature	Refs.
MgAl	pEGFP-N1 (plasmid DNA)	Self-assembly	Delivery and expression of gene-encoding green fluorescence protein (GFP)	Internalization into cytoplasm and subsequent enrichment in cellular nucleus of 20 nm LDH nanoparticles	[114]
MgAl	5-FU and siRNA	Intercalation and self-assembly	Co-delivery system of drug and gene	Enhanced therapeutic efficacy of combined treatment	[118]
MgAl	siRNA, folic acid, and FITC	Size control, silane coupling, and self-assembly	Theranostic nanohybrids system for targeted gene-based cancer therapy and imaging function	Selective tumor targeting and siRNA-based mRNA degradation in vitro and in vivo	[119]
MgAl	Ovalbumin (antigen) and CpG (ligand)	Self-assembly	Protein-based vaccination	Significant induction of both cell-mediated (Th1) and antibody-mediated (Th2) immune responses	[120]
MgAl	Disuccinatocisplatin (Pt (IV) cisplatin prodrug)	Intercalation	Cancer cell-specific delivery of Pt-based anticancer drugs	Reduction of high toxicity, severe side effects, and drug resistance of Pt (IV) complex cisplatin prodrug in normal cells	[123]
NiAl	Doxorubicin, FITC, folic acid, and magnetic supraparticles	Self-assembly	Selective drug delivery system with fluorescence cellular imaging	Controlled release of anticancer drug in only acidic cytoplasm of cancer cells	[124]
MgAl	Doxorubicin, Gd metal, and Au NP	Substitution and self-assembly	Diagnosis and simultaneous drug delivery in vitro and in vivo	Dual modal CT-MR imaging with selectively targeted cancer therapy in vivo through EPR effect	[45]

Acknowledgments This work was supported by a National Research Foundation of Korea (NRF) Grant funded by the Korean Government (MSIP) (2005-0049412).

References

1. Whittingham MS, Jacobson AJ (eds) (1982) Intercalation chemistry. Academic Press, New York, NY
2. Bruce DW, O'Hare D (eds) (1997) Inorganic materials. Wiley, New York, NY
3. Ogawa M, Kuroda K (1995) Chem Rev 95:399
4. Choy JH (2004) J Phys Chem Solids 65:373
5. Paek SM, Oh JM, Choy JH (2011) Chem Asian J 6:324
6. Park DH, Hwang SJ, Oh JM, Yang JH, Choy JH (2013) Prog Polym Sci 38:1442
7. Khan AI, O'Hare D (2002) J Mater Chem 12:3191
8. Auerbach SM, Carrado KA, Dutta PK (eds) (2004) Handbook of layered materials. CRC Press/Marcel Dekker, Boca Raton, FL/New York, NY
9. Williams GR, O'Hare D (2006) J Mater Chem 16:3065
10. Choy JH, Kwak SY, Park JS, Jeong YJ, Portier J (1999) J Am Chem Soc 121:1399
11. Choy JH, Kwak SY, Jeong JJ, Park JS (2000) Angew Chem Int Ed 39:4041
12. Evans DG, Duan X (2006) Chem Commun 5:485
13. Park DH, Kim JE, Oh JM, Shul YG, Choy JH (2010) J Am Chem Soc 132:16735
14. Oh JM, Park DH, Choy JH (2011) Chem Soc Rev 40:583
15. Park DH, Han CJ, Shul YG, Choy JH (2014) Sci Rep 4:4879
16. Choy JH, Oh JM, Park M, Sohn KM, Kim JW (2004) Adv Mater 16:1181
17. Oh JM, Biswick TT, Choy JH (2009) J Mater Chem 19:2553
18. Ruiz-Hitzky E, Aranda P, Darder M, Rytwo G (2010) J Mater Chem 20:9306
19. Bruce DW, O'Hare D, Walton RI (eds) (2010) Low-dimensional solids. Wiley, Chichester
20. Choy JH, Choi SJ, Oh JM, Park T (2007) Appl Clay Sci 36:122
21. Kim KM, Kang JH, Vinu A, Choy JH, Oh JM (2013) Curr Top Med Chem 13:488
22. Cavani F, Trifirò F, Vaccari A (1991) Catal Today 11:173
23. Duan X, Evans DG (eds) (2006) Layered double hydroxides. Springer, Heidelberg
24. Rives V (ed) (2001) Layered double hydroxides: present and future. Nova Science, New York, NY
25. Forano C, Hibino T, Leroux F, Taviot-Gueho C (2006) Handbook of Clay Science. Elsevier, Amsterdam
26. Miyata S (1985) US Patent 5,514,389
27. Jettka W, Gajdos B, Benedikt MD (1996) Eur. Patent EP 715846, to Rhone-Poulenc Rorer Gmbh, Germany
28. TALCID® (2000) Bayer Healthcare. http://www.meppo.com/pdf/drugs/1676-TALCID-1328540040.pdf
29. Oh JM, Park DH, Choi SJ, Choy JH (2012) Recent Pat Nanotechnol 6:200
30. Choy JH, Kwak SY, Park JS, Jeong YJ (2001) J Mater Chem 11:1671
31. Hwang SH, Han YS, Choy JH (2001) Bull Korean Chem Soc 22:1019
32. Khan AI, Lei L, Norquist AJ, O'Hare D (2001) Chem Commun 22:2342
33. Oh JM, Choi SJ, Kim ST, Choy JH (2006) Bioconjug Chem 17:1441
34. Choi SJ, Choi G, Oh JM, Oh YJ, Park MC, Choy JH (2010) J Mater Chem 20:9463
35. Chung HE, Park DH, Choy JH, Choi SJ (2012) Appl Clay Sci 65:24
36. Oh JM, Choi SJ, Lee GE, Kim JE, Choy JH (2009) Chem Asian J 4:67
37. Oh JM, Choi SJ, Lee GE, Han SH, Choy JH (2009) Adv Funct Mater 19:1617
38. Choi SJ, Oh JM, Choy JH (2009) J Ceram Soc Jpn 1175:543
39. Choi G, Kim SY, Oh JM, Choy JH (2012) J Am Ceram Soc 95:2758

40. Choi SJ, Oh JM, Chung HE, Hong SH, Kim IH, Choy JH (2013) Curr Pharm Des 19:7196
41. Choi G, Kwon OJ, Oh Y, Yun CO, Choy JH (2014) Sci Rep 10:1038
42. Choi SJ, Oh JM, Choy JH (2008) J Mater Chem 18:615
43. Choi SJ, Choy JH (2011) J Mater Chem 21:5547
44. Choi SJ, Choy JH (2011) Nanomedicine 6:803
45. Wang X, Li J-G, Zhu Q, Li X, Sun X, Sakka Y (2014) J Alloys Compd 603:28
46. Zerhouni E (2003) Science 302:63
47. Martin CR (2006) Nanomedicine 1:5
48. MIT Technology Review (2006) 10 breakthrough technologies. http://www2.technologyreview.com/tr10/?year=2006
49. Wagner V, Dullaart A, Bock AK, Zweck A (2006) Nat Biotechnol 24:1211
50. Kelkar SS, Reineke TM (2011) Bioconjug Chem 22:1879
51. Lee DE, Koo H, Sun IC, Ryu JH, Kim K, Kwon IC (2012) Chem Soc Rev 41:2656
52. Ferrari M (2005) Nat Rev Cancer 5:161
53. Kievit FM, Zhang M (2011) Adv Mater 23:H217
54. Thyveetil MA, Coveney PV, Greenwell HC, Suter JL (2008) J Am Chem Soc 130:4742
55. Thyveetil MA, Coveney PV, Greenwell HC, Suter JL (2008) J Am Chem Soc 130:12485
56. Oh JM, Kwak SY, Choy JH (2006) J Phys Chem Solids 67:1028
57. Desigaux L, Belkacem MB, Richard P, Cellier J, Léone P, Cario L, Leroux F, Taviot-Guého C, Pitard B (2006) Nano Lett 6:199
58. Ariga K, Ji O, McShane MJ, Lvov YM, Vinu A, Hill JP (2012) Chem Mater 24:728
59. Zhang L, Lia Y, Yu JC (2014) J Mater Chem B 2:452
60. Yhee JY, Lee S, Kim K (2014) Nanoscale 6:13383
61. Oh JM, Hwang SH, Choy JH (2002) Solid State Ion 151:285
62. Kim TH, Lee WJ, Lee JY, Paek SM, Oh JM (2014) Dalton Trans 43:10430
63. Yoon YS, Lee KS, Im GH, Lee BI, Byeon SH, Lee JH, Lee IS (2009) Adv Funct Mater 19:3375
64. Huang J, Gou G, Xue B, Yan Q, Sun Y, Dong LE (2013) Int J Pharm 450:323
65. Yan M, Zhang Z, Cui S, Lei M, Zeng K, Liao Y, Chu W, Deng Y, Zhao C (2014) Int J Nanomedicine 9:4867
66. Shu YQ, Yin PG, Liang B, Wang S, Gao L, Wang H, Guo L (2012) J Mater Chem 22:21667
67. Chen C, Yee LK, Gong H, Zhang Y, Xu R (2013) Nanoscale 5:4314
68. Shao M, Ning F, Zhao J, Wei M, Evans DG, Duan X (2012) J Am Chem Soc 134:1071
69. Posati T, Melucci M, Benfenati V, Durso M, Nocchetti M, Cavallini S, Toffanin S, Sagnella A, Pistone A, Muccini M, Ruani G, Zamboni R (2014) RSC Adv 4:11840
70. Wei PR, Cheng SH, Liao WN, Kao KC, Weng CF, Lee CH (2012) J Mater Chem 22:5503
71. Xu ZP, Stevenson GS, Lu CQ, Lu GQ, Bartlett PF, Gray PP (2006) J Am Chem Soc 128:36
72. Xu ZP, Stevenson G, Lu CQ, Lu GQ (2006) J Phys Chem B 110:16983
73. Choi SJ, Oh JM, Choy JH (2008) J Nanosci Nanotechnol 8:5297
74. Choy JH, Jung JS, Oh JM, Park M, Jeong J, Kang YK, Han OJ (2004) Biomaterials 25:3059
75. Oh JM, Park M, Kim ST, Jung JY, Kang YK, Choy JH (2006) J Phys Chem Solids 67:1024
76. Bibby MC (2004) Eur J Cancer 40:852
77. Kura AU, Hussein-Al-Ali SH, Hussein MZ, Fakurazi S (2014) Sci World J 8:47
78. Dong L, Gou G, Jiao L (2013) Acta Pharm Sin B 3:400
79. Bégu S, Pouëssel AA, Polexe R, Leitmanova E, Lerner DA, Devoisselle JM, Tichit D (2009) Chem Mater 21:2679
80. Li D, Xu X, Xu J, Hou W (2011) Colloid Surf A 384:585
81. Li B, He J, Evans DG, Duan X (2004) Int J Pharm 287:89
82. Choi G, Lee JH, Oh YJ, Choy YB, Park MC, Chang HC, Choy JH (2010) Int J Pharm 402:117
83. Chakraborti M, Jackson JK, Plackett D, Brunette DM, Burt HM (2011) Int J Pharm 416:305
84. Chakraborti M, Jackson JK, Plackett D, Gilchrist SE, Burt HM (2012) J Mater Sci Mater Med 7:1705
85. Park AY, Kwon H, Woo AJ, Kim SJ (2005) Adv Mater 17:106

86. Hu H, Wang XB, Xu SL, Yang WT, Xu FJ, Shen J, Mao C (2012) J Mater Chem 22:15362
87. Yan L, Chen W, Zhu X, Huang L, Wang Z, Zhu G, Roy VAL, Yu KN, Chen X (2013) Chem Commun 49:10938
88. Yan L, Wang Y, Li J, Kalytchuk S, Susha AS, Kershaw SV, Yan F, Rogach AL, Chen X (2014) J Mater Chem 2:4490
89. Tao Q, He H, Li T, Frost RL, Zhang D, He Z (2014) J Solid State Chem 213:176
90. Hu H, Xiu KM, Xu SL, Yang WT, Xu FJ (2013) Bioconjug Chem 24:968
91. Tao Q, Yuan J, Frost RL, He H, Yuan P, Zhu J (2009) Appl Clay Sci 45:262
92. Kotal M, Srivastava SK (2011) J Mater Chem 21:18540
93. Tao Q, Zhu J, Wellard RM, Bostrom TE, Frost RL, Yuan P, He P (2011) J Mater Chem 21:10711
94. Singha S, Sahoo M, Parida KM (2011) Dalton Trans 40:7130
95. Li C, Wang G, Evans DG, Duan X (2004) J Solid State Chem 177:4569
96. Kickelbick G (ed) (2007) Hybrid materials: synthesis, characterization, and applications. Wiley, Weinheim
97. Musumeci AW, Schiller TL, Xu ZP, Minchin RF, Martin DJ, Smith SV (2009) J Phys Chem C 114:734
98. Xu ZP, Kurniawan ND, Bartlett PF, Lu GQ (2007) Chem Eur J 13:2824
99. Kim SY, Oh JM, Lee JS, Kim TJ, Choy JH (2008) J Nanosci Nanotechnol 10:5181
100. Yoo SS, Razzak R, Bédard E, Guo L, Shaw AR, Moore RB, Roa WH (2014) Nanotechnology 25:425102
101. Lee BI, Lee KS, Lee JH, Lee IS, Byeon SH (2009) Dalton Trans 14:2490
102. Chen H, Zhang WG (2010) J Am Ceram Soc 93:2305
103. Chen YF, Zhou SH, Li F, Chen YW (2010) J Mater Sci 45:6417
104. Dominguez M, Perez-Bernal ME, Ruano-Casero RJ, Barriga C, Rives V, Ferreira RAS, Carlos LD, Rocha J (2011) Chem Mater 23:1993
105. Yan D, Lu J, Wei M, Evans DG, Duan X (2011) J Mater Chem 21:13128
106. Zhang Z, Chen G, Liu J (2014) RSC Adv 4:7991
107. Butturini E, Dolcet P, Casarin M, Speghini A, Pedroni M, Benetti F, Motta A, Badocco D, Pastore P, Diodati S, Pandolfo L, Gross S (2014) J Mater Chem B 2:6639
108. Zhang H, Pan D, Zou K, He J, Duan X (2009) J Mater Chem 19:3069
109. Liu J, Harrison R, Zhou JZ, Liu TT, Yu CZ, Lu GZ, Qiao SZ, Xu ZP (2011) J Mater Chem 21:10641
110. Liang R, Tian R, Shi W, Liu Z, Yan D, Wei M, Evans DG, Duan X (2013) Chem Commun 49:969
111. Sun J, Li J, Fan H, Ai S (2013) J Mater Chem B 1:5436
112. Adam D (2000) Nature. doi:10.1038/news001123-1
113. Sokolova V, Epple M (2008) Angew Chem Int Ed 47:1382
114. Li S, Li J, Wang CJ, Wang Q, Cader MZ, Lu J, Evans DG, Duan X, O'Hare D (2013) J Mater Chem B 1:61
115. Tyner KM, Roberson MS, Berghorn KA, Li L, Gilmour RF, Batt CA, Giannelis EP (2004) J Control Release 100:399
116. Ladewig K, Niebert M, Xu ZP, Gray PP, Lu GQ (2009) Biomaterials 10:1821
117. Dong H, Chen M, Rahman S, Parekh HS, Cooper HM, Xu ZP (2014) Appl Clay Sci 100:66
118. Li L, Gu W, Chen J, Chen W, Xu ZP (2014) Biomaterials 3:761
119. Park DH, Cho J, Kwon OJ, Yun CO, Choy JH (2014) submitted
120. Yan S, Rolfe BE, Zhang B, Mohammed YH, Gu W, Xu ZP (2014) Biomaterials 35:9508
121. Rives V, Del Arco M, Martín C (2013) J Control Release 169:28
122. Bi X, Zhang H, Dou L (2014) Pharmaceutics 6:298
123. Ma R, Ang Z, Li Y, Chen X, Zhu G (2014) J Mater Chem B 2:4868
124. Li D, Zhang YT, Yu M, Guo J, Chaudhary D, Wang CC (2013) Biomaterials 32:7913

Photochromic Intercalation Compounds

Tomohiko Okada, Minoru Sohmiya, and Makoto Ogawa

Contents

1   Introduction 178
2   General Background 180
    2.1   Photochromic Reactions in Solids 180
    2.2   Host–Guest Systems 181
    2.3   Forms of Intercalation Compounds 183
    2.4   Photochromic Reactions 184
3   Photochromic Reaction of Adsorbed Dyes: Fundamental Studies 184
    3.1   Spectroscopic and Photochemical Characteristics of the Adsorbed
          Photochromic Dyes 184
    3.2   Controlled Reaction Paths by Host–Guest Interaction 190
    3.3   Some Relevant Studies on the Optical Application of Intercalation Compounds .. 192
4   Photoresponsive Intercalation Compounds 193
    4.1   Photoinduced Structural and Morphological Changes 193
    4.2   Photoinduced Adsorption 197
    4.3   Photoresponse of Magnetic Properties 198
5   Photochromism Based on Photoinduced Electron Transfer 200
6   Conclusions and Future Perspectives 205
References 206

T. Okada (✉)
Faculty of Engineering, Department of Chemistry and Material Engineering, Shinshu University, Wakasato 4-17-1, Nagano 380-8553, Japan
e-mail: tomohiko@shinshu-u.ac.jp

M. Sohmiya
Department of Earth Sciences, Waseda University, Nishiwaseda 1-6-1, Shinjuku-ku, Tokyo 169-8050, Japan
e-mail: minoru.sohmiya@aoni.waseda.jp

M. Ogawa (✉)
Department of Earth Sciences, Waseda University, Nishiwaseda 1-6-1, Shinjuku-ku, Tokyo 169-8050, Japan

Graduate School of Creative Science and Engineering, Waseda University, Nishiwaseda 1-6-1, Shinjuku-ku, Tokyo 169-8050, Japan

Institute of Molecular Science and Engineering, Vidyasirimedhi Institute of Science and Technology, Rayong 21210, Thailand
e-mail: waseda.ogawa@gmail.com

Abstract Photochromism of intercalation compounds has been investigated so far. Starting from fundamental studies on the photochromic reactions of the dyes in the presence of layered materials, the precise design of the nanostructures of intercalation compounds toward controlled photochemical reactions and the creation of novel photoresponsive supramolecular systems based on layered solids have been a topic of interests. Various layered materials with different surface chemistries have been used as hosts for the controlled orientation, and aggregation of the intercalated dyes and the states of the intercalated guests affected photoresponses. Molecular design of the photochromic dyes has also been conducted in order to organize them on layered solids with the desired manner. On the other hand, layered solids with such functions as semiconducting and magnetic have been examined to host photochromic dyes for the photoresponsive changes in the materials' properties.

Keywords Electron transfer • Host–guest hybrid • Intercalation • Isomerization • Photochromism • Photoregulation

1 Introduction

Photochemical reactions in heterogeneous systems may differ significantly from analogous reactions in homogeneous solutions or in gas phases [1–6], and the possible roles of the reaction media to control the reaction rates and product selectivity have been recognized. In other words, one can tune the attractive properties, including photochemical reactions, by organizing functional species in nanospaces with appropriate geometry and chemical nature. Accordingly, photochemical reactions in various restricted geometries and chemical environments have been topics of interests of photochemistry and materials chemistry. Materials with nanospace have advantages in that the properties of the immobilized species can be discussed on the basis of their nanoscopic structures [7]. Their structure–property relationships will provide indispensable information on designing materials with controlled properties. Spectroscopic properties, which are very sensitive to the environment, of the immobilized species have given insights to the nanoscopic structures of the host–guest systems where conventional instrumental analysis does not have access [7–11]. By utilizing photoprocesses, one can obtain such information as distribution, orientation, and mobility of the guest species.

Layered materials offer a two-dimensional expandable interlayer space for organizing guest species, among available ordered or constrained nanoenvironments [7, 12–20]. The study on the intercalation reactions has been motivated by the facts that the optical and electronic properties of both guest and host can be altered by the reactions [21, 22]. If compared with other host–guest systems, the interlayer space of layered materials is characterized as the two-dimensional expandable nanospace, whose geometry and chemical nature can be tailored by

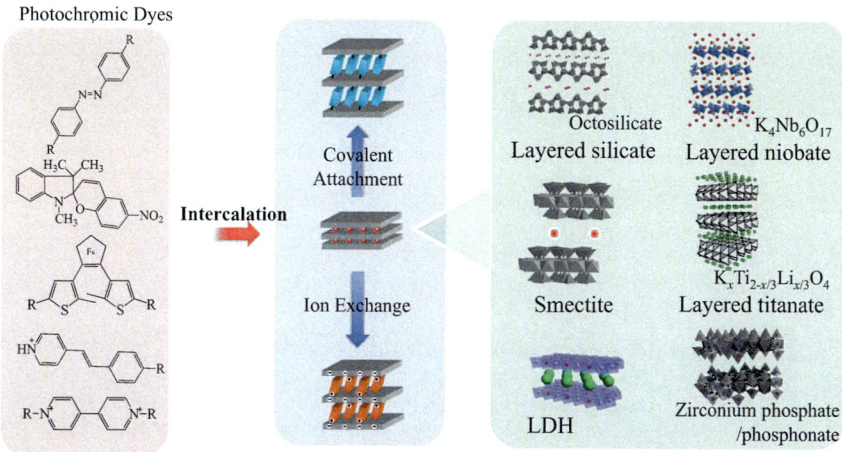

Fig. 1 Materials' variation in photochromic dyes to be intercalated into layered solids cited in the present review

selecting and designing both the guests and hosts and also by co-adsorption. From X-ray diffraction studies, interlayer distances are measured and the orientation of the intercalated species is estimated based on their size and shape. Moreover, some materials have been processed into single-crystal or oriented films, in which microscopic anisotropy can be converted into a macroscopic property [23]. The hierarchical anisotropy achieved for the single-crystal and oriented films and the detailed characterization of the orientation of the adsorbed species have been done [24, 25]. Such structural features make it possible to discuss the structure–property relationships in detail.

In this article, chemical and material investigations regarding the photochromic reactions of intercalation compounds are summarized (Fig. 1). Photochromism, which deals with photochemical reactions that are thermally or photochemically reversible (Scheme 1), has received considerable attention because of its actual and potential applications and for its paramount importance in biological phenomena [26–29]. Studies concerning photochromic reactions in solids have significance for practical applications such as optical recording. Accordingly, the interactions of photochromic dyes with layered solids have been investigated so far [30–42]. These studies have initially been done to characterize the interlayer nanospaces and the host–guest system and then to contribute to future practical applications of the intercalation compounds for optical recording and photoresponsive materials. The layered structures with the ability to accommodate a variety of guest species on the layer surface are very useful for organizing a variety of photoactive species to evaluate and to control photochromic reactions. Here, our attention will mainly be focused on the role of the nanostructures, which directly and indirectly correlate the photochromic reactions.

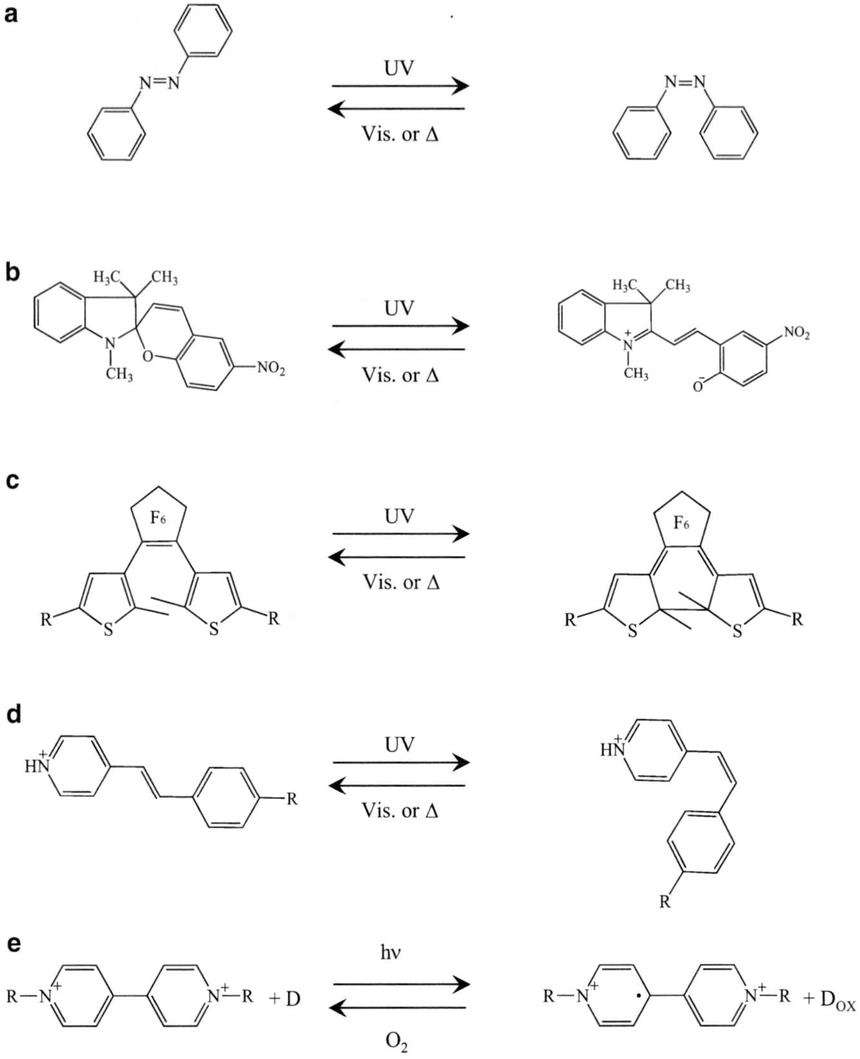

Scheme 1 Photochromism of representative photochromic dyes: (**a**) azobenzene, (**b**) spiropyran, (**c**) diarylethene, (**d**) stilbazolium, and (**e**) viologens (N,N'-bis(R)-4,4'-bipyridinium)

2 General Background

2.1 Photochromic Reactions in Solids

The photochromic reaction of dyes in solid matrices has been reported so far [29, 31]. In order to evaluate the photochemical reactions, solids that are photochemically inert and optically transparent in UV and visible light are useful.

Therefore, various polymers and silica gels have been used [26, 43–46]. The optical transparency has been achieved for amorphous systems, so that, in general, the concentration of the dye is low and the orientation and distribution of the photochromic moiety are random. In addition to the systems based on photochromic compounds in such transparent solid matrices as polymers and silicas, various supramolecular systems have been utilized in order to organize photochromic moiety in a controlled manner. Zeolites and mesoporous materials have been used as matrices for constructing photofunctional hybrids [1, 2, 4, 5, 47–51]. Chemical interactions between matrices and photochromic dyes have been designed to construct host–guest photochromic materials [52–55]. As a recent example of a solid-state photochromic host–guest system, which is relevant to the intercalation compounds, a metal–organic framework (MOF or a porous coordination polymer PCP) with an azobenzene group introduced to the organic linker has been developed for reversible change in CO_2 uptake upon external stimuli [56]. The positional changes of the dangling benzene group in the cubic cavity of an MOF upon reversible *trans*-to-*cis* photoisomerization have been a possible reason to the increased and decreased CO_2 uptake. Photoresponsive adsorption of N_2 has also been realized by photoisomerization of the azobenzene group in the material [57].

Photochromic moieties have been covalently immobilized into organic supramolecular systems such as surfactant assemblies and polymers. Surfactant assemblies often take membrane-like two-dimensional structures, therefore relevant to layered compounds and their intercalation compounds [6]. Chemical and structural stability, mechanical property, optical quality, preparation, and processing of different supramolecular systems should be discussed for practical applications. In order to achieve photoresponsive materials, the introduction of photochromic moiety into microheterogeneous systems has also been conducted extensively. The supramolecular photochemistry concepts have also been motivated to the biological application of photochemical switches [26]. The nanoparticles of layered solids can be a possible candidate for such purposes.

2.2 Host–Guest Systems

Smectite is a group of 2:1 clay minerals consisting of negatively charged silicate layers and readily exchangeable interlayer cations [16, 24, 58, 59]. The isomorphous substitution of framework metal cations with similar size and lower valency generates a net negative charge for layers, and to compensate for the negative charges and thus metal cations as sodium and calcium occupy the interlayer space. The amount and the site of the isomorphous substitution influence the surface and colloidal properties of smectites. Impurities are present both within the structure and on the particle surface, and their amounts vary depending on the source of the clay minerals. Synthetic analogs of smectites, i.e., hectorite (Laponite, Rockwood Ind. Co.) [60], saponite (Sumecton SA, Kunimine Ind. Co.) [61], and swelling mica (sodium-fluor-tetrasilicic mica, Na-TSM, Topy Ind. Co.) [62, 63], have advantages

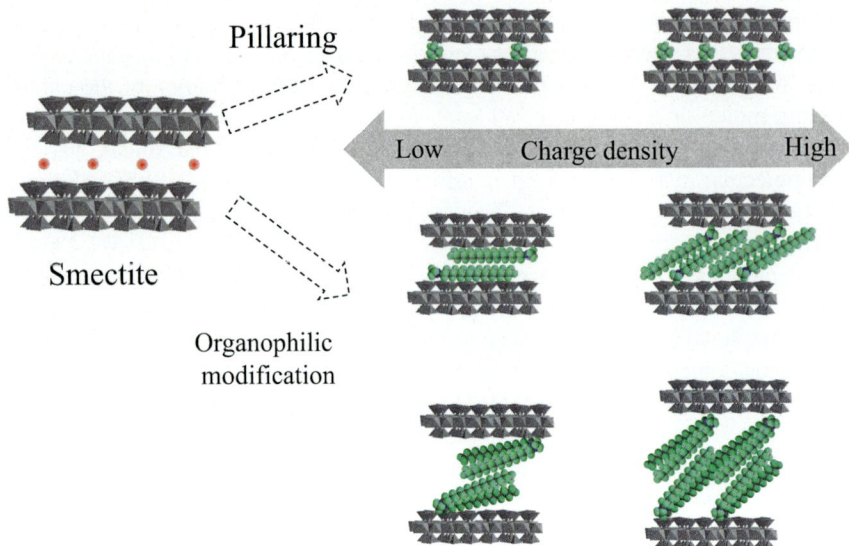

Fig. 2 Surface modification of smectites with organoammonium ions

for the photochemical studies since natural clay minerals contain impurities, which gave colored clays. Synthetic analogs of smectite have been prepared in the laboratory and used for the adsorption of dyes [19, 64–67].

The mechanism of the intercalation for smectite can be classified into two [16, 24]: the cation exchange with interlayer exchangeable cations and the adsorption of polar molecules by ion–dipole interactions with interlayer cations and/or hydrogen bonding with the surface oxygen atom of the silicate sheets. One of the characteristic features of smectites is the possible surface modification. Nanoporous pillared smectites have been obtained using inorganic particles and small organic cations as pillars [68, 69]. Organophilic modification has also been conducted by cation exchange with cationic surfactants of various structures [70–75] (Fig. 2).

Besides smectites, a large variety of layered solids with the ability to accommodate guest species in the interlayer space is available [7, 15, 17, 21, 22]. Layered alkali silicates are capable of incorporating guest species (organoammonium ions and polar molecules) in the interlayer space to form intercalation compounds [72, 76–79]. Organosilane-grafted derivatives have been obtained previously [80–91]. Compared with smectites, the series of layered alkali silicates possess such useful properties for organizing guest species as the high layer charge density and well-defined particle morphology. Metal phosphates and phosphonates and layered transition metal oxides have also been used for constructing photochromic intercalation compounds [89–93].

Due to the variation of the layer charge density, particle morphology, and electronic properties, host–guest systems with unique microstructures and properties have been obtained. On the other hand, the intercalation of guest species is not as easy as that is for smectites. In order to introduce bulky organic species in the interlayer spaces with designed manners, organoammonium-exchanged forms, which are prepared by conventional ion exchange reactions in aqueous media, have been used as the intermediates [25, 94–96].

Layered double hydroxides (LDHs) are composed of positively charged brucite-type layers of mixed-metal hydroxides and exchangeable anions located at the interlayer spaces, which compensate for the positive charge of the brucite-type layers [97–101]. Due to the structural and compositional characteristics, the application of LDHs in such areas as adsorption, catalyst, medical, and biochemical application has been proposed so far [100, 101]. The chemical composition of the LDHs is generally expressed as $[M(II)_{1-x}M'(III)_x(OH)_2][A^{n-}{}_{x/n}]^{x-}$ where $M(II) =$ Mg, Co, Ni, etc., $M(III) =$ Al, Cr, Fe, etc., and A is an interlayer anion such as CO_3^{2-} and Cl^-. Anionic species have been introduced into the interlayer spaces of LDHs by three methods. Anion exchange reactions using an aqueous solution of guest species have been investigated widely. Compared with the cation exchange of smectites, the ion exchange reaction for LDHs is more difficult because of their high selectivity to carbonate anions. Therefore, CO_2 should be excluded during the sample preparation. Intercalation compounds have also been prepared via a direct synthesis in which an LDH phase precipitates in the presence of a guest species [102]. The reaction of a mixed-metal oxide solid solution, which was obtained by the thermal decomposition of LDH-carbonate, with an aqueous solution of guest species results in the formation of the LDH intercalation compound [reconstruction method] [103].

In addition to the crystalline structures, the particle size and its distributions of layered solids are key issues in order to achieve optimum performance of layered solids and their intercalates, and accordingly, attention has been paid for the powder morphology [104]. Recently, monodispersed particles of LDHs have been obtained [105–108]. The effects of synthetic temperature on the particle size of layered transition metal oxides, which affected the photocatalytic ability, have also been reported recently [109, 110].

Thanks to the variation of available layered solids with designable chemical nature, one can investigate photochromic reactions of cationic, anionic, and non-ionic photochromic molecular species. The location, distribution, and orientation of the photochromic moieties can be designed by selecting host and host–guest complexation, in order to control the photochromic reactions.

2.3 Forms of Intercalation Compounds

For the evaluation of the photoprocesses, samples have been obtained as powders, suspensions, and thin films. One of the most unique and attractive properties of smectites is their spontaneous swelling in water. By dispersing in water, smectites form a stable thixotropic gel or suspension. The careful preparation of suspension occasionally led to liquid crystalline phases for smectites, layered transition metal oxides, arsenate, and so on [111–116]. One of the advantages of using exfoliated layered solids (nanosheet) suspension if compared with microheterogeneous systems composed of surfactant and polymer is the stable layered structures in the wide range of temperature, composition, and solvents. In microheterogeneous systems composed of soluble surfactants and polymers, the nanostructures vary significantly depending on the experimental conditions.

When the suspension is evaporated on a flat plate, platy particles pile up with their *ab* plane parallel to the substrate to form a film [117–119]. The preparation of thin films by the Langmuir–Blodgett technique from exfoliated platelets of clays has also been reported [120–122]. Inorganic–organic multilayered films have also been prepared via alternate adsorption of a cationic species and an anionic sheet of an exfoliated layered solid (layer-by-layer deposition technique, hereafter abbreviated as LbL technique) [123–130]. In order to apply LbL for the fabrication of thin films, the swelling of layered solids into the nanodimension is a basic prerequisite. Therefore, the exfoliation of various layered solids has been a topic of active research during these two decades after the successful preparation of the so-called nanosheet suspension and thin-film formation [126, 131–134]. The exfoliation and the film preparation will be described in other article of this volume by Sasaki et al.

2.4 Photochromic Reactions

The "E-Z" isomerization of azobenzenes, cyclization of spiropyrans and spirooxazines, cycloaddition of stilbenes, and reduction of viologen have been investigated so far. The involved chemical reactions are shown in the scheme (Scheme 1):

3 Photochromic Reaction of Adsorbed Dyes: Fundamental Studies

3.1 Spectroscopic and Photochemical Characteristics of the Adsorbed Photochromic Dyes

Amphiphilic cationic azobenzene derivatives (Scheme 2a, b) have been intercalated into layered silicates (magadiite and montmorillonite) [135–138]. The dye orientation in the interlayer spaces has been discussed from the spectral shifts and the gallery heights of the products. The intermolecular interactions of chromophores give aggregated states and the dye–dye interactions cause both bathochromic and hypsochromic spectral shifts depending on the nanostructures of aggregates [139]. The spectral shifts reflect the orientation of the dipoles in the aggregates; smaller spectral red shifts are expected for the aggregates with larger tilt angles of

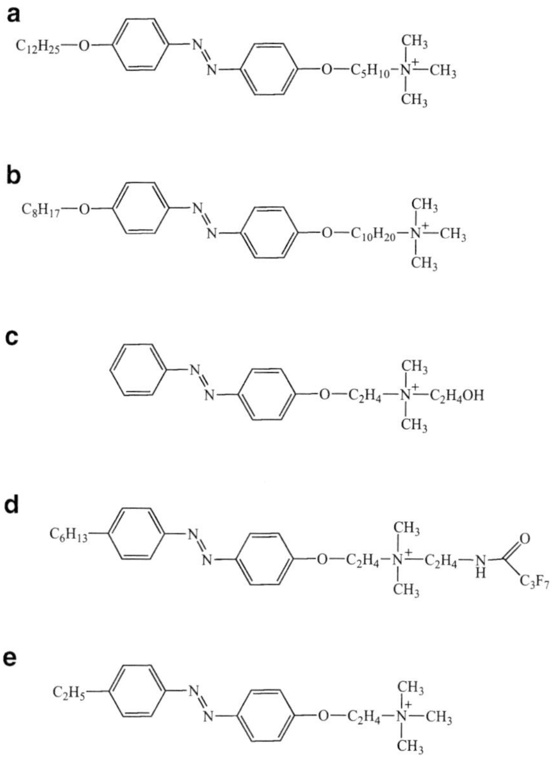

Scheme 2 Molecular structures of cationic azobenzene derivatives cited in this review. (**a**) π-(ω-Trimethylammoniodecyloxy)-π′-(octyloxy)azobenzene, (**b**) π-(ω-trimethylammonio heptyloxy)-π′-(dodecyloxy)azobenzene, (**c**) p-[2-(2-hydroxyethyldimethylammonio)ethoxy] azobenzene, (**d**) [2-(2,2,3,3,4,4,4-heptafluorobutylamino)ethyl]-{2-[4-(4-hexyphenylazo)-phenoxy]-ethyl}dimethylammonium, and (**e**) 2-[4-(4-ethylphenylazo)phenoxy]ethyl(trimethyl) ammonium

the dipoles. Depending on the layer charge density (cation exchange capacity, CEC) of host materials and the molecular structures of the amphiphilic azo dyes, aggregates (*J*- and *H*-aggregates) with different microstructures (tilt angle) formed in the interlayer spaces of layered silicates.

The intercalation of ionic photochromic dyes into the interlayer space of ion-exchangeable layered solids has been investigated extensively [30, 140–142]. Katsuhiko Takagi et al. reported the intercalation of 1′,3′,3′-trimethylspiro [2*H*-1-benzopyran-2,2′-indoline](H-SP) and its 6-nitro (NO_2-SP) and 6-nitro-8-(pyridinium)-methyl (Py^+-SP) derivatives into montmorillonite, and their photochromic behavior has been studied for colloidal suspension [32]. The effects of the intercalation on the rate of thermal coloration and decoloration have been compared with those in other systems such as colloidal silica, aqueous micellar solution of hexadecyltrimethylammonium bromide ($C_{16}3C_1N^+$ Br), or sodium dodecyl sulfate (SDS). Besides the results on the photochemical studies, one of the advantages of using layered solids is the robust layered structure existing in the wide range of temperature, composition, and solvents. In microheterogeneous systems composed of soluble surfactants and polymers, the nanostructures vary significantly depending on the experimental conditions.

Py^+-SP was intercalated into montmorillonite quantitatively as an equilibrium mixture with the corresponding merocyanine (MC) with the ratio of Py^+-SP:Py^+-MC of 35:65 and exhibited reverse photochromism. It is known that thermal equilibria between SP and MC are dependent on the polarity of the molecular surroundings; MC becomes the major product under the increased polarity environments. The reverse photochromism observed for the montmorillonite systems has been explained in terms of the polar nature of the interlayer space of montmorillonite. The thermal isomerization of Py^+-SP intercalated in aqueous colloidal montmorillonite suspension exhibited a linear combination of two components of first-order kinetics, indicating the presence of two different states of the intercalated Py^+-SP; one is a molecularly separated species and the other is an aggregated species.

In contrast, a preferential adsorption as Py^+–SP was observed for the hydrophobic montmorillonite co-adsorbing $C_{16}3C_1N^+$. Normal photochromism has been observed in these systems. A single first-order kinetic has been observed for the Py^+-SP-$C_{16}3C_1N^+$-montmorillonite. The effects of the co-adsorbing $C_{16}3C_1N^+$ on the photochromic behavior showed that the $C_{16}3C_1N^+$ surrounds Py^+-SP to create a hydrophobic environment for each Py^+-SP molecule.

The photochromism of a cationic diarylethene, 1,2-bis(2-methyl-3-thiophenyl) perfluorocyclopentene bearing two pyridinium substituents at each thiophenyl ring, intercalated in montmorillonite was reported so far [143]. The product was prepared as oriented films by casting and the dye orientation was deduced from the basal spacing and spectroscopic behavior, which was determined by using polarized light. The photochromic reaction was efficient and smooth, while the efficiency decreased upon repeated irradiation. The decrease was attributed to the formation of photo-inactive species (degradation). The degradation was successfully suppressed by the co-adsorption of dodecylpyridinium cations with the cationic diarylethene.

The intercalation of another cationic diarylethene, 1,2-bis(2′-methyl-5′-(1-″-methyl-3″-pyridinio)thiophen-3′-yl)-3,3,4,4,5,5-hexafluorocyclopentene (**1**, the molecular structure is shown in Fig. 3a), into montmorillonite was reported [33, 144]. As schematically shown in Fig. 3b, from XRD and UV–Vis polarized spectroscopy, the cationic diarylethene **1** was shown to be anisotropically accommodated with DMF in a gelatin film involving the dye-montmorillonite hybrid. The photochemical interconversion between a colorless **1** and blue-colored **2** was achieved with good repeatability (Fig. 3a, c). The interconversion was also observed for diarylethene, which was covalently immobilized to the surface silanol groups on magadiite [145, 146].

On the other hand, nonionic photochromic dyes have been intercalated to the long-chain organoammonium-modified silicates [36, 37, 41, 42, 135, 147]. The role of the surfactant is not only for producing hydrophobic interlayer spaces but also for

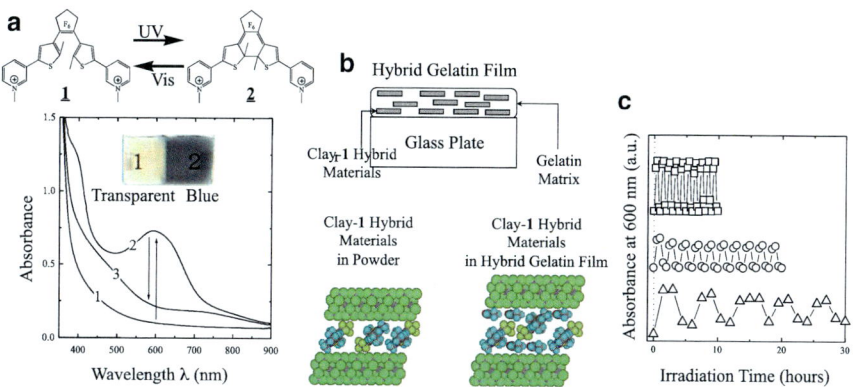

Fig. 3 (a) Absorption spectra of a clay-**1** hybrid film recorded (1) before and (2) after UV irradiation and (3) after subsequent visible light irradiation. The inserted photograph is the film after the UV irradiation. The left-half area was not exposed to the light using a black screen paper; (b) schematic drawing of proposed models of (*top*) clay-**1** hybrid film, (*bottom, left*) intercalated **1** in the powder samples, and (*bottom, right*) intercalated **1** in clay dispersed in the optically transparent film; (c) changes in the absorbance at 600 nm with alternative irradiation of UV and visible light: (*squares*) **1** in methanol, (*circles*) powder sample of the clay-**1** hybrid, (*triangles*) gelatin-clay-**1** hybridfilm

controlling the states of the adsorbed dyes. Seki and Ichimura have investigated the thermal isomerization kinetics of photoinduced merocyanine (MC) to spiropyran (SP) in solid films having multi-bilayer structures, which consist of ion complexes between the cationic bilayer forming amphiphiles (dioctadecyldimethylammonium cation, abbreviated as $2C_{18}2C_1N^+$) and polyanions (montmorillonite and poly(styrene sulfonate), abbreviated as PSS) [34]. 1,3,3-Trimethyl-6′-nitrospiro[indoline-2,2′-2′H-benzopyran](SP) was incorporated into the thin films of the polyion complexes, which were prepared by casting the chloroform solution or suspension of the polyion complexes.

X-ray diffraction studies showed that the cast films were composed of a multi-bilayer structure whose lamellar plane was oriented parallel to the film surface. Endothermic reactions were observed in the DSC (differential scanning calorimetry) curves of the films to show the phase transition at the temperatures of 54.5 and 48.5 °C for the $2C_{18}2C_1N^+$-montmorillonite and $2C_{18}2C_1N^+$-PSS films, respectively. Both DSC and X-ray diffraction results indicated that the $2C_{18}2C_1N^+$-montmorillonite film had a more ordered structure than $2C_{18}2C_1N^+$-PSS film. Annealing the film at 60–70 °C at the relative humidity of ca. 100 % for a few hours resulted in the improved ordering of $2C_{18}2C_1N^+$-PSS film, and the photochromic behavior was investigated for the annealed films. The difference in the film structure influenced the kinetics of the thermal decay of MC embedded in the films. The incorporated SP exhibited photochromism in both of the immobilized bilayer complexes with montmorillonite and PSS. The decoloration reaction rate was dependent on the mobility of the surroundings and, in polymer matrices, was influenced by the glass transition. It was found that the reaction rates abruptly increased near the gel to liquid-crystal phase transition temperature (54 °C) of the immobilized bilayer due to increased matrix mobility in this system. The film prepared with montmorillonite gave more homogeneous reaction environments for the chromophore than those with PSS. This led to the drastic changes in the reaction rate at the crystal to liquid-crystal phase transition of the bilayer, showing the effect of the phase transition of bilayers immobilized on layered solids to be more pronounced than that of the transition of bilayers immobilized on amorphous polymers.

The formation of H- (parallel type) and J- (head-to-tail type) aggregates of photomerocyanines upon adsorption in didodecyldimethylammonium($2C_{12}2C_1N^+$)-montmorillonite has been suggested for a series of 1'-alkyl-3',3'-dimethyl-6-nitro-8-alkanoyloxymethylspiro($2H$-1-benzopyran-2,2'-indoline) derivatives with different lengths of alkylchains [35]. A cast film consisting of the SP incorporated in the $2C_{12}2C_1N^+$-montmorillonite was prepared on a glass plate by slowly evaporating the suspension of the SP and $2C_{12}2C_1^+$-montmorillonite. When longer alkylchains were introduced, new very sharp absorption peak appeared at around 500 nm upon UV light irradiation in addition to the absorption at 570 nm due to the monomeric photomerocyanine (abbreviated as PMC). New sharp absorption bands appeared at a longer wavelength region (around 610 nm) when SPs bearing longer alkylchains were exposed to UV light. These new absorption bands are attributed to aggregates of PMCs which are reported to form occasionally in organized molecular assemblies [35]. The absorption bands at around 500 and 610 nm have been ascribed to H- and J-aggregates of PMC, respectively. A high activation energy and highly positive activation entropy for J- and H-aggregates of PMCs, which directly correlate with the thermal stability of these aggregates (which led the slow decoloration), have been observed. Thus, the kinetics of the thermal decoloration of the MCs were determined by the aggregation, which was controlled by the host–guest interactions. From this viewpoint, layered materials are quite useful as matrices of dye and dye aggregate (aggregation and isolation) because of the expandable interlayer space (geometrically adaptable size for both isolated molecules and dye aggregates) and the variable layer charge density, which directly correlate with the intermolecular distance between adjacent ionic guest species.

Photochemical *trans*-to-*cis* isomerization of azobenzenes intercalated in the hydrophobic interlayer space of alkylammonium-montmorillonite and TSMs has also been investigated [36, 37, 41]. Intercalation compounds were obtained by a solid–solid reaction between the organophilic hosts and nonionic azobenzene. The intercalated azobenzene showed reversible *trans*-to-*cis* photoisomerization upon UV irradiation and *cis*-to-*trans* isomerization by subsequent thermal treatment (or visible light irradiation). The hydrophobic interlayer space of the alkylammonium-smectites serves as reaction media for the immobilization and efficient photochemical isomerization of the azo dye. The degree of the isomerization at the photostationary states varied, suggesting the possible design toward optimized photochemical reactions by selecting host–guest systems.

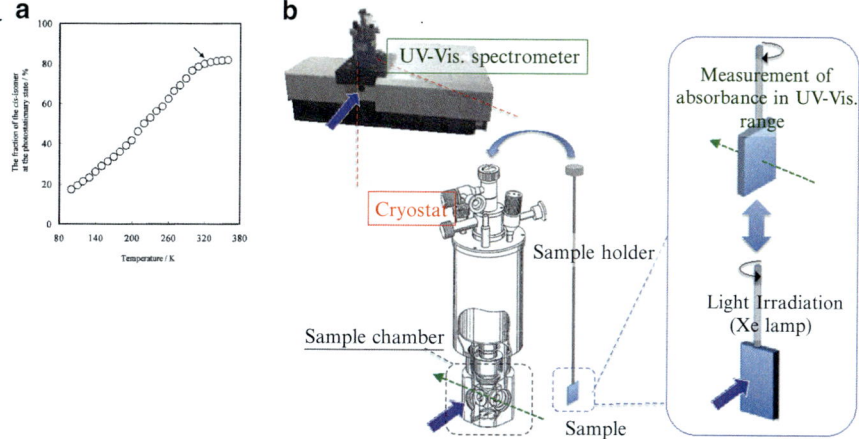

Fig. 4 (a) The temperature dependence of the fraction of the photochemically formed *cis*-isomer at the photostationary state for the $2C_{18}2C_1N^+$–TSM–azobenzene intercalation compound. *Arrow* indicates the phase transition temperature. (b) A photograph of experimental setup of recording absorption spectrum with UV and/or visible light irradiations under varied temperature using a cryostat

For detailed evaluation of the photochemical reactions, dialkyldimethylammonium($2C_n2C_1N^+$)-TSM-azobenzene intercalation compounds were fabricated as films by casting the suspension of the $2C_n2C_1N^+$-TSM and azobenzene on a flat substrate [37]. The intercalated azobenzene also exhibits reversible photochromic reactions. The fraction of the photochemically formed *cis*-isomer in photostationary states depends on the reaction temperature (Fig. 4a; the experimental setup is shown in Fig. 4b), suggesting that a change in the states of the interlayer $2C_n2C_1N^+$ occurs. The phase transition temperatures estimated from the photochemistry of azobenzene were in good agreement with the values determined by other techniques [34, 148, 149].

The d values for the $2C_n2C_1N^+$-smectites varied [34, 37, 148–150] and the difference in the basal spacings corresponds to the difference in the orientation of the alkylchains. It has been known that excess guest species can be accommodated in the interlayer spaces of smectites as a salt, and this phenomenon is referred to as "intersalation." The large d values reported in the literature may be due to the intersalation as well as the difference in the surface layer charge density.

The introduction of retinal, which is the chromophore of rhodopsin, into a surfactant (dimethyloctadecylammonium)-modified clay was investigated in order to mimic the properties of rhodopsin [38, 39]. The spectroscopic and photochemical properties of retinal in vitro are of interest in studying the primary chemical process of vision and in developing novel photoresponsive materials. The modified clay interlayer offers environments for retinal similar to rhodopsin in two respects: color regulation and efficient isomerization at a cryogenic temperature. Protein environments have the ability to tune the color of retinal Schiff bases; however, the color regulation in artificial systems was not satisfactory. In rhodopsins, a retinal molecule forms a Schiff base linkage with a lysine residue and the retinal Shiff base is protonated. It was proposed that a proton was supplied from dimethyloctadecylammonium to retinal Shiff base. The *trans*-to-*cis* isomerization of a protonated retinal Shiff base occurs even at 77 K as revealed by visible and infrared spectroscopy. On the other hand, azobenzene isomerization in dialkyldimethylammonium-TSM was reported to be suppressed at lower temperature. The efficient isomerization of retinal at 77 K was worth mentioning as a mimic of the primary photochemical reaction in rhodopsin. The difference is worth investigating systematically using similar materials (clay minerals, surfactants, and dyes). The adsorption of retinal onto smectites with different origin was examined to find the color development of retinal that depends on the nature of clay minerals [40]. The isomerization of the retinal adsorbed on various clay minerals is worth investigating further [40].

Amphiphilic cationic azobenzene derivatives (Scheme 2a, b) have been intercalated into the layered silicates magadiite and montmorillonite to show the controlled orientation by the host–guest interactions [135–139]. Layer charge density determines the orientation (tilt angle) of the intercalated dye when quantitative ion exchange was achieved. The azobenzene chromophore photoisomerized effectively in the interlayer space of silicates, despite the fact that the azobenzene chromophore is aggregated there. These unique characteristics have been used for the construction of photoresponsive intercalation compounds, which will be described in the following section of this article.

3.2 Controlled Reaction Paths by Host–Guest Interaction

The interlayer spaces of clay minerals have been shown to provide a stable and characteristic reaction field suitable for stereochemically controlled photochemical reactions. Regioselective photocycloaddition of stilbazolium cations, intercalated in the interlayer space of saponite, has been reported [147, 151–157]. There are four possible photochemical reaction paths of the stilbazolium ion Fig. 5a. Upon irradiation of UV light to a stilbazolium-saponite suspension, *syn*-head-to-tail dimers (**2**, in Fig. 5a) were predominantly formed at the expense of *cis*-to-*trans* isomerization (**4**, in Fig. 5a), which is a predominate path in homogeneous solution. The selective formation of head-to-tail dimers suggests that the intercalation occurs in an antiparallel fashion, as shown in Fig. 5b. Since the dimer yields were only slightly dependent on the loading amount, stilbazolium ions were adsorbed inhomogeneously and form aggregates with antiparallel alternative orientation even at very low loading (1 % of C.E.C.). This aggregation was supported by the fluorescence spectrum of the dye adsorbed on saponite, in which excimer fluorescence was observed at 490–515 nm at the expense of the monomer fluorescence at ca. 385–450 nm. The selective formation of *syn*-head-to-tail dimers indicates the

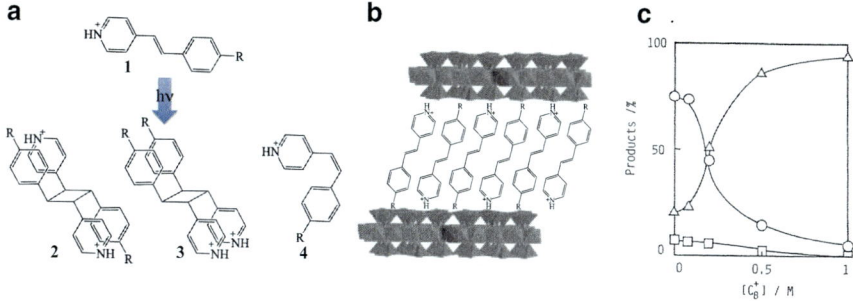

Fig. 5 (a) Possible photochemical reaction paths of stilbazolium. (b) A schematic drawing of the stilbazolium-intercalated saponite. (c) Effect of C_8N^+ on the photoreactivity of the pre-intercalated S-type monomer on saponite; syn head-to-tail dimer (*circle*), syn-head-to-head dimer (*square*), and Z-type monomer (*triangle*). Pre-intercalated stilbazolium-saponite was suspended in aqueous solution of C_8N^+

formation of the aggregates with antiparallel alternative orientation owing to the hydrophobic interaction between the adsorbate ions.

The aggregation state of γ-stilbazolium [4-(2-phenylvinyl)pyridinium] ion on saponite was changed by the co-adsorption of alkylammonium ions (C_nN^+), where *n* indicates the carbon number in the alkylchain [153]. Figure 5c shows the effect of C_8N^+ on photoreactivity of the pre-intercalated stilbazolium ions on the synthetic saponite. On co-adsorbing C_nN^+ with the alkyl group longer than the stilbazolium ion, the major photochemical reaction was changed from cyclodimerization to *trans*-to-*cis* isomerization, and the excimer emission of the intercalated stilbazolium ions was dramatically reduced.

Photochemical cycloaddition for several unsaturated carboxylates has been studied in the presence of a layered double hydroxide, hydrotalcite [152, 154, 155]. In addition to the antiparallel packing of the guest, the intermolecular distances of two double bonds of adjacent carboxylates were found to affect the stereoselectivity of the photochemical reactions. While cinnamate yielded head-to-head dimers exclusively, stilbene carboxylates gave a significant amount of head-to-tail dimer in addition to a head-to-head dimer. This difference was explained by the molecular packing of the dye anions in the interlayer space of hydrotalcite. Being similar to the effects of the organoammonium ions on the photochemistry of stilbazolium ions in smectite as mentioned before [153], the addition of *p*-phenethyl benzoate, a photochemically inactive co-adsorbate, affected significantly the product distribution of *p*-(2-phenylethenyl) benzoate upon irradiation. The series of pioneering investigations on the roles of host–guest interactions on the selectivity of the reactions paths shows that the organization of organic species into the interlayer space of layered materials is a way of crystal engineering in which reaction selectivity can be determined. In other words, the relation between the selectivity of reactions and the interlayer spacing is a method of probing the orientation and aggregation of the intercalated species. Thus, the photochemical reactions of the adsorbed photochromic dyes are useful for the evaluation of the orientation and packing in the interlayer space as well as for controlling the reaction paths and efficiency. More recently, Shinsuke Takagi and his coworkers reported the photochemical behaviors of a dicationic azobenzene on saponite and in aqueous solution to find an important role of the saponite surface to control the relative stability of two isomers (*trans*- and *cis*-forms). The quantum yield of *trans*-to-*cis* photoisomerization of the azo dye cation on the saponite was much smaller than that in water, while *cis*-to-*trans* isomerization was accelerated. Almost 100 % *trans*-isomer was successfully obtained after the visible light irradiation thanks to the interactions between the dye and saponite surface [142].

3.3 Some Relevant Studies on the Optical Application of Intercalation Compounds

Photochemical hole burning (PHB) is another phenomena relevant to the photochromism, which is a quite sensitive tool to probe the host–guest interactions. PHB is the site-selective and persistent photobleaching of an inhomogeneously broadened absorption band, induced by resonant laser light irradiation at cryogenic temperatures [158]. PHB has attracted attention due in part to its possible applicability to a high-density frequency domain optical storage, in which more than 10^3 times more storage density than present optical disk systems would in principle be available. Since the phenomena occurred at a cryogenic temperature using laser, naked eye observation of the color change is almost impossible. A search for new materials has been done because the hole formation processes depend significantly on the nature of host–guest systems. For the PHB materials, the host–guest system is essential. Since hole formation depends significantly on the structures of host–guest systems, it can be used as high-resolution solid-state spectroscopy. PHB reaction of intercalation compounds using synthetic saponite and 1,4-dihydroxyanthraquinone (abbreviated as DAQ) and cationic porphines, both of which are typical PHB dyes, has been reported [159, 160].

We have prepared the TMA–saponite–DAQ intercalation compound and investigated its PHB reaction to show the merits of nanoporous saponite. The PHB reaction of DAQ is 1 due to the breakage of internal hydrogen bond(s) and the subsequent formation of external hydrogen bond(s) to proton acceptor(s) within a matrix. A molecularly isolated chromophore is another basic prerequisite for efficient PHB materials, in order to avoid line broadening due to energy transfer. For this purpose, saponite was modified by tetramethylammonium (abbreviated as TMA) ions to obtain independent micropores in which a DAQ molecule was incorporated without aggregation. Taking into account the molecular size and shape of DAQ and the geometry of the micropore of TMA–saponite, DAQ was intercalated with the molecular plane nearly perpendicular to the silicate sheet.

A persistent spectral zero-phonon hole was obtained at liquid helium temperatures by Kr^+ laser light irradiation (520.8 nm). In spite of the high concentration of

DAQ (ca. 1.5 mol kg^{-1}), a narrow hole with an initial width of 0.25 cm^{-1} (at 4.6 K) was obtained. The width was narrower than those (e.g., 0.4–0.8 cm^{-1}) of DAQ doped in ordinary polymers and organic glasses (e.g., PMMA, ethanol/methanol mixed glass) obtained under similar experimental conditions. A narrow line width is desirable for the optical recording application, since the number of holes to be made in an inhomogeneously broadened band increases for narrow hole. The width is related mainly to the dephasing but contributions from spectral diffusion cannot be neglected. On the other hand, a broad *pseudo*-phonon sidehole, whose shift from the zero-phonon hole is 25 cm^{-1}, appears only after irradiation stronger than 1,500 mJ cm^{-2}. The burning efficiency was high, being similar to or higher than the typical one observed in ordinary dye-dispersed in amorphous solids. The microporous structure of the TMA–saponite intercalation compound therefore apparently leads to some desirable characteristics regarding hole formation.

The nonlinear optic (NLO) of organic compounds is another relevant phenomenon where nanostructures of intercalation compounds are designed by host–guest interactions. In the NLO application, photochemical reactions should be avoided to achieve durable performances. Nonlinear optics comprises the interaction of light with matter to produce a new light field that is different in wavelength or phase [161, 162]. Examples of nonlinear optical phenomena are the ability to alter the frequency (or wavelength) of light and to amplify one source of light with another, switch it, or alter its transmission characteristics throughout the medium, depending on its intensity. Since nonlinear optical processes provide key functions for photonics, the activity in many laboratories has been directed toward understanding and enhancing second- and third-order nonlinear effects. Second harmonic generation (SHG) is a nonlinear optical process that converts an input optical wave into an outwave of twice the input frequency. Large molecular hyperpolarizabilities of certain organic materials lead to anomalously large optical nonlinearities. Research efforts had been done toward (a) identifying new molecules possessing large nonlinear polarizability and (b) controlling molecular orientation at nanoscopic level. Intercalation compounds have potential because the arrangements of the intercalated species are determined by the host–guest interactions [163, 164].

4 Photoresponsive Intercalation Compounds

4.1 Photoinduced Structural and Morphological Changes

There are some examples on XRD-detectable nanostructural changes induced by photochemical reactions of the interlayer dyes. The change in the basal spacings triggered by the photoisomerization of the intercalated azobenzenes has been observed for organophilic smectites-nonionic azobenzene intercalation compounds [36, 37, 41, 147]. However, the location of the intercalated azobenzene is difficult to determine in those systems, since the dyes were solubilized in the hydrophobic interlayer spaces surrounded by alkylchains of the interlayer alkylammonium ions. Even after the careful control of the orientation of the intercalated amphiphilic cationic azobenzenes, the photoresponse of the basal spacing has hardly been observed in clay-amphiphilic cationic azo dye systems [135–138]. Irreversible change in the basal spacing upon UV and visible light irradiations has been observed by using Li-fluor-taeniolite (cation exchange capacity (CEC): 1.57 meq/g) and cationic azo dyes without flexible alkylchains in the structures [165].

When a cationic azo dye, p-[2-(2-hydroxyethyldimethylammonio)ethoxy] azobenzene bromide (Scheme 2c), was intercalated in magadiite with the adsorbed amount of 1.90 meq/g silicate), the basal spacing changed after UV irradiation from 2.69 to 2.75 nm and the value came back to 2.69 nm upon visible light irradiation (Fig. 6a) [42]. The reversible change in the basal spacing has been observed repeatedly as shown in XRD patterns of Fig. 6a. The spectral properties as well as XRD results revealed that the intercalated azo dye cations form head-to-head aggregates (H-aggregate) in the interlayer space, as schematically shown in Fig. 6b. The fraction of cis-isomer at room temperature was ca. 50 %, which is lower than that (ca. 70 %) for the azo dye occluded in a mesoporous silica film [166, 167]. Upon UV irradiation, half of the $trans$-form isomerized to cis-form and coexists with $trans$-form in a same interlayer space as suggested by the single-phase X-ray diffraction patterns before and during the irradiation. The first-order plot for the thermal cis-to-$trans$ isomerization of the azo dye–magadiite at 360 K showed that the rate became slow after 20 min [168]. This fact suggests that the $trans$-azo dye cations are thought to form a densely packed aggregate in the interlayer space. However, the cationic azo dye, which intercalated as J-aggregate (Fig. 6c) in the interlayer space of montmorillonite (CEC: 1.19 meq/g), did not give

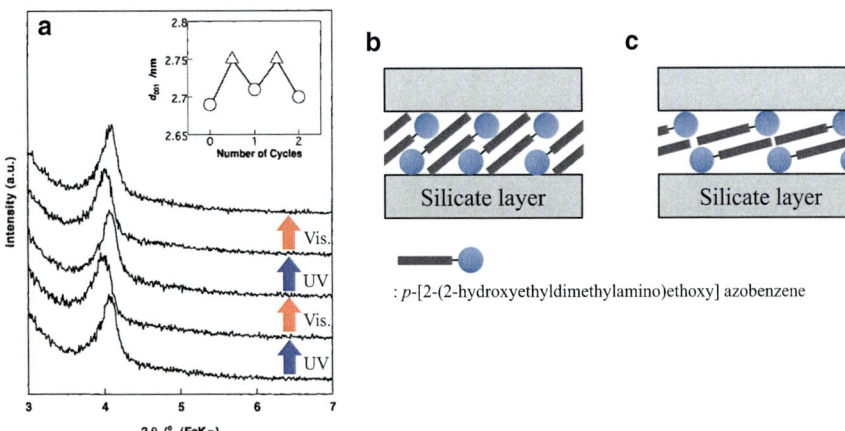

Fig. 6 (a) The reversible change in the XRD pattern of the magadiite modified with p-[2-(2-hydroxyethyldimethylammonio)ethoxy]azobenzene. Inset in this figure shows the change in the basal spacing from the photochemical reactions. Schematic drawing of possible arrangements of the cationic azobenzene in layered silicates as (**b**) H- and (**c**) J-aggregates

the photoresponses of the basal spacing during the photoisomerization [169]. It is postulated that a densely packed aggregate is difficult to form in the interlayer space of magadiite at the photostationary state due to the geometric difference of the two isomers, and this causes the change in the basal spacings. As mentioned previously, amphiphilic cationic azobenzene intercalated in layered silicates (smectites and magadiite) was photoisomerized while the basal spacing was hardly changed [136–138]. The cationic azo dye does not contain flexible units such as long alkylchains; therefore, the photoisomerization induced the change in the microstructure detectable by XRD. This demonstrated a type of photomechanical effect and larger volume change is expected.

XRD-detectable microstructural change has also been observed in a cationic spiropyran–montmorillonite system [170]. As schematically shown in Fig. 7, the introduction of a cationic spiropyran in montmorillonite has been conduced through two methods: cation exchange reactions with interlayer Na ions of the host ("ion exchange method") and partial exchange with pre-intercalated hexadecyltrimethylammonium ($C_{16}3C_1N^+$) ions ("guest exchange method"). Reversible conversions of spiropyran–merocyanine were observed upon UV and visible light irradiation for both hybrids. The basal spacing changed reversibly by the photoinduced conversion in the absence of $C_{16}3C_1N^+$, while the basal spacing did not change in the latter case due to the interlayer expansion by bulkier co-intercalated $C_{16}3C_1N^+$.

Inoue and his coworkers have reported the intercalation of a polyfluorinate cationic azo dye (C3F-azo: Scheme 2d) into [171] a layered niobate [172] and a layered titanoniobate [173]. Because the intercalation of the azo dye into the niobate and the titanoniobate is difficult, if compared with the intercalation into smectite, due to their higher layer charge densities, the pre-intercalation of hexylamine and 1,1'-dimethyl-4,4'-bipyridinium and the subsequent guest exchange with the C3F-azo dye were conducted. The basal spacings of the niobate [174] and the titanoniobate intercalates [from *(020)* reflection] decreased by the irradiation of a light with shorter wavelength than ca. 370 nm and increased to the value the same before the photoirradiations upon subsequent light irradiation at ca. 460 nm. Reversible *trans*-to-*cis* photoisomerization of the intercalated azobenzene moiety accompanied the change in the basal spacing and the change can be repeated. The direction of the interlayer distance upon each isomerization was opposite to that observed for the magadiite system as described above.

Morphological change in the C3F-azo-layered niobate hybrid film with response to light irradiation has been reported [175]. Right side of Fig. 8 shows an AFM image of the C3F-azo-layered niobate film, and the height profiles (along the white line perpendicular to the dashed line) are shown in the left side of this figure. After the irradiation of ca. 370 nm light, the bottom edge (point A in Fig. 8) of the nanosheet stack slid out from the interior of the layered film, while the height of the top edge (point B) maintained. Subsequent higher wavelength light (ca. 460 nm) irradiation resulted in the protruded bottom edge (point C) slid back to the original position before the photoirradiation. The sliding distance of the bottom edge of the film reached ca. 1,500 nm. The photoinduced nanosheet sliding back and forth on such a giant scale is interesting from the viewpoint of the driving mechanism in the

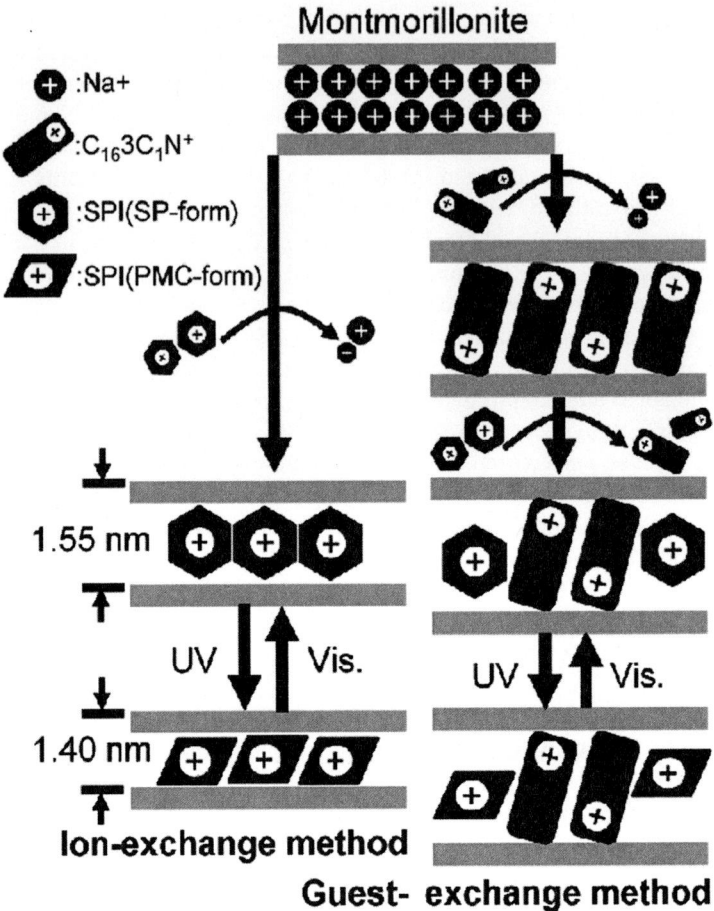

Fig. 7 Schematic illustration of mechanism for the intercalation of the cationic spiropyran through the ion and guest exchange methods in montmorillonite, followed by the conformational change in the spiropyran in the interlayer space upon the photoirradiations

morphological change. Photoresponsive morphological changes have also been reported in the hybrid film composed of an azobenzene-containing polymer and an LDH obtained by using a layer-by-layer self-assembly (LbL) technique [176]. AFM observations of the resulting film exhibited changes in the surface roughness with responses to UV and visible light irradiation.

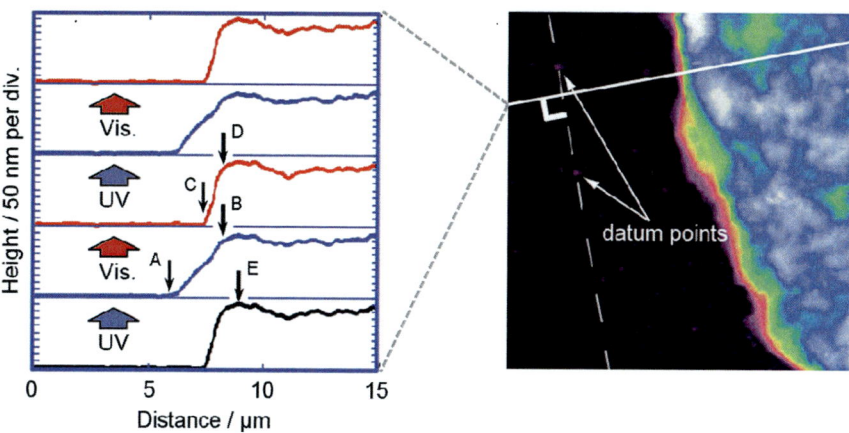

Fig. 8 Change in the height profile of in the C3F-azo-layered niobate hybrid film upon the repeated photoirradiations

4.2 Photoinduced Adsorption

We have reported reversible change in the basal spacing by phenol adsorption triggered by photoirradiations. This was achieved by the complexation of smectite with the cationic azo dye, *p*-[2-(2-hydroxyethyldimethylammonio)ethoxy] azobenzene [177]. Phenol was intercalated into montmorillonite (CEC: 1.19 meq/g clay) modified with the cationic azo dye to expand the interlayer space by mechanical mixing. From the change in the XRD pattern (Fig. 9), photoinduced intercalation of phenol was observed by the UV irradiation, and subsequent visible light irradiation indicated phenol deintercalation. It was assumed to the intercalation and deintercalation of phenol induced by reversible *trans*-to-*cis* isomerization of the azobenzene chromophore. On the contrary, both of the intercalation and the photoinduced intercalation were not observed for the azo dye Sumecton SA (the CEC of 0.71 meq/g clay). The dye orientation (with the molecular long axis inclined to the silicate layer) plays an important role in the photoinduced intercalation of phenol; the phenol intercalation before the irradiation is prerequisite to induce the photoresponsed intercalation [178]. Studies toward the following goals are worth investigating: no intercalation at ground state, larger amount of intercalation by irradiation, complete deintercalation by subsequent irradiation, and efficiencies of all the repeatable phenomena.

We have deduced that photoisomerization to *cis*-form of azobenzene chromophore leads to the intercalation of atmospheric water [42] as well as phenol [177]. If azo dye without hydroxyl group is used, the interactions of water with intercalation compounds are expected to be weaker. In order to confirm the idea, another type of azo dye, 2-[4-(4-ethylphenylazo)phenoxy]ethyl(trimethyl)ammonium (Scheme 2e), was synthesized and intercalated into the interlayer space of smectite to compare the adsorptive properties of the resulting intercalation compounds for phenol

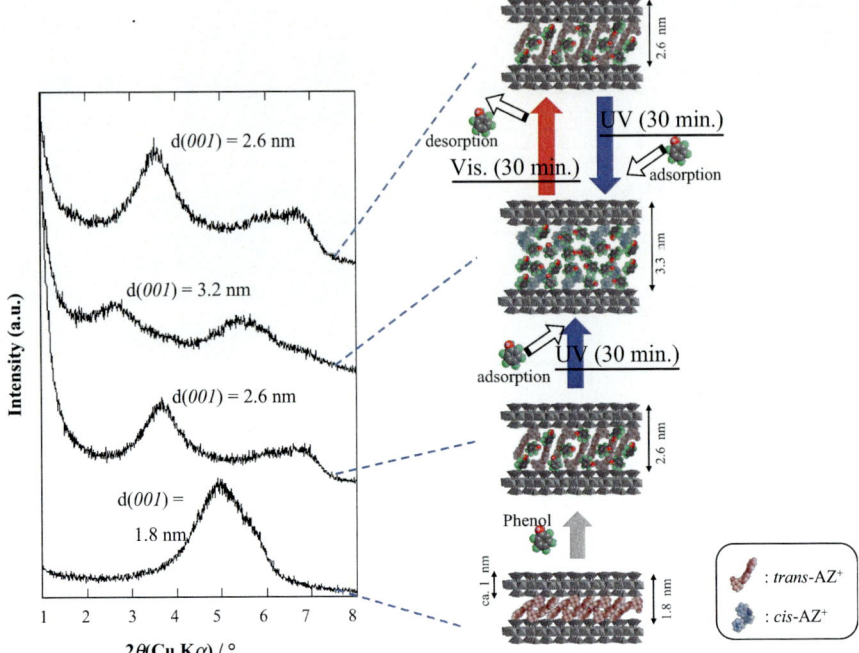

Fig. 9 Change in the XRD patterns of p-[2-(2-hydroxyethyldimethylammonio)ethoxy] azobenzene–montmorillonite by the reaction with phenol and by photoirradiations. Photoinduced phenol intercalation behavior is also schematically summarized in this figure

[179]. Due to the absence of the hydroxyl group, the hybrids are hydrophobic and the larger amount of phenol was intercalated before the irradiation. Then the amount of the intercalated phenol by the UV irradiation was smaller compared to azo dye with a hydroxyl group.

Recently, molecular dynamics (MD) simulation of the layered silicates intercalating a series of cationic azo dyes has been investigated by Heinz and his coworkers [180]. The simulation results indicate that the reversible change in the basal spacing upon UV and visible light irradiation may be improved through the (1) presence of a co-intercalated species, (2) conformational rigidity of azo dye, and (3) upright orientation of the dye. A moderate-to-high CEC, the absence of flexible alkyl spacers in the cationic azo dye, and the use of rigid macrocyclic "pedestals" support this objective.

4.3 Photoresponse of Magnetic Properties

Magnetic properties of transition metal-layered hydroxides and double hydroxides have been investigated and the photoresponses of magnetic properties have been reported. Such magnetism as ferromagnetism, ferrimagnetism, and antiferromagnetism depend on the interlayer distance [163, 181]. Fujita and Awaga reported the intercalation of an anionic dye, 8-((p-(phenylazo)phenyl)oxy-octanoate, into the interlayer space of $[Cu_2(OH)_3]^-$ and investigated the magnetic properties of the intercalation compound [182]. While the interlayer distance of the layered hydroxide was 2.07 nm in methanol, that in acetonitrile was increased to 3.87 nm, which is almost twice as long as the molecular height of the 8-((p-(phenylazo)phenyl)oxy-octanoate anion, forming a membrane-like bilayer. Reversible mono-to-bilayer phase transition of the anion layer in the hybrid resulted in the drastic change in magnetic properties. The monolayer phase was paramagnetic down to 3 K, while the bilayer phase became a weak ferromagnet with critical temperature of $T_c = 10.8$ K. The origin of the magnetic variety was thought to be due to the sensitivity of the magnetic interactions to the Cu–OH–Cu bridging angles in the $[Cu_2(OH)_3]^-$ network. Although the photoisomerization of azobenzene chromophore was not linked to the magnetic properties [183], this is a successful example on controllable magnetic properties induced by using the structural change of the intercalation compounds.

Recently, Abellián et al. reported the photoresponses of magnetization and critical temperature (T_c) of ferromagnetic Co^{II}–Al^{III}-layered double hydroxide (LDH) induced by *trans*-to-*cis* photoisomerization of the intercalated azobenzene dianion. The interlayer spacing changed reversibly by the isomerization. UV irradiation (*trans*-to-*cis* isomerization of the azobenzene chromophore) led to the increased T_c from 4.5 to 5.2 and the decreased magnetization (external magnetic field of 2 T at 2 K) from 0.66 to 0.48 emu/mol. The magnetic properties in T_c and magnetization came back to the values close to that observed before the UV irradiation. It was explained that the observed reversibility was concerned with antiferromagnetic dipolar interactions and magnetic correlation length with the aid of flexible LDH layers [184].

Reversible photoresponsive changes in magnetization have been achieved by hybridizing amphiphilic cationic azobenzene, montmorillonite, and Prussian blue (CN–Fe^{III}–NC–Fe^{II}–O) [185, 186]. A magnetic film has been fabricated through a Langmuir–Blodgett technique (azobenzene-clay film), cation exchange reactions with $FeCl_2$, and subsequent reactions with $K_3[Fe(CN)_6]$. Under an external magnetic field of 10 G, the hybrid multilayer films exhibited ferromagnetic properties with a critical temperature (T_c) of 3.2 K. Upon repeated light irradiation of UV and Vis, the magnetization of the hybrid film at 2 K repeatedly decreased and increased, respectively, and the photoinduced changes were estimated to be ca. 11 % in total. The change in the values has been explained by electrostatic field driven by the photoisomerization of the azobenzene chromophore. Such changes are necessary to transfer an electron from the $[Fe(CN)_6]^{3-}$ to Fe^{III}, and this might affect the superexchange interaction between the spins in the Prussian blue. In a separated paper, the photoinduced electron transfer from Fe^{II} to Co^{III} in a Co–Fe Prussian blue clay oriented film has been reported to vary the magnetization depending on the direction of the applied magnetic field [187].

Reversible change in π-conjugation in the dye assemblies is also useful for controlled magnetic interactions between layers. Kojima and his coworkers reported the intercalation of a photochromic diarylethene divalent anion into a layered cobalt(II) hydroxide, which contains both tetrahedrally and octahedrally coordinated cobalt(II) ions [188]. In the dark and under UV-irradiated (313 nm) conditions, open and closed forms of the interlayer diarylethene anion were obtained, respectively. The photochemical reactions forming a closed form led to the change in Curie temperature from $T_c = 9$ to 20 K. The enhancement has been explained by the delocalization of the π-electrons in the closed form, which correlates the enhancement of the interlayer magnetic interactions.

Bénard et al. have investigated the intercalation of a cationic spiropyran into layered $MnPS_3$ [189] and the variation of the magnetic properties in response to spiropyran–merocyanine conversion upon UV irradiation. Intercalation via cation exchange reactions into $MnPS_3$ often results in the appearance of a spontaneous magnetization derived from generation of intralayer vacancies by losing Mn^{2+} ions. Upon repeating UV irradiation and thermal treatment, a reversible change in the basal spacing by 0.01 nm was observed repeatedly in one of the film samples. The merocyanine form (J-aggregate) was quite stable over several months in dark. While the critical temperature (T_c) did not change by the UV irradiation, remanence and coercivity increased. After the subsequent thermal treatment, the VSM curves returned to the shape close to that before UV irradiation.

5 Photochromism Based on Photoinduced Electron Transfer

Viologens (N,N'-bis(R)-4,4′-bipyridinium, or di(R)viologen, Scheme 1e) are photoreduced reversibly in the presence of an electron donor to form blue radical cations, which is a kind of photochromic reaction between colorless and blue [190]. The structural variation of viologens using the organic moiety at 4,4′-positions is a merit for the use as a building block of hybrid materials. The color development by electrochemical reduction and UV light irradiation is shown in Scheme 1e. Kakegawa et al. reported that some synthetic smectites (Sumecton SA, and Laponite XLG) play a role as an electron donor for the photoinduced reduction of the intercalated dimethylviologen [191]. Proposed electron donating sites are bridging Si–O–Al (in a silicate layer of Sumecton SA) and crystal edges of smectites. From the aspect of designing adsorbents, dimethylviologen has been used as a scaffold to create nanospace in the interlayer spaces of smectites to accommodate 2,4-dichlorophenol [66, 192]. It was shown that charge-transfer interactions with dimethylviologen–smectites are a driving force for the adsorption.

Miyata et al. reported the photochromism of dipropyl and di-n-heptylviologens intercalated in the interlayer space of montmorillonite with co-intercalating

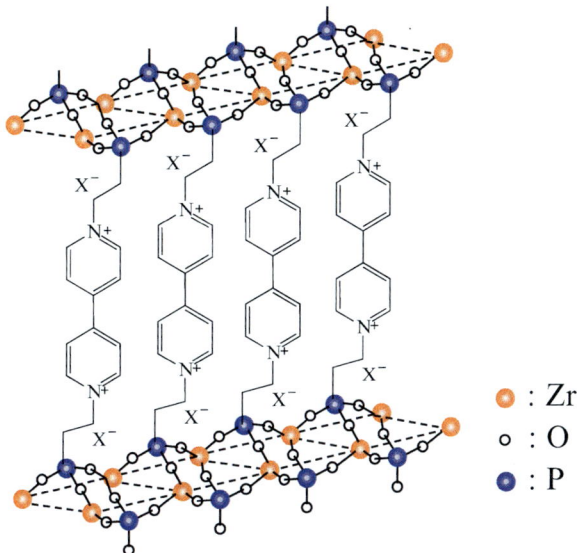

Fig. 10 Proposed structure of ZrPV(X)

● : Zr
○ : O
● : P

polyvinylpyrrolidone (PVP) [193]. The viologen dications were intercalated in the interlayer space of pre-synthesized montmorillonite–PVP intercalation compound by cation exchange. Photochemical studies were conducted for cast films of viologen–montmorillonite–PVPs by the UV irradiation with Hg lamp. Upon UV irradiation, viologen radical cations formed as shown by blue color, characteristic absorption bands at 610 and 400 nm, and ESR signal. Reversible color development by UV irradiation and color fading were observed. In this system, co-intercalated PVP was assumed to act as the electron donor for the reduction of the viologens. The color fading required a longer period than that in a pure PVP matrix. Since the color-fading process in the PVP matrix was an oxidation caused by oxygen in air, the slow color-fading reaction observed for the viologen–montmorillonite–PVP system was explained by the prevention of contact between the intercalated viologen radical cations and the atmospheric oxygen.

Thompson and his coworkers have reported stable photoinduced charge separation in layered zirconium phosphonate containing viologen moiety in both powder and thin-film samples [194, 195]. In order to overcome the scattering problems associated with powder, transparent multilayer thin films of zirconium phosphonate viologen salt [ZrPV(X)] were grown directly onto fused-silica substrates from aqueous solution. The sequential growth method has been applied in the preparation of ZrPV(X) films (Fig. 10). In this case, fused-silica slides are treated with $(et)_3SiCH_2CH_2CH_2NH_2$, followed by treatment with $POCl_3$. This procedure leads to a phosphonate-rich surface suitable for treatment with $ZrOCl_2$. The slides are then allowed to react with $H_2O_3P\text{-}CH_2CH_2\text{-bipyridinium-}CH_2CH_2\text{-}PO_3H_2X_2$.

Photolysis of $Zr(O_3P\text{-}CH_2CH_2(\text{bipyridinium})\ CH_2CH_2\text{-}PO_3)X_2$ (X = Cl, Br, I), ZrPV(X), resulted in the formation of blue radical cations of viologen which were stable in air. The photoreduction of viologen in these thin-film samples was very efficient (quantum yields = 0.15), showing simple isosbestic behavior in the electronic spectra. Contrary to bulk solids, photoreduced thin films are very air sensitive. The mechanism for the formation of charge-separated states in these materials involves both irreversible and reversible components. An irreversible component is proposed to involve hydrogen atom abstraction by photochemically formed halide radicals, followed by structural rearrangements. Optimization of the reversible process may make it possible to use these materials for the efficient conversion and storage of photochemical energy.

The photochromic behavior of dimethylviologen intercalated into a series of layered transition metal oxides has been reported [196]. $K_2Ti_4O_9$ [197], $HTiNbO_5$ [198], $K_4Nb_6O_{17}$ [199, 200], HNb_3O_8 [200], and $HA_2Nb_3O_{10}$ (A = Ca, Sr) [201] were used as host materials and dimethylviologen was intercalated by cation exchange. The photochemical studies were conducted for powdered samples by irradiation with a Hg lamp, and the reactions were monitored by diffuse reflectance spectra. Semiconducting host layers acted as electron donors for the reduction of viologen to form radical cations of the intercalated dimethylviologen in the interlayer space. The stability of the photochemically formed blue radical cations has been discussed with respect to their microscopic structures.

The photochemistry of intercalation compounds formed between layered niobates $K_4Nb_6O_{17}$ and HNb_3O_8 with dimethylviologen can be controlled by changing the interlayer structures [200]. Two types of dimethylviologen-intercalated compounds with different structures have been prepared for each host. In the $K_4Nb_6O_{17}$ system, two intercalation compounds were obtained by changing the reaction conditions. In both of the intercalation compounds, dimethylviologen ions are located only in the interlayer I. HNb_3O_8 also gave two different intercalation compounds; one was prepared by the direct reaction of HNb_3O_8 with dimethylviologen and the other was obtained by using propylammonium-exchanged HNb_3O_8 as an intermediate. In the latter compound, propylammonium ions and dimethylviologen were located in the same layer. All the intercalation compounds formed dimethylviologen radical cation in the interlayers by host–guest electron transfer upon UV irradiation. The presence of co-intercalated K^+ and propylammonium ion in the $K_4Nb_6O_{17}$ and HNb_3O_8 systems, respectively, significantly affected the decay of the viologen radical cation. This difference was explained by the guest–guest interactions with the co-intercalated photo-inactive guests (K^+ and propylammonium ion).

Viologens on a layered solids have also been used as electron mediator. The photoreduction of dipropyl and di-n-heptylviologens from incorporated PVP is an example [193]. Electron transfer quenching of tris(2,2'-bipyridine)ruthenium (II) ($[Ru(bpy)_3]^{2+}$) by dimethylviologen in aqueous suspension of smectites (Sumecton SA, Laponite XLG, and ME-100) in the presence of PVP was also investigated [202]. It is known that dimethylviologen strongly interacts with clay surfaces [191, 203, 204] and does not quench the excited state of $[Ru(bpy)_3]^{2+}$ on clay due to segregation, which is a phenomenon occasionally observed for the intercalation as schematically shown in Fig. 11 [205]. The concept has been proposed in 1980s [205] and more recently discussed in our recent review article

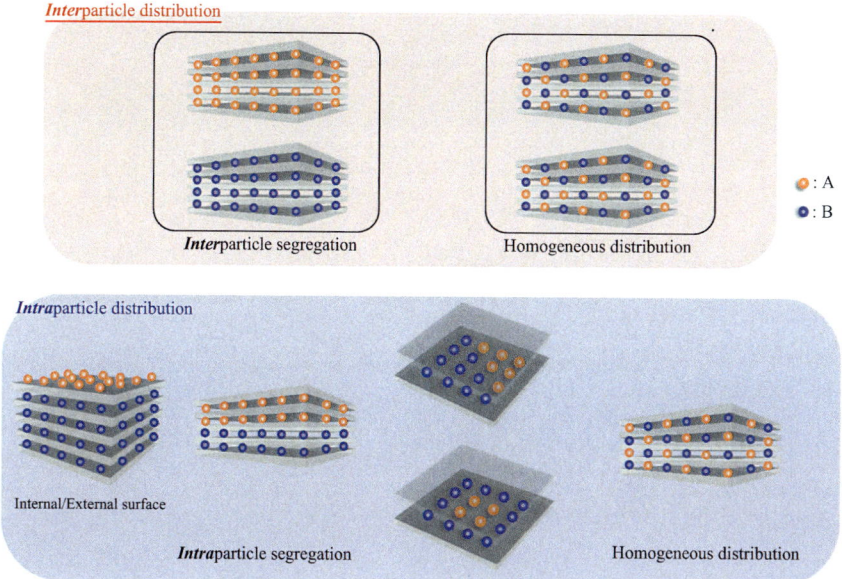

Fig. 11 Variation in the spatial distribution of guest species in/on layered materials

[17]. On the contrary, the adsorption of PVP on clay resulted in the co-adsorption of [Ru(bpy)$_3$]$^{2+}$ and dimethylviologen without segregation and the photoluminescence study on the [Ru(bpy)$_3$]$^{2+}$/dimethylviologen in PVP–smectite indicated the homogeneous distribution of the adsorbed dyes without segregation.

The intraparticle electron transfer on smectite in the absence of PVP was revealed for the system containing co-intercalation of dihexadecylviologen cation and tetraphenylporphine through hydrophobic interactions (Fig. 12) [206]. Energy-transfer among naphthyl-, anthryl-, and pyrenylalkylammonium bound to zirconium phosphate and photoinduced electron transfer from 5,10,15,20-tetrakis (4-phosphonophenyl)porphyrin to N,N'-bis(3-phosphonopropyl)-4,4'-bipyridinium organized in Zr phosphonate-based self-assembled multilayers have been reported [207–209]. The layered structure plays an important role in organizing reactants to control the reaction. Recently, Shinsuke Takagi and his coworkers estimated the size of aggregate (island) of a cationic porphine around N,N'-bis(2,4-dinitrophenyl) viologen cations on Sumecton SA by using time-resolved fluorescent measurement [210]. The segregation structure was found to be different depending on the molecular structure of porphyrin and the island size is a key factor responsible for enhancing fluorescence quenching by electron transfer from viologen to porphyrin. The photoinduced electron transfer through interparticle electron hopping from [Ru (bpy)$_3$]$^{2+}$-intercalated clays (Kunipia F, Sumecton SA, Laponite, and fluorohectorite) to the dimethylviologen counterparts in the presence of ethylenediamine tetraacetate (sacrificial electron donor) was shown by Nakato et al. [211]. The photoreduction occurred depending on the lateral size of the clay

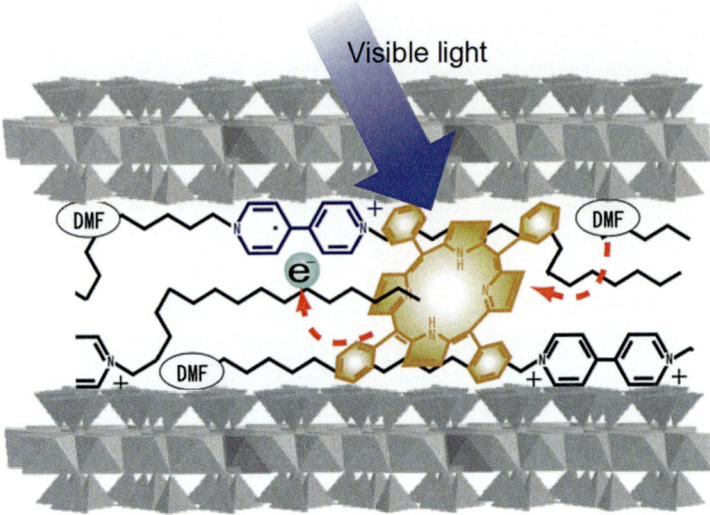

Fig. 12 Schematic drawing of intraparticle electron transfer reactions from tetraphenylporphine to dihexadecylviologen in the interlayer space of smectite

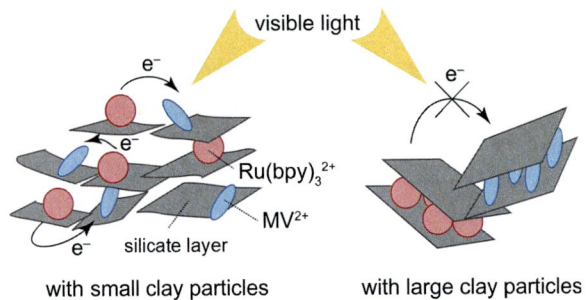

Fig. 13 Schematic representation of the interparticle visible light-induced electron transfer from [Ru(bpy)$_3$]$^{2+}$- to dimethylviologen clays

mineral particles; electron transfer was observed in the case of smaller sized clay minerals [synthetic saponite (Sumecton SA) and synthetic hectorite (Laponite)] with appropriate aggregation, while the electron transfer did not proceed in large clay mineral particles (natural montmorillonite and synthetic swelling mica) irrespective of the flocculation (Fig. 13).

Spatial separation of acceptor and donor species has been optimized to stabilize a long-lived charge-separated state as a result of suppressed back electron transfer [212]. K$_4$Nb$_6$O$_{17}$ nanosheets are excited upon UV irradiation to generate electron–hole pairs, and then, dimethylviologen (divalent cation) adsorbed on hectorite nanosheets accepted photoexcited electrons from the niobate nanosheets. The stable photoinduced charge separation has been explained by the slower diffusion of the

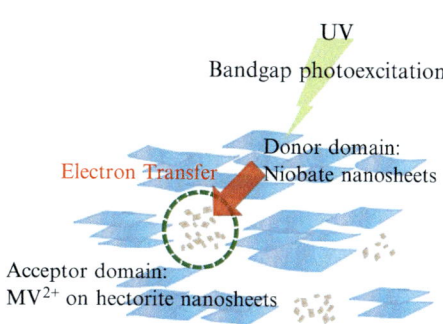

Fig. 14 Schematic illustration of spatial separation of acceptor (dimethylviologen on hectorite nanosheets) and donor (niobate nanosheets) to stabilize a long-lived charge-separated state

nanosheets that form the microdomains (Fig. 14) than that of the molecular species. As a result, the generation of the radical cation was slower, followed by slow oxidation of the formed radical cation to the dication. Thus, the efficiency of the reduction and the stability of the formed charge-separated states have been designed through the nanostructural design and the interparticle level controlled spatial distribution of donor and acceptors in and on layered solids. Since the roles of viologens in photocatalytic applications have long been recognized, the organization (and the characterization of them) of viologens achieved so far will bring useful information and hint to understand and modify the artificial photosynthesis based on layered solids [213].

6 Conclusions and Future Perspectives

Examples and progresses of photochromic reactions of intercalation compounds are summarized. Interactions of various photochromic dyes with layered solids have been investigated to control the photochromic reactions (selectivity, yield, and dynamics). The photochromic reaction of the intercalated dyes have also been used to probe the microenvironments of interlayer space, to which conventional instrumental analyses are not useful. Not only the variation of layered materials but also possible chemical modification of the chemical nature of interlayer spaces makes the materials' diversity. Photoirradiation triggered responses of such properties of intercalation compounds as adsorptive and magnetic ones have successfully been achieved, which motivates further study on the photoresponsive functional intercalation compounds.

Developments regarding controlled solid-state forms of intercalation compounds from nano (nanoparticle with designed shape, size, and size distribution) to macroscopic ones (large single-crystal and oriented film) represent milestones for the application of photochromic intercalation compounds. Not only for the solid-state, the structure of suspension (or dispersion) of layered materials have been characterized and controlled, and the possible role of the suspension for the controlled photochromic reaction has been proposed. Combining the materials' diversity and the hierarchical control of the forms of intercalation compounds, the materials performances will be optimized further to lead future practical applications. The preparation of layered solids with novel structure and chemical nature, the modification of the interlayer space, and the complexation of newly developed photochromic dyes with appropriate layered solids will be conducted to find novel photoresponsive phenomena as well as to control photochromic reactions precisely.

References

1. Anpo M, Matsuura T (eds) (1989) Photochemistry on solid surfaces. Elsevier, Amsterdam
2. Anpo M (ed) (1996) Surface photochemistry. Wiley, Chichester
3. Klafter J, Drake JM (eds) (1989) Molecular dynamics in restricted geometries. Wiley, New York, NY
4. Ramamurthy V (ed) (1991) Photochemistry in organized & constrained media. VCH, New York, NY
5. Ramamurthy V, Schanze KS (eds) (2000) Solid state and surface photochemistry. Marcel Dekker, New York, NY
6. Fendler JH (ed) (1994) Membrane-mimetic approach to advanced materials. Springer, Heidelberg
7. Alberti G, Bein T (eds) (1996) Solid-state supramolecular chemistry: two- and three-dimensional inorganic networks. Pergamon, Oxford
8. Thomas JK (1987) J Phys Chem 91:267–276
9. Thomas JK (1993) Chem Rev 93:301–320
10. Turro NJ, Gratzel M, Braun AM (1980) Angew Chem Int Ed 19:675–696
11. Ramamurthy V (1986) Tetrahedron 42:5753–5839
12. Thomas JK (1988) Acc Chem Res 21:275–280
13. Ogawa M, Kuroda K (1995) Chem Rev 95:399–438
14. Ogawa M (1998) Annu Rep Prog Chem Sect C Phys Chem 94:209–257
15. Auerbach SM, Carrado KA, Dutta PK (eds) (2004) Handbook of layered materials. Taylor & Francis, Boca Raton, FL
16. Bergaya F, Theng BKG, Lagaly G (eds) (2006) Handbook of clay science. Elsevier, Amsterdam
17. Ogawa M, Saito K, Sohmiya M (2014) Dalton Trans 43:10340–10354
18. Takagi K, Shichi T (2000) Photophysics and photochemistry in clay materials. In: Ramamurthy V, Schanze KS (eds) Solid state and surface photochemistry, vol 5. Marcel Dekker, New York, NY (Chapter 2)
19. Okada T, Ide Y, Ogawa M (2012) Chem Asian J 7:1980–1992
20. Takagi S, Shimada T, Ishida Y, Fujimura T, Masui D, Tachibana H, Eguchi M, Inoue H (2013) Langmuir 29:2108–2119
21. Whittingham MS, Jacobson AJ (eds) (1982) Intercalation chemistry. Academic Press, New York, NY
22. Müller-Warmuth W, Schöllhorn R (eds) (1994) Progress in intercalation research. Kluwer Academic, Dordrecht
23. Inadomi T, Ikeda S, Okumura Y, Kikuchi H, Miyamoto N (2014) Macromol Rapid Commun 35:1741–1746
24. Theng BKG (1974) The chemistry of clay-organic reactions. Wiley, London
25. Kaito R, Miyamoto N, Kuroda K, Ogawa M (2002) J Mater Chem 12:3463–3468
26. Morrison H (ed) (1993) Biological applications of photochemical switches. Wiley, New York, NY

27. Crano JC, Guglielmetti RJ (1990) Main photochromic families. Organic photochromic and thermochromic compounds, vol 1. Plenum, New York, NY
28. Crano JC, Guglielmetti RJ (1999) Physicochemical studies, biological applications, and thermochromism. Organic photochromic and thermochromic compounds, vol 2. Kluwer Academic/Plenum, New York, NY
29. Dürr H, Bouas-Laurent H (eds) (2003) Photochromism: molecules and systems. Elsevier, Amsterdam
30. Adams JM, Gabbutt AJ (1990) J Inclus Phenom Mol 9:63–83
31. Irie M, Yokoyama Y, Seki T (eds) (2013) New frontiers in photochromism. Springer, Japan
32. Takagi K, Kurematsu T, Sawaki Y (1991) J Chem Soc Perkin Trans 2:1517–1522
33. Sasai R, Ogiso H, Shindachi I, Shichi T, Takagi K (2000) Tetrahedron 56:6979–6984
34. Seki T, Ichimura K (1990) Macromolecules 23:31–35
35. Tomioka H, Itoh T (1991) J Chem Soc Chem Commun 532–533
36. Ogawa M, Kimura H, Kuroda K, Kato C (1996) Clay Sci 10:57–65
37. Ogawa M, Hama M, Kuroda K (1999) Clay Miner 34:213–220
38. Sasaki M, Fukuhara T (1997) Photochem Photobiol 66:716–718
39. Kandori H, Ichioka T, Sasaki M (2002) Chem Phys Lett 354:251–255
40. Furutani Y, Ido K, Sasaki M, Ogawa M, Kandori H (2007) Angew Chem Int Ed 46:8010–8012
41. Ogawa M, Fujii K, Kuroda K, Kato C (1991) Mater Res Soc Symp Proc 233:89–94
42. Ogawa M, Ishii T, Miyamoto N, Kuroda K (2001) Adv Mater 13:1107
43. Tran-Thi TH, Dagnelie R, Crunaire S, Nicole L (2011) Chem Soc Rev 40:621–639
44. Lebeau B, Innocenzi P (2011) Chem Soc Rev 40:886–906
45. Naito T, Kunishige M, Yamashita T, Horie K, Mita I (1991) React Polym 15:185–192
46. Naito T, Horie K, Mita I (1993) Polymer 34:4140–4145
47. Yoon KB (2003) Photoinduced electron transfer in zeolites. In: Auerbach SM, Carrado KA, Dutta PK (eds) Handbook of zeolite science and technology. CRC Press, New York, NY (Chapter 13)
48. Yoon KB (1993) Chem Rev 93:321–339
49. Hashimoto S (2003) J Photochem Photobiol C Photochem Rev 4:19–49
50. Ogawa M (2002) J Photochem Photobiol C Photochem Rev 3:129–146
51. Ogawa M, Saito K, Sohmiya M. (2015) Eur J Inorg Chem: 1126–1136
52. Oshita S, Matsumoto A (2003) Chem Lett 32:712–713
53. Oaki Y, Imai H (2009) Bull Chem Soc Jpn 82:613–617
54. Guo S, Sugawara-Narutaki A, Okubo T, Shimojima A (2013) J Mater Chem C 1:6989
55. Liu N, Yu K, Smarsly B, Dunphy DR, Jiang YB, Brinker CJ (2002) J Am Chem Soc 124:14540–14541
56. Park J, Yuan D, Pham KT, Li JR, Yakovenko A, Zhou HC (2012) J Am Chem Soc 134:99–102
57. Yanai N, Uemura T, Inoue M, Matsuda R, Fukushima T, Tsujimoto M, Isoda S, Kitagawa S (2012) J Am Chem Soc 134:4501–4504
58. Grim RE (1953) Clay mineralogy. McGraw-Hill, New York, NY
59. Van Olphen H (1977) An introduction to clay colloid chemistry, 2nd edn. Wiley, New York, NY
60. Thompson DW, Butterworth JT (1992) J Colloid Interface Sci 151:236–243
61. Ogawa M, Nagafusa Y, Kuroda K, Kato C (1992) App Clay Sci 7:291–302
62. Kitajima K, Daimon N (1974) Nippon Kagaku Kaishi 1:685
63. Soma M, Tanaka A, Seyama H, Hayashi S, Hayamizu K (1990) Clay Sci 8:1–8
64. Ogawa M, Matsutomo T, Okada T (2008) J Ceram Soc Jpn 116:1309–1313
65. Ogawa M, Matsutomo T, Okada T (2009) Bull Chem Soc Jpn 82:408–412
66. Okada T, Matsutomo T, Ogawa M (2010) J Phys Chem C 114:539–545
67. Egawa T, Watanabe H, Fujimura T, Ishida Y, Yamato M, Masui D, Shimada T, Tachibana H, Yoshida H, Inoue H, Takagi S (2011) Langmuir 27:10722–10729

68. Barrer RM (1978) Zeolites and clay minerals as sorbents and molecular sieves. Academic Press, London
69. Mitchell IV (ed) (1990) Pillared layered structures: current trends and applications. Elsevier, London
70. Lagaly G (1981) Clay Miner 16:1–21
71. Lagaly G (1986) Solid State Ion 22:43–51
72. Lagaly G, Beneke K (1991) Colloid Polym Sci 269:1198–1211
73. Ogawa M, Kuroda K (1997) Bull Chem Soc Jpn 70:2593–2618
74. Okada T, Ogawa M (2011) Clay Sci 15:103–110
75. Okada T, Seki Y, Ogawa M (2014) J Nanosci Nanotechnol 14:2121–2134
76. Lagaly G (1979) Adv Colloid Interface Sci 11:105–148
77. Schwieger W, Lagaly G (2004) Alkali silicates and crystalline silicic acids. In: Auerbach SM, Carrado KA, Dutta PK (eds) Handbook of layered materials. Taylor & Francis, Boca Raton, FL (Chapter 11)
78. Ide Y, Ochi N, Ogawa M (2011) Angew Chem Int Ed 50:654–656
79. Takahashi N, Kuroda K (2011) J Mater Chem 21:14336–14353
80. Ruiz-Hitzky E, Rojo JM (1980) Nature 287:28–30
81. Ruiz-Hitzky E, Rojo JM, Lagaly G (1985) Colloid Polym Sci 263:1025–1030
82. Ogawa M, Okutomo S, Kuroda K (1998) J Am Chem Soc 120:7361–7362
83. Ogawa M, Miyoshi M, Kuroda K (1998) Chem Mater 10:3787
84. Isoda K, Kuroda K, Ogawa M (2000) Chem Mater 12:1702–1707
85. Fujita I, Kuroda K, Ogawa M (2005) Chem Mater 17:3717–3722
86. Ide Y, Fukuoka A, Ogawa M (2007) Chem Mater 19:964–966
87. Ide Y, Iwasaki S, Ogawa M (2011) Langmuir 27:2522–2527
88. Nakamura T, Ogawa M (2012) Langmuir 28:7505–7511
89. Clearfield A, Constantino U (1996) Layered metal phosphates and their intercalation chemistry. In: Alberti G, Bein T (eds) Solid-state supramolecular chemistry: two- and three-dimensional inorganic networks, vol 7. Pergamon, Oxford (Chapter 4)
90. Raveau B (1987) Rev Inorg Chem 9:37–64
91. Ide Y, Sadakane M, Sano T, Ogawa M (2014) J Nanosci Nanotechnol 14:2135–2147
92. Kumar CV, Bhambhani A, Hnatiuk N (2004) Layered alpha-zirconium phosphates and phosphonates. In: Auerbach SM, Carrado KA, Dutta PK (eds) Handbook of layered materials. Taylor & Francis, Boca Raton, FL (Chapter 7)
93. Alberti G (1996) Layerd metal phosphonates and covalently pillared diphosphonates. In: Alberti G, Bein T (eds) Solid-state supramolecular chemistry: two- and three-dimensional inorganic networks, vol 7. Pergamon, Oxford (Chapter 5)
94. Ogawa M, Maeda N (1998) Clay Miner 33:643–650
95. Ogawa M, Takizawa Y (1999) J Phys Chem B 103:5005–5009
96. Ogawa M, Takizawa Y (1999) Chem Mater 11:30
97. Evans DG, Slade RCT (2006) Structural aspects of layered double hydroxides. In: Duan X, Evans DG (eds) Layered double hydroxides. Springer, Heidelberg
98. Reichle WT (1986) Chemtech 16:58–63
99. Trifirò F, Vaccari A (1996) Hydrotalcite-like anionic clays (layered double hydroxides). In: Alberti G, Bein T (eds) Solid-state supramolecular chemistry: two- and three-dimensional inorganic networks, vol 7. Pergamon, Oxford (Chapter 8)
100. Rives V, Ulibarri MA (1999) Coord Chem Rev 181:61–120
101. Li F, Duan X (2006) Applications of layered double hydroxides. In: Duan X, Evans DG (eds) Layered double hydroxides. Springer, Heidelberg
102. Park IY, Kuroda K, Kato C (1990) J Chem Soc Dalton Trans 3071–3074
103. Chibwe M, Pinnavaia TJ (1993) J Chem Soc Chem Commun 278–280
104. Ogawa M, Inomata K (2011) Clay Sci 15:131–137
105. Li L, Ma RZ, Ebina Y, Iyi N, Sasaki T (2005) Chem Mater 17:4386–4391

106. Liu Z, Ma R, Osada M, Iyi N, Ebina Y, Takada K, Sasaki T (2006) J Am Chem Soc 128:4872–4880
107. Kayano M, Ogawa M (2006) Bull Chem Soc Jpn 79:1988–1990
108. Arai Y, Ogawa M (2009) Appl Clay Sci 42:601–604
109. Igarashi S, Sato S, Takashima T, Ogawa M (2013) Ind Eng Chem Res 52:3329–3333
110. Ogawa M, Morita M, Igarashi S, Sato S (2013) J Solid State Chem 206:9–13
111. Gabriel JC, Camerel F, Lemaire BJ, Desvaux H, Davidson P, Batail P (2001) Nature 413:504–508
112. Nakato T, Furumi Y, Okuhara T (1998) Chem Lett 27:611–612
113. Nakato T, Miyamoto N (2002) J Mater Chem 12:1245–1246
114. Nakato T, Miyamoto N, Harada A (2004) Chemical Commun 78–79
115. Fossum JO, Gudding E, Fonseca DDM, Meheust Y, DiMasi E, Gog T, Venkataraman C (2005) Energy 30:873–883
116. Fonseca DM, Meheust Y, Fossum JO, Knudsen KD, Parmar KP (2009) Phys Rev E Stat Nonlin Soft Matter Phys 79:021402
117. Ogawa M, Takahashi M, Kato C, Kuroda K (1994) J Mater Chem 4:519–523
118. Isayama M, Sakata K, Kunitake T (1993) Chem Lett 22:1283–1286
119. Ogawa M (1998) Langmuir 14:6969–6973
120. Inukai K, Hotta Y, Taniguchi M, Tomura S, Yamagishi A (1994) J Chem Soc Chem Commun 959–959
121. Hotta Y, Taniguchi M, Inukai K, Yamagishi A (1997) Clay Miner 32:79–88
122. Suzuki Y, Tenma Y, Nishioka Y, Kawamata J (2012) Chem Asian J 7:1170–1179
123. Kleinfeld ER, Ferguson GS (1994) Science 265:370–373
124. Kleinfeld ER, Ferguson GS (1996) Chem Mater 8:1575
125. Lvov Y, Ariga K, Ichinose I, Kunitake T (1996) Langmuir 12:3038–3044
126. Keller SW, Kim HN, Mallouk TE (1994) J Am Chem Soc 116:8817–8818
127. Sasaki T, Watanabe M, Hashizume H, Yamada H, Nakazawa H (1996) Chem Commun 229–230
128. Sasaki T, Nakano S, Yamauchi S, Watanabe M (1997) Chem Mater 9:602–608
129. Lotsch BV, Ozin GA (2008) Adv Mater 20:4079
130. Ariga K, Ji Q, McShane MJ, Lvov YM, Vinu A, Hill JP (2012) Chem Mater 24:728–737
131. Guang C, Hong HG, Mallouk TE (1992) Acc Chem Res 25:420–427
132. Sasaki T, Watanabe M, Hashizume H, Yamada H, Nakazawa H (1996) J Am Chem Soc 118:8329–8335
133. Osada M, Sasaki T (2009) J Mater Chem 19:2503–2511
134. Osada M, Sasaki T (2012) Adv Mater 24:210–228
135. Ogawa M (1996) Chem Mater 8:1347
136. Ogawa M, Ishikawa A (1998) J Mater Chem 8:463–467
137. Ogawa M, Yamamoto M, Kuroda K (2001) Clay Miner 36:263–266
138. Ogawa M, Goto R, Kakegawa N (2000) Clay Sci 11:231–241
139. Shimomura M, Aiba S (1995) Langmuir 11:969–976
140. Iyi N, Kurashima K, Fujita T (2002) Chem Mater 14:583–589
141. Sudo H, Hatano B, Kadokawa JI, Tagaya H (2007) J Ceram Soc Jpn 115:901–904
142. Umemoto T, Ohtani Y, Tsukamoto T, Shimada T, Takagi S (2014) Chem Commun 50:314–316
143. Sasai R, Shichi T, Gekko K, Takagi K (2000) Bull Chem Soc Jpn 73:1925–1931
144. Sasai R, Itoh H, Shindachi I, Shichi T, Takagi K (2001) Chem Mater 13:2012–2016
145. Shindachi I, Hanaki H, Sasai R, Shichi T, Yui T, Takagi K (2004) Chem Lett 33:1116–1117
146. Shindachi I, Hanaki H, Sasai R, Shichi T, Yui T, Takagi K (2007) Res Chem Intermed 33:143–153
147. Fujita T, Iyi N, Klapyta Z (1998) Mater Res Bull 33:1693–1701
148. Ahmadi MF, Rusling JF (1995) Langmuir 11:94–100
149. Okahata Y, Shimizu A (1989) Langmuir 5:954–959

150. Hu NF, Rusling JF (1991) Anal Chem 63:2163–2168
151. Takagi K, Usami H, Fukaya H, Sawaki Y (1989) J Chem Soc Chem Commun 1174–1175
152. Shichi T, Takagi K, Sawaki Y (1996) Chem Lett 25:781–782
153. Usami H, Takagi K, Sawaki Y (1992) J Chem Soc Faraday Trans 88:77–81
154. Usami H, Takagi K, Sawaki Y (1991) Bull Chem Soc Jpn 64:3395–3401
155. Shichi T, Takagi K, Sawaki Y (1996) Chem Commun 2027–2028
156. Usami H, Takagi K, Sawaki Y (1990) J Chem Soc Perkin Trans 2:1723–1728
157. Takagi K, Shichi T, Usami H, Sawaki Y (1993) J Am Chem Soc 115:4339–4344
158. Moerner WE (ed) (1988) Persistent spectral hole-burning: science and application. Springer, Berlin
159. Sakoda K, Kominami K (1993) Chem Phys Lett 216:270–274
160. Ogawa M, Handa T, Kuroda K, Kato C, Tani T (1992) J Phys Chem 96:8116–8119
161. Prasad PN, Karna SP (1994) Int J Quant Chem 52:395–410
162. Munn RW, Ironside CN (eds) (1993) Principles and applications of nonlinear optical materials. Springer, Netherlands
163. Rabu P, Drillon M (2003) Adv Eng Mater 5:189–210
164. Delahaye É, Eyele-Mezui S, Bardeau J-F, Leuvrey C, Mager L, Rabu P, Rogez G (2009) J Mater Chem 19:6106
165. Iyi N, Fujita T, Yelamaggad CV, Arbeloa FL (2001) Appl Clay Sci 19:47–58
166. Ogawa M, Kuroda K, Mori J (2000) Chem Commun 2441–2442
167. Ogawa M, Kuroda K, Mori J (2002) Langmuir 18:744–749
168. Ogawa M (2002) J Mater Chem 12:3304–3307
169. Ogawa M, Ishii T, Miyamoto N, Kuroda K (2003) Appl Clay Sci 22:179–185
170. Kinashi K, Kita H, Misaki M, Koshiba Y, Ishida K, Ueda Y, Ishihara M (2009) Thin Solid Films 518:651–655
171. Yui T, Yoshida H, Tachibana H, Tryk DA, Inoue H (2002) Langmuir 18:891–896
172. Tong ZW, Takagi S, Shimada T, Tachibana H, Inoue H (2006) J Am Chem Soc 128:684–685
173. Tong ZW, Sasamoto S, Shimada T, Takagi S, Tachibana H, Zhang XB, Tryk DA, Inoue H (2008) J Mater Chem 18:4641–4645
174. Nabetani Y, Takamura H, Hayasaka Y, Sasamoto S, Tanamura Y, Shimada T, Masui D, Takagi S, Tachibana H, Tong Z, Inoue H (2013) Nanoscale 5:3182–3193
175. Nabetani Y, Takamura H, Hayasaka Y, Shimada T, Takagi S, Tachibana H, Masui D, Tong Z, Inoue H (2011) J Am Chem Soc 133:17130–17133
176. Han J, Yan D, Shi W, Ma J, Yan H, Wei M, Evans DG, Duan X (2010) J Phys Chem B 114:5678–5685
177. Okada T, Watanabe Y, Ogawa M (2004) Chem Commun 320–321
178. Okada T, Watanabe Y, Ogawa M (2005) J Mater Chem 15:987–992
179. Okada T, Sakai H, Ogawa M (2008) Appl Clay Sci 40:187–192
180. Heinz H, Vaia RA, Koerner H, Farmer BL (2008) Chem Mater 20:6444–6456
181. Hornick C, Rabu P, Drillon M (2000) Polyhedron 19:259–266
182. Fujita W, Awaga K (1997) J Am Chem Soc 119:4563–4564
183. Fujita W, Awaga K, Yokoyama T (1999) Appl Clay Sci 15:281–303
184. Abellan G, Coronado E, Marti-Gastaldo C, Ribera A, Jorda JL, Garcia H (2014) Adv Mater 26:4156–4162
185. Yamamoto T, Umemura Y, Sato O, Einaga Y (2004) Chem Mater 16:1195–1201
186. Yamamoto T, Umemura Y, Sato O, Einaga Y (2006) Sci Technol Adv Mater 7:134–138
187. Yamamoto T, Saso N, Umemura Y, Einaga Y (2009) J Am Chem Soc 131:13196
188. Shimizu H, Okubo M, Nakamoto A, Enomoto M, Kojima N (2006) Inorg Chem 45:2006
189. Benard S, Leaustic A, Riviere E, Yu P, Clement R (2001) Chem Mater 13:3709–3716
190. Monk PMS (1998) The viologens: physicochemical properties, synthesis and applications of the salts of 4,4′-bipyridine. Wiley, Chichester
191. Kakegawa N, Kondo T, Ogawa M (2003) Langmuir 19:3578–3582
192. Okada T, Ogawa M (2003) Chem Commun 1378–1379

193. Miyata H, Sugahara Y, Kuroda K, Kato C (1987) J Chem Soc Faraday Trans 183:1851–1858
194. Vermeulen LA, Thompson ME (1992) Nature 358:656–658
195. Vermeulen LA, Snover JL, Sapochak LS, Thompson ME (1993) J Am Chem Soc 115:11767–11774
196. Nakato T, Kuroda K (1995) Eur J Sol State Inorg 32:809–818
197. Miyata H, Sugahara Y, Kuroda K, Kato C (1988) J Chem Soc Faraday Trans 184:2677–2682
198. Nakato T, Miyata H, Kuroda K, Kato C (1988) React Solids 6:231–238
199. Nakato T, Kuroda K, Kato C (1989) J Chem Soc Chem Commun 1144–1145
200. Nakato T, Kuroda K, Kato C (1992) Chem Mater 4:128–132
201. Nakato T, Ito K, Kuroda K, Kato C (1993) Microporous Mater 1:283–286
202. Kakegawa N, Ogawa M (2004) Langmuir 20:7004–7009
203. Weber JB, Perry PW, Upchurch RP (1965) Soil Sci Soc Am J 29:678–688
204. Raupach M, Emerson WW, Slade PG (1979) J Colloid Interface Sci 69:398
205. Ghosh PK, Bard AJ (1984) J Phys Chem 88:5519–5526
206. Kakegawa N, Ogawa M (2003) Clay Sci 12:153–158
207. Snover JL, Thompson ME (1994) J Am Chem Soc 116:765–766
208. Ungashe SB, Wilson WL, Katz HE, Scheller GR, Putvinski TM (1992) J Am Chem Soc 114:8717–8719
209. Kumar CV, Chaudhari A (1994) J Am Chem Soc 116:403–404
210. Konno S, Fujimura T, Otani Y, Shimada T, Inoue H, Takagi S (2014) J Phys Chem C 118:20504–20510
211. Nakato T, Watanabe S, Kamijo Y, Nono Y (2012) J Phys Chem C 116:8562–8570
212. Miyamoto N, Yamada Y, Koizumi S, Nakato T (2007) Angew Chem Int Ed 46:4123–4127
213. Abe R, Shinmei K, Koumura N, Hara K, Ohtani B (2013) J Am Chem Soc 135:16872–16884

Photochemistry of Graphene

Liming Zhang and Zhongfan Liu

> *Carbon has this genius of making a chemically stable two-dimensional, one-atom-thick membrane in a three-dimensional world. And that, I believe, is going to be very important in the future of chemistry and technology in general [1].*
> —Richard Smalley (*from Nobel Lecture in 1996*)

Contents

1 Introduction 214
2 Free-Radical-Induced Photochemistry of Graphene 218
   2.1 Generation of Free Radicals 218
   2.2 Photohalogenation 219
   2.3 Photoarylation and Photoalkylation 220
   2.4 Photocatalytic Oxidation 220
   2.5 Addition of Other Functional Groups 222
3 Geometry-Correlated Effects in Graphene Photochemistry 222
   3.1 Single- and Double-Sided Additions 223
   3.2 Janus Graphene from Asymmetric 2D Chemistry 224
   3.3 Edge Effect 225
   3.4 Interlayer Coupling Effect 226
4 Applications 228
   4.1 Band Structure Engineering 229
   4.2 Surface Property Modulation 229
   4.3 Graphene Derivatives and Its 2D Hybrids 230
5 Summary and Outlook 232
References 233

Abstract As a two-dimensional (2D) giant polycyclic aromatic molecule, graphene provides a great opportunity for studying the behaviors of chemical reactions in two dimensions. However, the chemistry of graphene is challenging because of its extreme inertness due to the highly delocalized π electron system.

L. Zhang • Z. Liu (✉)
Center for Nanochemistry (CNC), Beijing Science and Engineering Center for Nanocarbons, Beijing National Laboratory for Molecular Sciences, College of Chemistry and Molecular Engineering, Peking University, Beijing 100871, People's Republic of China
e-mail: zfliu@pku.edu.cn

Recently, photogenerated free radicals have been demonstrated to effectively activate the chemical reactions of graphene. Such kinds of graphene photochemistry provide a new route for the covalent functionalization and band structure engineering of zero-gap graphene. In this chapter, the graphene photochemistry based on photogenerated free radicals is reviewed, including photohalogenation, photoarylation, photoalkylation, and photocatalytic oxidation. Although most photochemical reactions on graphene can be inspired by its small organic analogues, graphene photochemistry is of particular attraction due to its infinite 2D geometry, which offers a platform for studying geometry-correlated covalent chemistry, including single- and double-sided covalent addition reactions, asymmetric chemistry on two faces of monolayer graphene, edge-selective chemistry, and interlayer coupling-dependent few-layer graphene chemistry. In addition to the modulation of surface properties and the band structure engineering, new 2D derivatives and superlattices with fascinating features beyond mother graphene can be built by graphene photochemistry, which greatly expands the graphene family and its attraction. This chapter also summarizes the potential applications of graphene photochemistry with a specific focus being laid on the general consideration and understanding of graphene photochemistry towards electronic/optoelectronic devices and materials science. At the end, a brief discussion on the future directions, challenges, and opportunities in this emerging area is provided.

Keywords Graphene • Photochemistry • Free-radical addition • Janus graphene • Band structure engineering • Graphene superlattice

1 Introduction

Human beings have been using carbon for thousands of years; however, acquaintance with atomic layers of carbon atoms arose only when A. Geim and K. Novoselov successfully isolated graphene in 2004 [2]. To obtain deep insight into the chemistry of graphene, it is required to know the conventional hybridization of the atomic orbitals of carbon atoms first. Elemental carbon exists dominantly in three bonding states—sp^3, sp^2, and sp hybridization, and the corresponding three carbon allotropes with an integer degree of carbon bond hybridization are diamond, graphite, and carbyne [3]. Interestingly, there are also intermediate carbon forms with a non-integer degree of carbon bond hybridization, sp^n. For example, some closed-shell carbon structures, such as fullerene, carbon onion, and carbon nanotube, have carbon bonding hybridization states with $2 < n < 3$ [3–5]. The newly known carbon allotrope, graphyne, has a hybridization state in the range of $1 < n < 2$ [6–8]. In addition, some carbon allotropes have multiple hybridized states, such as amorphous carbon, which comprises carbon atoms with sp^2 and

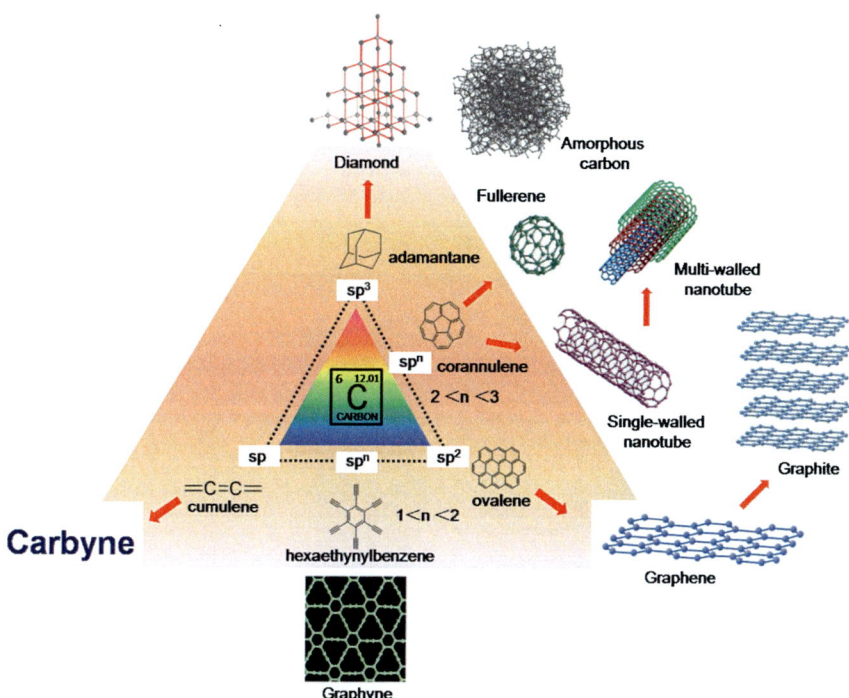

Fig. 1 A carbon allotrope diagram based on carbon valence bond hybridization

sp^3 hybridization simultaneously, each state with short-range ordering. The different carbon allotropes can be visualized in a carbon allotrope diagram based on carbon valence bond hybridization, as shown in Fig. 1. Starting with small organic molecules (the inner circle in Fig. 1), the hierarchy of carbon materials can be described as an extension of organic molecular species to the bulk inorganic all-carbon materials. Comparing with other crystalline carbon allotropes, carbon atoms in graphene are densely packed in a perfect hexagonal pattern with an infinite two-dimensional (2D) structure [9]. All of the carbon atoms in graphene are bonded with three other carbon atoms via sp^2 hybridization, namely, there are three σ orbitals placed in the graphene plane with angle 120° and one π orbital along Z-axis in the perpendicular direction. This unique structure gives graphene extraordinary mechanical, optical, and electronic properties. It is the thinnest material in the universe but the strongest ever measured [10]; it has excellent optical transparency (~97.7 %) [11]; its charge carriers exhibit a giant intrinsic carrier mobility (~200,000 cm^2/Vs), have zero effective mass, and can travel for micrometers without scattering at room temperature [12–16]. The promising applications of graphene in the future microelectronics, nanoelectronics, and optoelectronics have already motivated intense researches in various aspects ranging from theoretical predictions [17–25] and fundamental properties [10–12, 26–31] to large-scale synthesis [32–50] and practical applications [51–69].

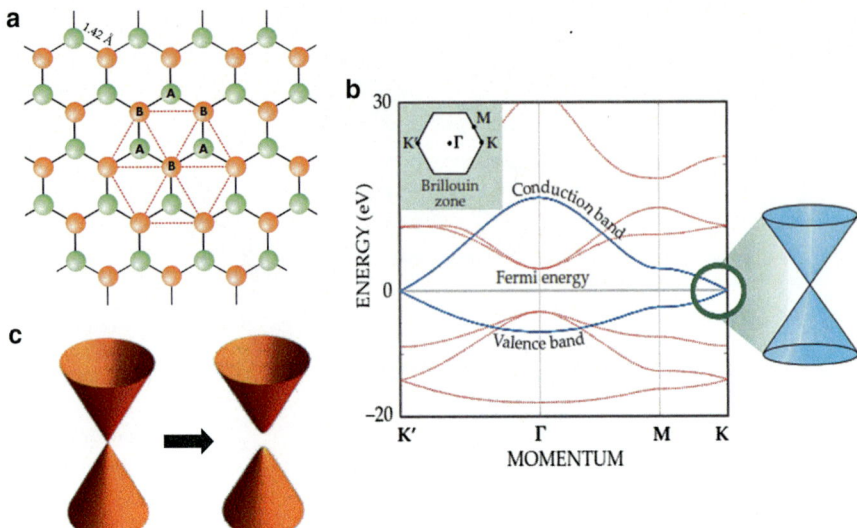

Fig. 2 (**a**) Hexagonal honeycomb lattice of graphene with two carbon atoms (A and B) per unit cell. The sites of one sub-lattice (A) are at the centers of triangles defined by the other (B). The length of –C=C– bond is 1.42 Å. (**b**) Energy-momentum dispersion in graphene. The π and π* bands (*blue curves*) are decoupled from the σ and σ* bands (*red curves*) due to the inversion symmetry and are closer to the Fermi energy as they participate less in bonding. The bands form conical valley that touch at two of the high symmetry points, conventionally labeled as K and K′, in the Brillouin zone (inset, *right*). (**c**) Band structure change of monolayer graphene near K point of the Brillouin zone after chemical modification. (**a**) and (**b**) are adapted from [70]

As early as the 1940s, graphene is endowed with a unique band structure by the special 2D atomic arrangement with two sp^2 hybridized carbon atoms in one unit cell (Fig. 2a) [71]. Each atom has one 2p orbital oriented perpendicular to the molecular plane, hybridizing into π and π* to form the valence and conduction bands, respectively. These two bands are only allowed to touch at six points, named Dirac points, at the Fermi level with no overlap [70]. This means that graphene can be considered as a semiconductor with zero bandgap or, alternatively, as a metal with zero density of states (DOS) at the Fermi level. In the vicinity of a Dirac point, the valence and conduction bands form a double cone (Dirac cone) and approach the Dirac point with zero curvature along certain symmetry-determined directions (Fig. 2b) [70]. The lack of bandgap is one big challenge for the practical application of graphene, since it leads to a high leakage of currents and power dissipation in the logic circuits. To open up a bandgap, it is necessary to break the symmetry of carbon atoms in one unit cell, making the potential on the two atoms different. Several strategies have been proposed to break the symmetry of graphene. For instance, formation of nanoribbon [19, 72–74]/nanomesh [23, 75–77] can generate a large bandgap by confining the electrons and holes to a "quantum box." Additionally, a tunable bandgap up to 250 mV can be opened in Bernal-stacked bilayer graphene by an external electric field [78, 79]. The limited edge controllability and/or lack of scalability as well as the complicated and time-consuming procedures of these approaches call for more efforts and breakthroughs on graphene band structure engineering for electronic applications.

Single-layer graphene is a giant 2D molecule with all the atoms accessible from surface. Graphene chemistry provides a powerful pathway to break the symmetry of graphene through covalent attachment of chemical groups, which leads to the transition of carbon hybridization from sp^2 to sp^3, thus breaking the conducting π bands and opening a tunable bandgap in graphene (Fig. 2c) [80]. In addition to the bandgap modification, graphene chemistry also provides some other attractive possibilities. For instance, novel crystalline graphene derivatives can be created by controlling the stoichiometry. A. Geim and his colleagues pioneered such graphene chemistry and achieved two crystalline graphene derivatives, graphane (hydrogen atoms are bonded to each carbon atom) [81] and fluorographene [82]. Both 2D derivatives are insulators with a large bandgap (more than 3 eV) and are stable at ambient temperature. In addition, surface chemical properties of graphene can be drastically changed by the attached decorations. Homogeneous dispersion of graphene in aqueous [44] and organic solvents [83] can be obtained by selectively attaching different functional groups on its surface, which facilitates various solution chemistry of graphene and practical applications. Different chemical groups can also be attached onto two faces of 2D graphene asymmetrically, creating an anisotropic wetting property on a single layer of carbon atoms [84]. Moreover, the in-plane hybrids of graphene, such as graphene/n-graphene [85, 86], graphene/hexagonal boron nitride (h-BN) [87, 88], graphene/graphane [89, 90], graphene/chlorinated graphene [91], graphene/graphene oxide [92], etc., can be fabricated by periodic chemical functionalization, providing an ideal platform to study the rich physics expected in 2D electronic system. Furthermore, chemical modification can be employed to modulate the optical, chemical, and mechanical properties of graphene for the purpose of fabricating sensors and actuators for biochemical detection and energy storage [57, 60, 63, 66, 68, 93].

Graphene is built with an infinite number of benzene rings two-dimensionally. Similar to benzene, all the sp^2 carbon atoms in graphene contribute a $2p_z$ electron for π bonding; thus, various reactions occurring on –C=C– of benzene can be expected for graphene. However, graphene chemistry is more difficult as a result of the giant-conjugation system, little structural curvature, and absence of dangling bonds. Thus, highly reactive chemical species are needed to activate the reaction. So far, only a few examples, such as graphene hydrogenation [81, 89], fluorination [82, 94], oxidation [95], diazonium salt reactions [83, 96], and some cycloaddition reactions [97–99], have been successfully demonstrated though the reaction efficiency is very limited.

Free radicals are known to be highly reactive chemical species, which can be easily generated by photochemical or thermochemical processes [100]. The high reactivity of free radicals makes it possible to overcome the large energy barriers in graphene chemistry and break the inert –C=C– bonds. In this chapter, we present a review on the photochemical modification of graphene using a variety of photoinduced free radicals [101]. As schematically illustrated in Fig. 3, the free radicals X · are photogenerated under illumination from neutral molecules. These radicals can

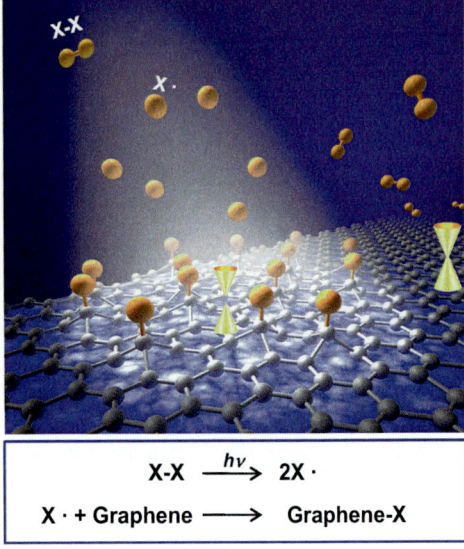

Fig. 3 Schematic illustration of graphene photochemistry based on photogenerated high energy free radicals. Adapted from [101]

react with the exposed graphene surface, leading to the light-guided covalent modification of graphene.

In the following section, we summarize –C=C– addition/oxidation reactions of graphene with various photogenerated radicals. Emphasis will be laid on the gas–solid interface reactions rather than liquid phase dispersions, targeting the carbon-based high-performance electronic devices. In the third section, we show various geometry-correlated effects in graphene photochemistry, including the single- and double-sided covalent addition reactions, asymmetric chemistry on two faces of monolayer graphene, edge-selective chemistry, and interlayer coupling-determined few-layer graphene chemistry. Finally, we demonstrate the applications of graphene photochemistry together with a brief discussion on the future directions, challenges, and opportunities in this fascinating research area.

2 Free-Radical-Induced Photochemistry of Graphene

2.1 Generation of Free Radicals

The electrons in molecules are paired as bonding/lone electron pairs. However, free radical is a chemical species (atom, molecule, or ion) that has single unpaired electron. The unpaired electron in free radicals has the highest energy among all bonding and nonbonding electrons in a molecule, which makes free radicals highly reactive towards other molecules. The free radicals are typically generated by photochemical or thermal homolytic bond cleavage, which requires significant amount of energy [100]. In particular, the photochemical approach is applied widely since it is clean and easy to operate. By tuning the wavelength of illumination, photons can well satisfy the varied energy requirements in the radical generation.

The required energy for generating free radicals varies with different radicals. For example, splitting H_2 into $2H\cdot$ has a $\Delta H^{\ominus}$ of +435 kJ/mol, and that of Cl_2 has a $\Delta H^{\ominus}$ of +243 kJ/mol [102]. This is known as the homolytic *bond dissociation energy* (BDE). Radicals requiring more energy to form are less stable than those requiring less energy. The reactivity of radical is dependent on its stability and lifetime. As a general role, the more stable the radical is, the less active it is. Overall, we can get a rough trend of the chemical reactivity of free radicals [100, 103]:

$F\cdot > HO\cdot > F_3C\cdot > H\cdot , Cl\cdot , MeO\cdot > Ph\cdot > Me\cdot > Et\cdot > Cl_3C\cdot , Br\cdot > PhCH_2\cdot > Ph_3C\cdot > I\cdot$

In the following, we will demonstrate a few examples of graphene photochemistry based on the free radicals with high chemical reactivity outlined above.

2.2 Photohalogenation

The chemical reactivity of halogen radicals essentially increases with the trend $I\cdot < Br\cdot < Cl\cdot < F\cdot$ [100]. In particular, $F\cdot$ and $Cl\cdot$ are very reactive and can easily break $-C=C-$ bonds. For example, $F\cdot$ can functionalize graphene in a very short time once generated and achieve stoichiometric fluorographene (C/F ratio 1:1) ultimately [82, 94, 104, 105]. In order to slow down the reaction and get better controllability, $F\cdot$ has to be generated via fluorination agents, such as XeF_2 and fluoropolymers, from which $F\cdot$ can be mildly released [94]. Recently, a facile photochemical approach has been developed to obtain fluorographene by laser irradiating fluoropolymer-covered graphene, in which the fluoropolymer molecules can release $F\cdot$ under the irradiation and functionalize graphene ultimately [104].

The photochlorination of graphene was inspired by the chlorine addition reaction of benzene in the presence of light [106]. Unlike the rapid destruction of graphene by fluorine, much slower reaction kinetics between chlorine and graphene were observed, allowing for controlled chlorination. The photogenerated chlorine radicals were first adsorbed onto graphene surface when exposing to Cl_2 gas under UV light, leading to a strong *p*-type doping effect and a significant increase of electrical conductivity. Addition reaction occurred afterwards (Fig. 4a), forming a chlorinated graphene with ~8 atom % Cl coverage at most [91]. The photochlorination reaction was highly efficient as shown in Fig. 4b, in which only a few minutes were required to reach the saturated coverage. Plasma treatment was also applied for graphene halogenation. For example, graphene can be successfully chlorinated under the chlorine plasma, achieving ~8.5 atom % Cl coverage in 1 min [108].

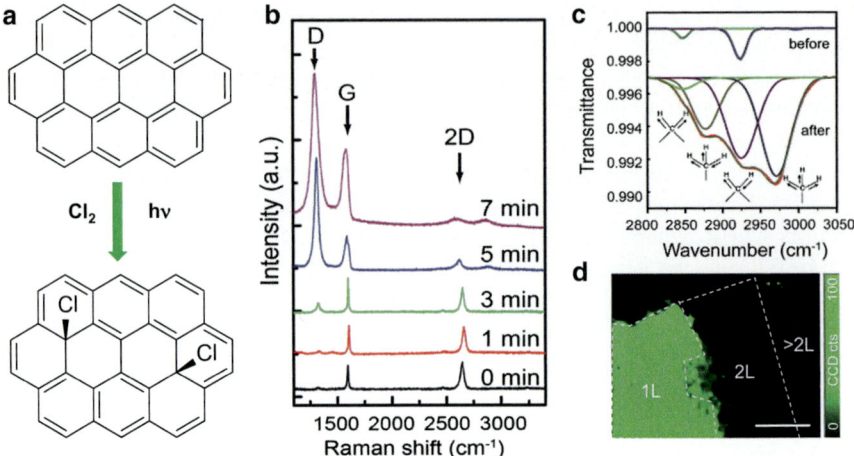

Fig. 4 (**a**) Schematic illustration of photochlorination of graphene. (**b**) Raman spectral evolution of graphene upon photochlorination, where the rapid increase of D band and decrease of 2D band intensity indicate the fast reaction kinetics. (**c**) IR spectra of pristine (*top*) and methylated graphene (*bottom*), confirming the successful addition of methyl groups. (**d**) Raman mapping of D band on graphene after photomethylation, showing the higher reactivity of monolayer graphene. (**a**) and (**b**) are adapted from [91]; (**c**) and (**d**) are adapted from [107]

2.3 Photoarylation and Photoalkylation

Aryl and alkyl groups can be attached onto graphene surface by radical chemistry. Such kinds of functional groups will neither introduce much doping effect nor change the surface wettability of graphene. Typically, the reactivities of these radicals are not very strong, thus providing a good opportunity to study the reaction selectivity on graphene 2D surface, such as interlayer coupling and edge-determined reaction behaviors. Brus et al. pioneered the photoinduced graphene arylation. Using benzoyl peroxide as the radical source, they introduced aryl groups onto the graphene scaffold via photo-assisted aryl radical addition [109]. Similar strategy also works on the photomethylation of graphene, using ditertbutyl peroxide as the methyl radical source [107]. As shown in Fig. 4c, the successful addition of methyl groups can be confirmed by infrared (IR) spectrum. Raman spectra showed that the addition preferably takes place on monolayer graphene surface as compared with few-layer graphene (Fig. 4d).

2.4 Photocatalytic Oxidation

Complementary to $-C=C-$ addition, oxidation is another strategy to achieve covalent chemical modification of graphene. In the early 1850s, graphene oxide,

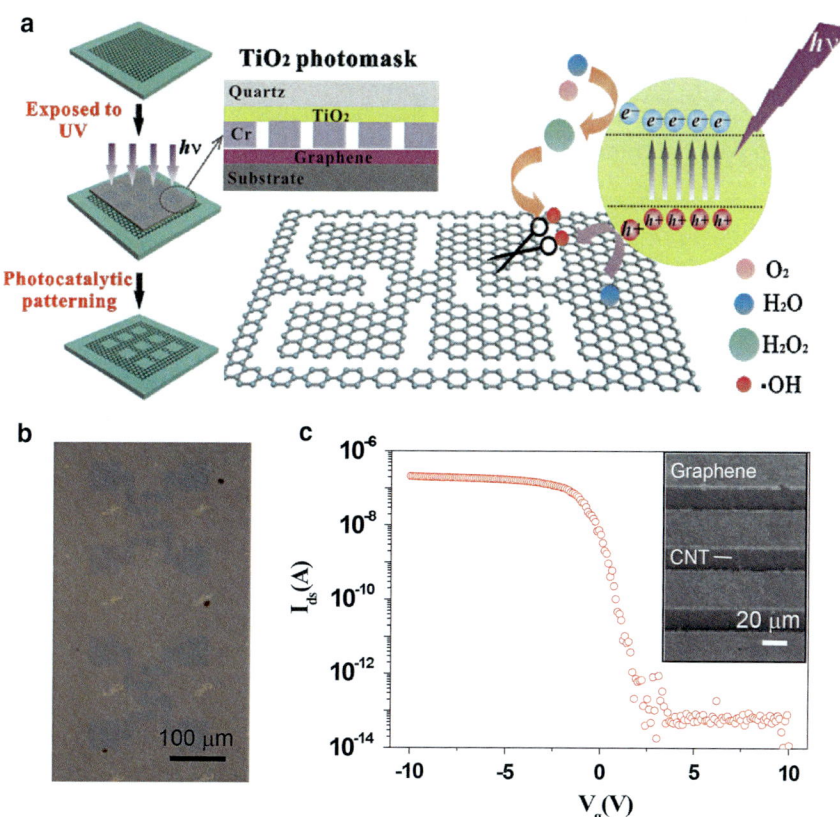

Fig. 5 (**a**) Schematic of TiO$_2$-based photocatalytic oxidation of graphene, where the photogenerated ·OH is employed as chemical scissors for graphene tailoring. (**b**) Optical microscope image of the obtained graphene electrode patterns by photocatalytic tailoring. (**c**) I_{ds}–V_g curve (at $V_{ds} = 0.1$ V) of an all-carbon FET with a semiconducting carbon nanotube serving as the conduction channel and the patterned graphene as the contacts. Inset, SEM image of the device. Adapted from [92]

another insulating derivative of graphene, has been obtained by the oxidation of graphite powder in strong oxidative solvent [95]. Graphene oxide can also be achieved through oxygen plasma [93] or O$_2$ treatment under high temperature [110]. As mentioned in Sect. 2.1, hydroxyl radical (·OH) possesses a strong oxidation capability, and it can be used for highly efficient oxidation of graphene. One successful example of controllable generation of ·OH is TiO$_2$-based photocatalytic strategy (Fig. 5a). TiO$_2$-based photocatalytic oxidation is a typical photochemical reaction and has been widely used to decompose various organic pollutants for environmental protection [111, 112]. With the aid of TiO$_2$, highly reactive ·OH can be generated under UV illumination and suitable humidity. These photogenerated radicals progressively react with the graphenic carbons, leading to the incorporation of oxy groups (e.g., epoxy, hydroxyl, carbonyl, and carboxyl groups) onto the basal plane of graphene [92]. By controlling the UV light intensity and irradiation time, the photogenerated ·OH can work as sharp chemical scissors to cut graphene into arbitrary patterns, allowing for the versatile design and fabrication of graphene-based devices and circuits (Fig. 5b) [92]. For example, an all-carbon field effect transistor (FET) array can be constructed with the patterned graphene as the electrode array and single-walled carbon nanotube as the conducting channel (Fig. 5c).

2.5 Addition of Other Functional Groups

A variety of other kinds of functional groups can also be decorated onto graphene surface via photochemistry, offering diverse surface properties of graphene derivatives. For instance, phenyl and alkyl azides react with the –C=C– bonds by the formation of reactive intermediate nitrene [113–116]. By photochemically activating a series of para-substituted perfluorophenylazides (PFPA), various functional groups were attached onto graphene via a three-membered aziridine ring [114]. These reactions result in varying solubility–dispersibility and surface energy of the modified graphenes.

In addition to the photoactivation, free radicals can also be initiated via thermal treatment. Upon heating diazonium salt, a highly reactive free radical is produced, which attacks the –C=C– of graphene, forming a covalent bond. This reaction has been used by J. Tour and coworkers to decorate graphene with nitrophenyls [83, 117]. Trifluoromethylation of graphene can be realized by applying highly reactive trifluoromethyl ($·CF_3$) free radicals. For example, $·CF_3$ was generated via copper-catalyzed process and grafted onto the basal plane of graphene, achieving ~4 atom % CF_3 coverage [118]. As CF_3 group is a very important moiety in medicines and agrochemicals, implanting CF_3 onto graphene may open up a gateway to graphene's applications in biotechnology.

3 Geometry-Correlated Effects in Graphene Photochemistry

Graphene is a 2D giant polycyclic aromatic molecule that provides opportunities for studying the behaviors of chemical reactions in two dimensions. For example, monolayer graphene has two faces and all of the atoms are accessible from the surface. This unique structure offers a platform to study the geometry-correlated chemical reactions on 2D surface, such as single- and double-sided covalent addition reactions, asymmetric chemistry on two faces of monolayer graphene, and edge-selective chemistry. In addition, layer numbers and stacking orientations provide a diversity to tune the chemical reactivity of few-layer graphene from the vertical direction; hence, probing the thickness- and stacking configuration-dependent graphene chemistry can give rich information to understand the interlayer coupling effect. All of the characteristics described above originate from the specific ultrathin 2D feature of graphene. In this section, we will provide more details on these geometric effects in the free-radical-based graphene photochemistry.

3.1 Single- and Double-Sided Additions

As a perfect 2D material, single-layer graphene has two faces. A number of theoretical works have shown that single-sided functionalization of graphene is not kinetically favorable relative to the double-sided functionalization [80, 119–121]. For example, calculation has shown that chemisorption of hydrogen by adjacent carbon atoms from both sides simultaneously is more energetically preferable [80]. This side-dependent reaction can be partially explained by the steric hindrance effect, especially when the functional groups are in a large size [120]. When the functional groups come from two sides, steric hindrance can be largely decreased, leading to a higher coverage of decorations (Fig. 6a). In addition, transformation of the sp^2 hybridization of carbon to the sp^3 hybridization will lead to the change of the bond lengths and angles, from which a surface deformation will be induced. For sp^3 hybridization, the standard value of –C–C– bond length is 1.54 Å, and the angle is 109.5° [80]. Therefore, the carbon atoms attached with functional groups will be stretched out from the plane, leading to the deformation of the neighboring carbon atoms to the other direction, which will subsequently lower the activation energy of the following attachment of functional groups coming from the other side.

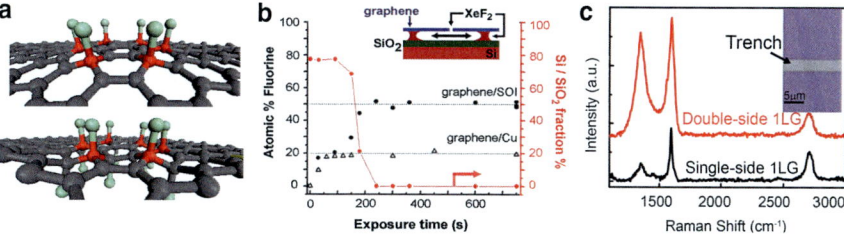

Fig. 6 (a) Schematic illustration of single- (*top*) and double-sided (*bottom*) functionalization of graphene. (b) Fluorine fraction of graphene for single (graphene/Cu)- and double [graphene/silicon-on insulator (SOI)]-side exposure to XeF_2 for different times. SOI can be etched by XeF_2; thus, graphene loading on SOI will be suspended ultimately. (c) Raman spectra of suspended (double-side) and supported (single-side) monolayer graphene after photochlorination under the same condition. Inset shows the optical image of a monolayer graphene, part of which has been suspended via loading on a trench. (**b**) and (**c**) are adapted from [94] and [122], respectively

Perfect graphene is an impermeable membrane in spite of one atomic layer thickness [123–128]; thus, it is possible to experimentally carry out the side-selective chemistry on monolayer graphene. In fact, free-standing graphene always exhibits higher chemical reactivity than the solid surface-supported graphene because of its higher accessibility to the reactants. For example, fluorination of as-grown graphene on copper is a single-sided reaction with a preferred C_4F stoichiometry, while fluorination of suspended graphene results in a double-sided reaction with a dominant CF stoichiometry (Fig. 6b) [94]. Free-standing graphene can be achieved by loading graphene on a substrate with trenches. As shown in Fig. 6c, the reaction behavior of suspended and supported graphene has been compared, using photochlorination reaction as a probe [122]. Raman spectra clearly show that the free-standing graphene exhibits a higher chlorination extent at the same experimental conditions, which is more than twice compared to the graphene supported on the substrate. This dramatic increase in reactivity indicates that the steric hindrance and surface deformation of graphene have a synergic effect on increasing the reaction kinetics of suspended graphene.

3.2 Janus Graphene from Asymmetric 2D Chemistry

Encouraged by the side-dependent graphene chemistry, asymmetric functionalization on two faces of monolayer graphene is highly desired to pursuit. *Janus* graphene can be obtained in such a way, which has a molecular-thick sandwich structure, X–G–Y, as shown in Fig. 7a, b. In addition to the unique structure, such kinds of asymmetrically modified graphene may provide additional attractions. For instance, the half-hydrogenated and half-fluorinated graphene was predicated to have a direct bandgap, which was larger than that of fluorographene but smaller than graphane [129]. A series of *Janus* graphene asymmetrically modified with two different species, e.g., H, F, Cl, Br, and phenyl, onto the different sides of monolayer graphene was theoretically studied [130]. It was found that the asymmetric functionalization was able to break the symmetric restrictions of graphene, leading to a robust nonzero bandgap, which was linearly correlated with the binding energy of two functionalities. Experimentally, the first *Janus* graphene was achieved by a two-step functionalization method [84]. A successful *Janus* graphene example is shown in Fig. 7c, in which one side of graphene was photochlorinated while the other side was photophenylated. A particularly interesting phenomenon is that, compared with pristine graphene, no discernible kinetic change for phenylation was detected when having phenyl groups on the opposite side, while a rapid increase of kinetics was found when having the same decoration extent of chlorine groups on the opposite side (Fig. 7d). This phenomenon was explained by the different cluster configurations of chlorine and phenyl groups attached on graphene surface [120, 121]. The chlorine groups can introduce more activation effect to the phenylation on the opposite side and thus lead to a dramatic increase in the reaction rate. This phenomenon indicates that the functionalities attached on the opposite

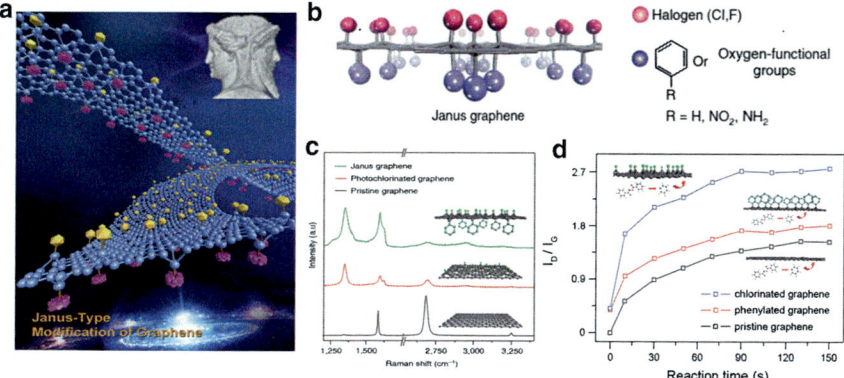

Fig. 7 (**a, b**) Schematic structure of a *Janus* graphene. The *red balls* in (**b**) represent the Cl or F atoms covalently attached on one side of the as-prepared *Janus* graphene, and the *blue balls* represent the oxygen functional groups or the different phenyl-based groups. (**c**) Raman spectral evolution of graphene following one- and two-sided asymmetric modifications. (**d**) Functionality dependence of phenylation kinetics on the other side, probed by in situ Raman spectroscopy. Change of I_D/I_G in response to the phenylation time with varied functionalities on the other side. Adapted from [84]

sides of graphene can "communicate" with each other through the single-layer carbon scaffold.

3.3 Edge Effect

Graphene confines its π-electrons in the space defined by the σ-bond network; thus, π-electron will respond to both the confinement and the shape of the peripheries. When one of the two dimensions of graphene is nano-sized, the bandgap enlarges exponentially with further shrinking [18, 131]. Nanographenes have different shapes and topologies, such as one-dimensional (1D) nanoribbons and zero-dimensional (0D) nanodots, in which the edge effect will be more pronounced. Depending on the cutting directions, two unique types of edges can be obtained: zigzag and armchair (Fig. 8a) [19]. By constructing an analytical model for the π-electron wave function [133], Nakada et al. showed that the zigzag edge in a semi-infinite graphene sheet results in a degenerate flat band near the Fermi level for the k vector between $2\pi/3$ and π. For $k = \pi$, the wave function is completely localized at the edge sites, leading to localized states. Comparing with zigzag edges, armchair edges are more stable and in most cases can be understood as double bonds. That is why most of the polycyclic aromatic hydrocarbons (PAHs) synthesized have armchair peripheries, while the synthesis of PAHs with consecutive zigzag edges remains challenging [133].

For edge effect study, the majority of published research stays in the theoretical field, which indicates the huge challenges in the controllable synthesis and

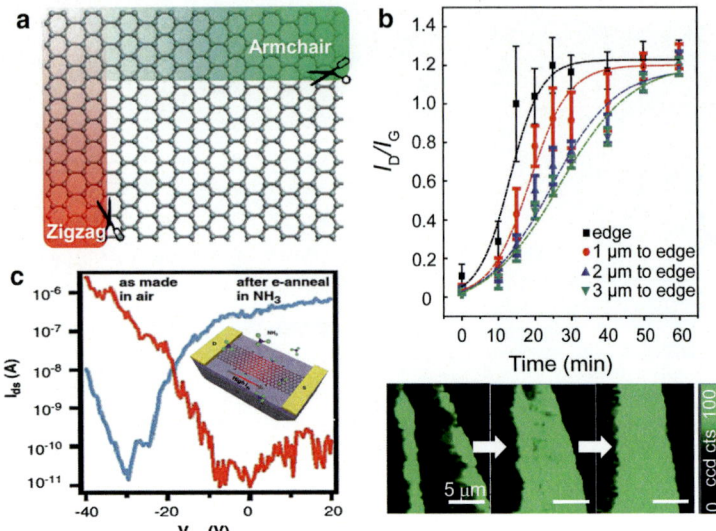

Fig. 8 (a) Cutting through an infinite graphene sheet to obtain zigzag and armchair edges. (b) Time evolution of photomethylation demonstrating that the reaction starts from graphene edges. *Bottom pictures* show the Raman mapping of I_D/I_G on single-layer graphene with the reaction time, revealing that the photomethylation starts from the edge sites of graphene. (c) $I_{ds}-V_g$ curves of the as-made graphene nanoribbon-based FET device in vacuum (*p*-type, *red*) and after e-annealing in NH_3 (*n*-type, *blue*). Inset shows the schematic of a device annealed under high current in NH_3. (**b**) and (**c**) are adapted from [107] and [132], respectively

characterization of the atomically flat edges. With the localized states, the edge sites of graphene are more chemically reactive than the basal plane atoms [96, 132, 134, 135]. Remarkably, the photomethylation of graphene usually starts from graphene edges (Fig. 8b) [107], which may offer a unique pathway to synthesize graphene nanoribbons simply by narrowing the electrically conductive pristine graphene areas. Similarly, hydrogen plasma has been used to decrease the lateral size of monolayer graphene nanoribbons with a rate of 0.27 ± 0.05 nm/min, without breaking the planar –C=C– bonds [135]. The higher chemical reactivity of edge atoms can be useful in the doping of graphene devices. For example, *n*-type molecules can be selectively attached onto the edge sites of graphene nanoribbon without damaging the conjugative carbon network, thus demonstrating a possibility to construct an *n*-type graphene-based electronic device with high carrier mobility (Fig. 8c) [132].

3.4 Interlayer Coupling Effect

Monolayer graphene is a zero-gap semiconductor with a linear Dirac-like spectrum around the Fermi energy, while graphite shows a semimetallic behavior with a band

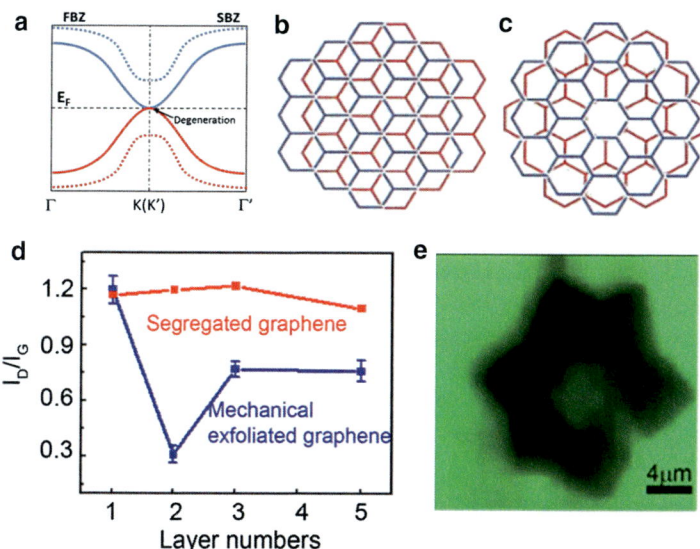

Fig. 9 (**a**) The electronic band structure of bilayer graphene with Bernal stacking orientation. (**b**, **c**) Schematic atomic structures of bilayer graphene with Bernal (**b**) and random (**c**) stacking configurations. (**d**) Statistical analysis of the I_D/I_G in Raman spectra as a function of layer numbers in photochlorination, in which an anomalous chemical inertness of bilayer graphene with Bernal stacking orientation (mechanically exfoliated graphene) was observed. The segregated graphene with random stacking orders does not show this anomalous phenomenon. (**e**) Raman mapping of D band on graphene grown by CVD after photochlorination. The bilayer region shows a uniformly distributed dark color, indicating a high homogeneity and low radical reactivity of AB stacked bilayer graphene domains. Adapted from [122]

overlap of about 41 meV [136]. In between is the few-layer graphene (FLG) whose electronic structure and physical/chemical properties strongly depend on the layer thickness and stacking configurations [9, 137, 138]. For example, bilayer graphene exhibits totally different band structure with monolayer graphene based on tight-binding theoretical model [136]. No linear dispersion was found at the Fermi energy of bilayer graphene; instead, a parabolic spectrum with a band overlap of 0.16 meV was observed (Fig. 9a). In addition to the thickness, stacking configuration offers another parameter to tune the properties of FLG. For instance, a tunable bandgap up to 250 mV can be opened in Bernal-stacked bilayer graphene by an external electric field, which cannot work successfully on the bilayer with random stacking configurations (Fig. 9b, c) [78, 79]. Similarly, trilayer graphenes with different stacking orientations, such as ABA and ABC stacking orders, exhibit distinct band structures and response differently under the vertical electric field [139].

The chemical reactivity of FLG can also be influenced by the layer thickness and stacking configurations. This influence partially comes from the band structure evolution, which determines the transfer rate of excited carriers at the Fermi level [140]. Another possible reason arises from the geometric distortion of chemical bonds. Inherent ripples and corrugations are observed on suspended graphene, which is believed to be the reason why monolayer graphene can stably exist [28, 141]. The associated strain and curvature may cause the distortion of –C=C– bonds; thus, the energy barriers of graphene chemistry can be reduced accordingly. Typically, graphene becomes stiffer as the layer number increases; therefore, the energy barriers for sp^2 to sp^3 structural change will increase, leading to a more chemical resistance to free radicals. It has been experimentally confirmed that the barrier for most chemical reactions on graphene will enlarge as the thickness increases, leading to a lower decoration coverage on thicker graphene films [96, 109, 110, 142]. For example, the reactivity of monolayer graphene is seven times higher than that of bilayer in photoarylation reaction with benzoyl peroxide [109]. Interestingly, an anomalous chemical inertness of bilayer graphene was observed in graphene photochlorination [122]. As shown in Fig. 9d, bilayer graphene exhibits the lowest I_D/I_G compared with the monolayer and multilayer counterparts, which was attributed to the inversion symmetry broken by the displacement field perpendicular to graphene surface when Cl atoms adsorbed onto the top layer. On the contrary, bilayer graphene with random stacking configuration did not show the anomalous chemical inertness. The different reactivities can be applied to rapidly identify the stacking orientations of bilayer graphene grown by chemical vapor deposition (CVD) (Fig. 9e).

4 Applications

Graphene chemistry provides a powerful pathway to break the symmetry of graphene through covalent attachment of functional groups, leading to the transition of carbon hybridization from sp^2 to sp^3. Band structure of graphene can be largely modulated, paving the ways for the application of graphene in electronics and optoelectronics. From the structural point of view, graphene has a very inert surface without dangling bonds; thus, it is hard to be dispersed in any solvents. However, surface properties of graphene can be modified by the attached decorations, which facilitate graphene to carry out chemical reactions in various solutions and practically apply in a large scale. Moreover, new 2D materials and superlattices with fascinating properties beyond mother graphene can be built from graphene scaffold, greatly expanding the graphene family and its attraction. In this section, we will briefly demonstrate the applications of graphene chemistry, and the specific focus is laid on the general consideration and understanding of graphene photochemistry towards electronic and optoelectronic devices and materials science.

4.1 Band Structure Engineering

The covalent modification of graphene can provide a chemical path to make the two carbon atoms in one unit cell different from each other and thus open the bandgap. The bandgap engineering can be confirmed by the changes of electronic and optical properties after chemical functionalization. For example, the aryl radicals generated from diazonium salts can break the –C=C– double bonds and introduce aryl groups onto graphene. Thus, decorated graphene supported on a Si/SiO$_2$ substrate behaved as a granular metal with a mobility gap of 100 meV at low temperature and a variable-range hopping transport. For the suspended graphene modified in both sides by aryl groups, it was found to behave as a semiconductor with 80 meV gap at room temperature [143]. From the device fabrication point of view, photochemical reaction is advantageous since light is particularly attractive as a clean, solvent-free, and easily operated tool. Calculation based on density functional theory (DFT) has shown that chlorination can tune the bandgap of graphene in a range of 0–1.3 eV by attaching Cl on both sides of monolayer graphene [121]. Experimentally, when graphene was exposed to Cl$_2$ gas under UV light, the photogenerated Cl· with high chemical reactivity will break graphenic carbon bonds, leading to the formation of chlorinated graphene, which has a transport gap of ca. 45 meV [91]. It is notable that the expected bandgap of chlorinated graphene should be much larger than the measured value of 45 meV due to the presence of localized states in the bandgap. Graphene can be made luminescent by reducing the connectivity of π-electron network and introducing a bandgap. For example, near-UV-to-blue photoluminescence (PL) has been observed in solution-processed graphene oxide, which originates from the recombination of electron–hole (e–h) pairs localized within small sp^2 carbon clusters, which are embedded within an sp^3 matrix [131, 144]. In the solid states, individual graphene flakes can be made brightly luminescent by photocatalytic oxidation and oxygen plasma treatment [93]. The graphene-based fluorescent organic compounds are of great importance to the development of low-cost optoelectronic devices [63, 93, 131, 145].

4.2 Surface Property Modulation

Homogeneous dispersion of graphene in aqueous and organic solvents can be obtained by selectively attaching different functional groups on the surface [44, 83, 114, 116]. The development of chemical reactions in various solutions largely improves the practical applications of graphene. For instance, introducing aziridine to the graphene surface through photochemical reaction was highlighted by Yan and coworkers [114]. Analogues of perfluorophenylazide containing long alkyl, ethylene oxide, or perfluoroalkyl chains were used to graft different moieties on graphene surface to modify the solubility of the hybrid graphene materials in aqueous and organic solvents. The hybrid graphene materials with alkyl and perfluoroalkyl chains were highly soluble in *o*-dichlorobenzene, while the hybrid graphene material with ethylene oxide chains was soluble in the aqueous solution. Interestingly, different chemical groups can also be attached onto two faces of monolayer graphene asymmetrically, creating the anisotropic wetting property on a single layer of carbon atoms [84]. For example, the half-chlorinated and half-phenylated *Janus* graphene shows different wettability on two sides. With only one atom thickness, the surface wettability can be remarkably altered by chemical modification on the other side.

4.3 Graphene Derivatives and Its 2D Hybrids

Graphene can serve as an excellent scaffold to build up new 2D materials with designed structures and properties since all the sp^2 hybridized carbon atoms in graphene are chemically accessible, offering a versatile choice for chemical designs. Therefore, graphene chemistry can be employed to create novel crystalline graphene derivatives, which is of great importance from the viewpoint of new 2D materials synthesis. Graphane, an extended 2D hydrocarbon derived from a single graphene sheet, is predicted to be stable with a bonding energy comparable to other hydrocarbons such as benzene, cyclohexane, and polyethylene [119]. Experimentally, graphane was first achieved by functionalizing suspended monolayer graphene with hydrogen plasma [81]. All of the carbons in graphane are in sp^3 hybridization, and the hydrogen atoms are bonded to carbon on both sides of the plane in an alternative manner (Fig. 10a). Another representative example of 2D materials derived from mother graphene is fluorographene, a fully fluorinated graphene via addition reaction of –C=C– double bonds [82, 147]. Theoretical simulation showed that both graphane and fluorographene have two stable conformations, in which the chair conformer is more stable than the boat conformer [80, 119, 148]. Their bandgaps are higher than 3.0 eV, and both are transparent at most visible frequencies [81, 82].

Although the formation of stoichiometric graphene derivatives can tune the bandgap of graphene in a large extent, the change of hybridization leads to a sharp decrease of carrier mobility, which sacrifices the major merits of graphene-based electronic devices. One way to overcome this issue is to perform site-selective periodic chemical modification. Graphene superlattice with charge carriers confined in the intact regions can be fabricated in such a way. A typical graphene superlattice structure is the mosaic structure composed of graphene and *h*-BN, which exhibits ubiquitous opened bandgap governed by the width of the wall between BN quantum dots (Fig. 10b) [146, 149]. One of the most important features of such superlattice structure is the extremely high carrier mobility together with the bandgap opening, which is essential for fabricating graphene-based high-performance electronic devices [150]. The uniform graphene/*h*-BN patchworks have been experimentally achieved by CVD technique on Cu [87] and Rh (111) [88] metal substrates. Another interesting mosaic graphene structure can be realized by

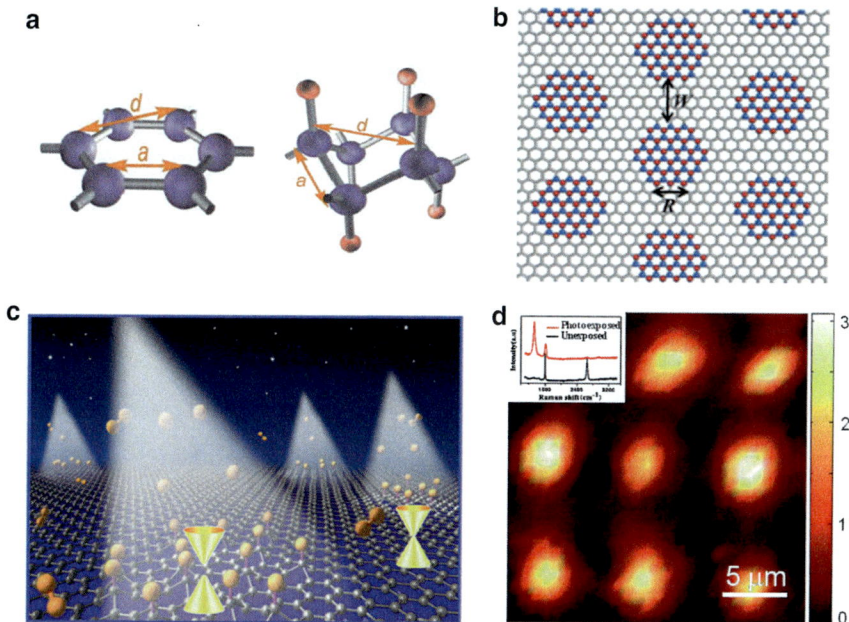

Fig. 10 (a) Schematic structures of graphene (*left*) and theoretically predicted graphane (*right*). (b) Schematic illustration of *h*-BN embedded graphene superlattice: R represents the radius of *h*-BN quantum dots and W represents the width of the wall. (c) Cartoon illustration of photoinduced free-radical-based site-selective photochemical modification of graphene. (d) Raman D/G intensity ratio mapping of graphene after photocatalytic oxidation using a circular TiO_2 photomask, the *red cycles* represent the modified area with larger D/G ratio and the *black* remain pristine. Inset shows the Raman spectra of exposed and unexposed regions after photocatalytic oxidation. (a), (b), and (d) are adapted from [81], [146], and [92], respectively

alternatively growing pristine and *n*-type graphene using methane and acetonitrile as the carbon sources, respectively [85, 86]. Remarkably, photochemistry of graphene is advantageous in fabricating 2D superlattice since light is attractive as an easily patternable tool (Fig. 10c). For instance, a 2D superlattice structure can be achieved under masked photocatalytic oxidation, where the UV light-exposed areas were selectively oxidized via oxidation reaction while the unexposed areas remained intact [92]. As shown in Fig. 10d, the Raman D band mapping clearly shows the formation of 2D graphene/graphene oxide superlattice. Similarly, the photochlorination under photomask could also be employed for fabricating graphene/chlorinated graphene superlattice [91]. The challenge for the photomask-derived graphene patterning is the sub-diffraction-limited transverse feature size, which still needs more technological breakthrough.

One promising way to fabricate the 2D graphene superlattices with the nanometer-scale feature size is to utilize the substrate-induced geometric/electronic effects. The chemical reactivity of graphene is strongly affected by the underlying substrate because of its atomically thin nature. For example, a moiré pattern can be observed on the as-grown monolayer graphene on Ir (111) surface, which is attributed to the geometric and electronic effects of the substrate. Such moiré superlattice can induce a patterned adsorption of hydrogen atoms on graphene [89]. Another interesting approach is the use of strain-enhanced reactivity of graphene. The nanoscale deformation may be introduced by designed surface nanostructures, e.g., nanosphere arrays, which would enable the nanometer-scale modifications of graphene by inducing highly localized chemical reactions.

5 Summary and Outlook

Graphene chemistry is a promising research area; however, one challenge remained along this direction is the enhancement of reaction efficiency. Photogenerated free radicals can effectively react with the chemically inert graphene, leading to the versatile covalent modification. Such kinds of free-radical-based photochemical approaches offer an alternative pathway for bandgap engineering of graphene towards electronic applications with advantages of high efficiency, easy patterning, and solvent-free process. Although most reactions can be inspired from those of small organic analogues, the chemistry of graphene has distinct characteristics due to the infinite 2D feature and offers a platform to study the geometry-induced selective chemistry on 2D surface. Remarkably, new 2D materials and superlattices with fascinating features can be created using the graphene scaffold based on a well-controlled covalent modification, which provides an attractive route to make graphene go beyond.

Another challenge in graphene chemistry is the carrier mobility drop accompanied by the covalent modification, which sacrifices the major merits of graphene-based electronic devices. The formation of graphene superlattices by site-selective modification may solve this problem. As an easily patternable technique, light-guided photochemistry of graphene provides more promising possibilities along this direction. However, the feature sizes of most periodic structures obtained so far are too large to tune the bandgap. To improve the electrical performance, a nanoscale feature size is highly desirable, which calls for more technological breakthroughs. One promising way is to utilize the substrate-induced strain effect, which is able to localize the reactive graphenic carbon in the nanoscale regions. Moreover, as a 2D atomic crystal model system, graphene's chemistry provides rich information for chemical studies in other 2D systems, such as h-BN, metal dichalcogenides, and graphene–boron nitride composite (BCN). Overall, the chemical research on graphene definitely needs more chemists to join in.

Acknowledgments The work was supported by the Natural Science Foundation of China (Grants 51432002, 50121091, 51290272) and the Ministry of Science and Technology of China (Grants 2013CB932603, 2012CB933404, 2011CB933003).

References

1. Smalley RE (1997) Discovering the fullerenes. Rev Mod Phys 69(3):723–730
2. Novoselov KS, Geim AK, Morozov S et al (2004) Electric field effect in atomically thin carbon films. Science 306(5696):666–669
3. Shenderova O, Zhirnov V, Brenner D (2002) Carbon nanostructures. Crit Rev Solid State Mater Sci 27(3–4):227–356
4. Kroto HW, Heath JR, Obrien SC, Curl RF, Smalley RE (1985) C_{60}- Buckminster fullerene. Nature 318(6042):162–163
5. Iijima S (1991) Helical microtubules of graphitic carbon. Nature 354(6348):56–58
6. Narita N, Nagai S, Suzuki S et al (1998) Optimized geometries and electronic structures of graphyne and its family. Phys Rev B 58(16):11009
7. Narita N, Nagai S, Suzuki S et al (2000) Electronic structure of three-dimensional graphyne. Phys Rev B 62(16):11146
8. Zhang H, Zhao M, He X et al (2011) High mobility and high storage capacity of lithium in sp–sp^2 hybridized carbon network: the case of graphyne. J Phys Chem C 115(17):8845–8850
9. Geim AK, Novoselov KS (2007) The rise of graphene. Nat Mater 6(3):183–191
10. Lee C, Wei XD, Kysar JW et al (2008) Measurement of the elastic properties and intrinsic strength of monolayer graphene. Science 321(5887):385–388
11. Nair RR, Blake P, Grigorenko AN et al (2008) Fine structure constant defines visual transparency of graphene. Science 320(5881):1308
12. Novoselov KS, Geim AK, Morozov SV et al (2005) Two-dimensional gas of massless Dirac fermions in graphene. Nature 438(7065):197–200
13. Avouris P, Chen ZH, Perebeinos V (2007) Carbon-based electronics. Nat Nanotechnol 2(10):605–615
14. Bolotin KI, Sikes K, Jiang Z et al (2008) Ultrahigh electron mobility in suspended graphene. Solid State Commun 146(9):351–355
15. Du X, Skachko I, Barker A et al (2008) Approaching ballistic transport in suspended graphene. Nat Nanotechnol 3(8):491–495
16. Schwierz F (2010) Graphene transistors. Nat Nanotechnol 5(7):487–496
17. Gusynin V, Sharapov S (2005) Unconventional integer quantum Hall effect in graphene. Phys Rev Lett 95(14):146801
18. Barone V, Hod O, Scuseria GE (2006) Electronic structure and stability of semiconducting graphene nanoribbons. Nano Lett 6(12):2748–2754
19. Son YW, Cohen ML, Louie SG (2006) Energy gaps in graphene nanoribbons. Phys Rev Lett 97(21):216803
20. Hwang E, Adam S, Sarma SD (2007) Carrier transport in two-dimensional graphene layers. Phys Rev Lett 98(18):186806
21. Pisana S, Lazzeri M, Casiraghi C et al (2007) Breakdown of the adiabatic Born–Oppenheimer approximation in graphene. Nat Mater 6(3):198–201
22. Nomura K, MacDonald A (2007) Quantum transport of massless Dirac fermions. Phys Rev Lett 98(7):076602
23. Park CH, Yang L, Son YW et al (2008) Anisotropic behaviours of massless Dirac fermions in graphene under periodic potentials. Nat Phys 4(3):213–217
24. Neto AC, Guinea F, Peres N et al (2009) The electronic properties of graphene. Rev Mod Phys 81(1):109–162
25. Abergel D, Apalkov V, Berashevich J et al (2010) Properties of graphene: a theoretical perspective. Adv Phys 59(4):261–482
26. Zhang YB, Tan YW, Stormer HL et al (2005) Experimental observation of the quantum Hall effect and Berry's phase in graphene. Nature 438(7065):201–204
27. Novoselov KS, Jiang Z, Zhang Y et al (2007) Room-temperature quantum hall effect in graphene. Science 315(5817):1379

28. Meyer JC, Geim AK, Katsnelson MI et al (2007) The structure of suspended graphene sheets. Nature 446(7131):60–63
29. Katsnelson M, Novoselov K, Geim A (2006) Chiral tunnelling and the Klein paradox in graphene. Nat Phys 2(9):620–625
30. Balandin AA, Ghosh S, Bao WZ et al (2008) Superior thermal conductivity of single-layer graphene. Nano Lett 8(3):902–907
31. Seol JH, Jo I, Moore AL et al (2010) Two-dimensional phonon transport in supported graphene. Science 328(5975):213–216
32. Kim KS, Zhao Y, Jang H et al (2009) Large-scale pattern growth of graphene films for stretchable transparent electrodes. Nature 457(7230):706–710
33. Li X, Cai W, An J et al (2009) Large-area synthesis of high-quality and uniform graphene films on copper foils. Science 324(5932):1312–1314
34. Reina A, Jia X, Ho J et al (2009) Large area, few-layer graphene films on arbitrary substrates by chemical vapor deposition. Nano Lett 9(1):30–35
35. Bae S, Kim H, Lee Y et al (2010) Roll-to-roll production of 30-inch graphene films for transparent electrodes. Nat Nanotechnol 5(8):574–578
36. Sun Z, Yan Z, Yao J et al (2010) Growth of graphene from solid carbon sources. Nature 468 (7323):549–552
37. Liu N, Fu L, Dai B et al (2010) Universal segregation growth approach to wafer-size graphene from non-noble metals. Nano Lett 11(1):297–303
38. Li XS, Magnuson CW, Venugopal A et al (2011) Large-area graphene single crystals grown by low-pressure chemical vapor deposition of methane on copper. J Am Chem Soc 133 (9):2816–2819
39. Yan K, Peng HL, Zhou Y et al (2011) Formation of bilayer bernal graphene: layer-by-layer epitaxy via chemical vapor deposition. Nano Lett 11(3):1106–1110
40. Dai B, Fu L, Zou Z et al (2011) Rational design of a binary metal alloy for chemical vapour deposition growth of uniform single-layer graphene. Nat Commun 2:522
41. Zou Z, Fu L, Song X et al (2014) Carbide-forming groups IVB-VIB metals: a new territory in the periodic table for CVD growth of graphene. Nano Lett 14(7):3832–3839
42. Sun J, Gao T, Song X et al (2014) Direct growth of high-quality graphene on high-κ dielectric $SrTiO_3$ substrates. J Am Chem Soc 136(18):6574–6577
43. Eda G, Fanchini G, Chhowalla M (2008) Large-area ultrathin films of reduced graphene oxide as a transparent and flexible electronic material. Nat Nanotechnol 3(5):270–274
44. Li D, Müller MB, Gilje S et al (2008) Processable aqueous dispersions of graphene nanosheets. Nat Nanotechnol 3(2):101–105
45. Park S, Ruoff RS (2009) Chemical methods for the production of graphenes. Nat Nanotechnol 4(4):217–224
46. Cai JM, Ruffieux P, Jaafar R et al (2010) Atomically precise bottom-up fabrication of graphene nanoribbons. Nature 466(7305):470–473
47. Sutter PW, Flege JI, Sutter EA (2008) Epitaxial graphene on ruthenium. Nat Mater 7 (5):406–411
48. Emtsev KV, Bostwick A, Horn K et al (2009) Towards wafer-size graphene layers by atmospheric pressure graphitization of silicon carbide. Nat Mater 8(3):203–207
49. Hernandez Y, Nicolosi V, Lotya M et al (2008) High-yield production of graphene by liquid-phase exfoliation of graphite. Nat Nanotechnol 3(9):563–568
50. Si Y, Samulski ET (2008) Synthesis of water soluble graphene. Nano Lett 8(6):1679–1682
51. Stankovich S, Dikin DA, Dommett GH et al (2006) Graphene-based composite materials. Nature 442(7100):282–286
52. Stoller MD, Park S, Zhu Y et al (2008) Graphene-based ultracapacitors. Nano Lett 8 (10):3498–3502
53. Wang X, Zhi L, Müllen K (2008) Transparent, conductive graphene electrodes for dye-sensitized solar cells. Nano Lett 8(1):323–327

54. Becerril HA, Mao J, Liu Z et al (2008) Evaluation of solution-processed reduced graphene oxide films as transparent conductors. ACS Nano 2(3):463–470
55. Ramanathan T, Abdala A, Stankovich S et al (2008) Functionalized graphene sheets for polymer nanocomposites. Nat Nanotechnol 3(6):327–331
56. Yoo E, Kim J, Hosono E et al (2008) Large reversible Li storage of graphene nanosheet families for use in rechargeable Li ion batteries. Nano Lett 8(8):2277–2282
57. Sun X, Liu Z, Welsher K et al (2008) Nano-graphene oxide for cellular imaging and drug delivery. Nano Res 1(3):203–212
58. Chen H, Müller MB, Gilmore KJ et al (2008) Mechanically strong, electrically conductive, and biocompatible graphene paper. Adv Mater 20(18):3557–3561
59. Blake P, Brimicombe PD, Nair RR et al (2008) Graphene-based liquid crystal device. Nano Lett 8(6):1704–1708
60. Lu CH, Yang HH, Zhu CL et al (2009) A graphene platform for sensing biomolecules. Angew Chem Int Ed 121(26):4879–4881
61. Xia F, Mueller T, Lin YM et al (2009) Ultrafast graphene photodetector. Nat Nanotechnol 4(12):839–843
62. Lin YM, Dimitrakopoulos C, Jenkins KA et al (2010) 100-GHz transistors from wafer-scale epitaxial graphene. Science 327(5966):662
63. Bonaccorso F, Sun Z, Hasan T et al (2010) Graphene photonics and optoelectronics. Nat Photonics 4(9):611–622
64. Qu L, Liu Y, Baek J-B et al (2010) Nitrogen-doped graphene as efficient metal-free electrocatalyst for oxygen reduction in fuel cells. ACS Nano 4(3):1321–1326
65. Wang H, Cui L-F, Yang Y et al (2010) Mn_3O_4-graphene hybrid as a high-capacity anode material for lithium ion batteries. J Am Chem Soc 132(40):13978–13980
66. Wang H, Casalongue HS, Liang Y et al (2010) $Ni(OH)_2$ nanoplates grown on graphene as advanced electrochemical pseudocapacitor materials. J Am Chem Soc 132(21):7472–7477
67. Li Y, Wang H, Xie L et al (2011) MoS_2 nanoparticles grown on graphene: an advanced catalyst for the hydrogen evolution reaction. J Am Chem Soc 133(19):7296–7299
68. Liang Y, Li Y, Wang H et al (2011) Co_3O_4 nanocrystals on graphene as a synergistic catalyst for oxygen reduction reaction. Nat Mater 10(10):780–786
69. Wu Y, Lin YM, Bol AA et al (2011) High-frequency, scaled graphene transistors on diamond-like carbon. Nature 472(7341):74–78
70. Geim AK, MacDonald AH (2007) Graphene: exploring carbon flatland. Phys Today 60(8):35–41
71. Wallace P (1947) The band theory of graphite. Phys Rev 71(9):622–634
72. Han M, Özyilmaz B, Zhang Y et al (2007) Energy band-gap engineering of graphene nanoribbons. Phys Rev Lett 98(20):206805
73. Jiao L, Zhang L, Wang X et al (2009) Narrow graphene nanoribbons from carbon nanotubes. Nature 458(7240):877–880
74. Kosynkin DV, Higginbotham AL, Sinitskii A et al (2009) Longitudinal unzipping of carbon nanotubes to form graphene nanoribbons. Nature 458(7240):872–876
75. Rosales L, Pacheco M, Barticevic Z et al (2009) Transport properties of antidot superlattices of graphene nanoribbons. Phys Rev B 80(7):073402
76. Bai JW, Zhong X, Jiang S et al (2010) Graphene nanomesh. Nat Nanotechnol 5(3):190–194
77. Kim M, Safron NS, Han E et al (2010) Fabrication and characterization of large-area, semiconducting nanoperforated graphene materials. Nano Lett 10(4):1125–1131
78. Ohta T, Bostwick A, Seyller T et al (2006) Controlling the electronic structure of bilayer graphene. Science 313(5789):951–954
79. Zhang Y, Tang TT, Girit C et al (2009) Direct observation of a widely tunable bandgap in bilayer graphene. Nature 459(7248):820–823
80. Boukhvalov D, Katsnelson M, Lichtenstein A (2008) Hydrogen on graphene: electronic structure, total energy, structural distortions and magnetism from first-principles calculations. Phys Rev B 77(3):035427

81. Elias DC, Nair RR, Mohiuddin TM et al (2009) Control of graphene's properties by reversible hydrogenation: evidence for graphane. Science 323(5914):610–613
82. Nair RR, Ren WC, Jalil R et al (2010) Fluorographene: a two-dimensional counterpart of teflon. Small 6(24):2877–2884
83. Lomeda JR, Doyle CD, Kosynkin DV et al (2008) Diazonium functionalization of surfactant-wrapped chemically converted graphene sheets. J Am Chem Soc 130(48):16201–16206
84. Zhang L, Yu J, Yang M et al (2013) Janus graphene from asymmetric two-dimensional chemistry. Nat Commun 4:1443
85. Yan K, Wu D, Peng H et al (2012) Modulation-doped growth of mosaic graphene with single-crystalline p–n junctions for efficient photocurrent generation. Nat Commun 3:1280
86. Wu D, Yan K, Zhou Y et al (2013) Plasmon-enhanced photothermoelectric conversion in chemical vapor deposited graphene p–n junctions. J Am Chem Soc 135(30):10926–10929
87. Ci L, Song L, Jin C et al (2010) Atomic layers of hybridized boron nitride and graphene domains. Nat Mater 9(5):430–435
88. Gao Y, Zhang Y, Chen P et al (2013) Toward single-layer uniform hexagonal boron nitride–graphene patchworks with zigzag linking edges. Nano Lett 13(7):3439–3443
89. Balog R, Jorgensen B, Nilsson L et al (2010) Bandgap opening in graphene induced by patterned hydrogen adsorption. Nat Mater 9(4):315–319
90. Sun ZZ, Pint CL, Marcano DC et al (2011) Towards hybrid superlattices in graphene. Nat Commun 2:559
91. Li B, Zhou L, Wu D et al (2011) Photochemical chlorination of graphene. ACS Nano 5(7):5957–5961
92. Zhang L, Diao S, Nie Y et al (2011) Photocatalytic patterning and modification of graphene. J Am Chem Soc 133(8):2706–2713
93. Gokus T, Bonetti A, Lombardo A et al (2009) Making graphene luminescent by oxygen plasma treatment. ACS Nano 3(12):3963–3968
94. Robinson JT, Burgess JS, Junkermeier CE et al (2010) Properties of fluorinated graphene films. Nano Lett 10(8):3001–3005
95. Hummers WS, Offeman RE (1958) Preparation of graphitic oxide. J Am Chem Soc 80(6):1339
96. Sharma R, Baik JH, Perera CJ et al (2010) Anomalously large reactivity of single graphene layers and edges toward electron transfer chemistries. Nano Lett 10(2):398–405
97. Quintana M, Spyrou K, Grzelczak M et al (2010) Functionalization of graphene via 1,3-dipolar cycloaddition. ACS Nano 4(6):3527–3533
98. Georgakilas V, Bourlinos AB, Zboril R et al (2010) Organic functionalisation of graphenes. Chem Commun 46(10):1766–1768
99. Devadoss A, Chidsey CED (2007) Azide-modified graphitic surfaces for covalent attachment of alkyne-terminated molecules by "click" chemistry. J Am Chem Soc 129(17):5370–5371
100. Mu G (1983) The reaction of free radicals, Higher Education Press, Beijing, pp 27–37
101. Zhang L, Zhou L, Yang M et al (2013) Photo-induced free radical modification of graphene. Small 9(8):1134–1143
102. Benson SW (1965) Bond energies. J Chem Educ 42:502
103. Bansal RK (1978) Organic reaction mechanisms, Tata McGraw Hill, New Delhi, Chapter 6.
104. Lee WH, Suk JW, Chou H et al (2012) Selective-area fluorination of graphene with fluoropolymer and laser irradiation. Nano Lett 12(5):2374–2378
105. Jeon KJ, Lee Z, Pollak E et al (2011) Fluorographene: a wide bandgap semiconductor with ultraviolet luminescence. ACS Nano 5(2):1042–1046
106. Bowen EJ, Hinshelwood CN, Sidgwick NV (1932) Annu Rep Progr Chem 29:13–73
107. Liao L, Song Z, Zhou Y et al (2013) Photoinduced methylation of graphene. Small 9(8):1348–1352
108. Wu J, Xie L, Li Y et al (2011) Controlled chlorine plasma reaction for noninvasive graphene doping. J Am Chem Soc 133(49):19668–19671

109. Liu H, Ryu S, Chen Z et al (2009) Photochemical reactivity of graphene. J Am Chem Soc 131 (47):17099–17101
110. Liu L, Ryu S, Tomasik MR et al (2008) Graphene oxidation: thickness-dependent etching and strong chemical doping. Nano Lett 8(7):1965–1970
111. Fujishima A, Zhang XT, Tryk DA (2008) TiO_2 photocatalysis and related surface phenomena. Surf Sci Rep 63(12):515–582
112. Tatsuma T, Tachibana S, Miwa T et al (1999) Remote bleaching of methylene blue by UV-irradiated TiO_2 in the gas phase. J Phys Chem B 103(38):8033–8035
113. Liu LH, Yan MD (2009) Simple method for the covalent immobilization of graphene. Nano Lett 9(9):3375–3378
114. Liu LH, Yan MD (2010) Derivatization of pristine graphene with well-defined chemical functionalities. Nano Lett 10(9):3754–3756
115. Liu LH, Zorn G, Castner DG et al (2010) A simple and scalable route to wafer-size patterned graphene. J Mater Chem 20(24):5041–5046
116. Liu LH, Yan MD (2011) Functionalization of pristine graphene with perfluorophenyl azides. J Mater Chem 21(10):3273–3276
117. Sinitskii A, Dimiev A, Corley DA et al (2010) Kinetics of diazonium functionalization of chemically converted graphene nanoribbons. ACS Nano 4(4):1949–1954
118. Zhou L, Zhou L, Wang X et al (2014) Trifluoromethylation of graphene. Appl Phys Lett 2 (9):092505
119. Sofo J, Chaudhari A, Barber G (2007) Graphane: a two-dimensional hydrocarbon. Phys Rev B 75(15):153401
120. Zhang YH, Zhou KG, Xie KF et al (2010) Tuning the electronic structure and transport properties of graphene by noncovalent functionalization: effects of organic donor, acceptor and metal atoms. Nanotechnology 21(6):065201
121. Yang M, Zhou L, Wang J et al (2012) Evolutionary chlorination of graphene: from charge-transfer complex to covalent bonding and nonbonding. J Phys Chem C 116(1):844–850
122. Zhou L, Zhou L, Yang M et al (2013) Free radical reactions in two dimensions: a case study on photochlorination of graphene. Small 9(8):1388–1396
123. Bunch JS, Verbridge SS, Alden JS et al (2008) Impermeable atomic membranes from graphene sheets. Nano Lett 8(8):2458–2462
124. Leenaerts O, Partoens B, Peeters FM (2008) Graphene: a perfect nanoballoon. Appl Phys Lett 93(19):193107
125. Garaj S, Hubbard W, Reina A et al (2010) Graphene as a subnanometre trans-electrode membrane. Nature 467(7312):190–193
126. Merchant CA, Healy K, Wanunu M et al (2010) DNA translocation through graphene nanopores. Nano Lett 10(8):2915–2921
127. Koenig SP, Boddeti NG, Dunn ML et al (2011) Ultrastrong adhesion of graphene membranes. Nat Nanotechnol 6(9):543–546
128. Min SK, Kim WY, Cho Y et al (2011) Fast DNA sequencing with a graphene-based nanochannel device. Nat Nanotechnol 6(3):162–165
129. Zhou J, Wu MM, Zhou X et al (2009) Tuning electronic and magnetic properties of graphene by surface modification. Appl Phys Lett 95(10):103108
130. Yang M, Zhao R, Wang J et al (2013) Bandgap opening in Janus-type mosaic graphene. J Appl Phys 113(8):084313
131. Eda G, Mattevi C, Yamaguchi H et al (2010) Blue photoluminescence from chemically derived graphene oxide. Adv Mater 22(4):505–509
132. Wang X, Li X, Zhang L et al (2009) N-doping of graphene through electrothermal reactions with ammonia. Science 324(5928):768–771
133. Nakada K, Fujita M, Dresselhaus G et al (1996) Edge state in graphene ribbons: nanometer size effect and edge shape dependence. Phys Rev B 54(24):17954
134. Wang X, Dai H (2010) Etching and narrowing of graphene from the edges. Nat Chem 2 (8):661–665

135. Xie L, Jiao L, Dai H (2010) Selective etching of graphene edges by hydrogen plasma. J Am Chem Soc 132(42):14751–14753
136. Partoens B, Peeters F (2006) From graphene to graphite: electronic structure around the K point. Phys Rev B 74(7):075404
137. Mak KF, Shan J, Heinz TF (2010) Electronic structure of few-layer graphene: experimental demonstration of strong dependence on stacking sequence. Phys Rev Lett 104(17):176404
138. Frank O, Bousa M, Riaz I et al (2011) Phonon and structural changes in deformed Bernal stacked bilayer graphene. Nano Lett 12(2):687–693
139. Lui CH, Li ZQ, Mak KF et al (2011) Observation of an electrically tunable band gap in trilayer graphene. Nat Phys 7(12):944–947
140. Goh MS, Pumera M (2010) The electrochemical response of graphene sheets is independent of the number of layers from a single graphene sheet to multilayer stacked graphene platelets. Chem Asian J 5(11):2355–2357
141. Fasolino A, Los J, Katsnelson MI (2007) Intrinsic ripples in graphene. Nat Mater 6(11):858–861
142. Luo ZQ, Yu T, Kim KJ et al (2009) Thickness-dependent reversible hydrogenation of graphene layers. ACS Nano 3(7):1781–1788
143. Zhang H, Bekyarova E, Huang JW et al (2011) Aryl functionalization as a route to band gap engineering in single layer graphene devices. Nano Lett 11(10):4047–4051
144. Mei Q, Guan G, Liu B et al (2010) Highly efficient photoluminescent graphene oxide with tunable surface properties. Chem Commun 46(39):7319–7321
145. Luo Z, Mele EJ, Johnson ATC et al (2009) Photoluminescence and band gap modulation in graphene oxide. Appl Phys Lett 94(11):111909
146. Zhao R, Wang J, Yang M et al (2012) BN-embedded graphene with a ubiquitous gap opening. J Phys Chem C 116(39):21098–21103
147. Zboril R, Karlicky F, Bourlinos AB et al (2010) Graphene fluoride: a stable stoichiometric graphene derivative and its chemical conversion to graphene. Small 6(24):2885–2891
148. Şahin H, Topsakal M, Ciraci S (2011) Structures of fluorinated graphene and their signatures. Phys Rev B 83(11):115432
149. Zhao R, Wang J, Yang M et al (2012) Graphene quantum dots embedded in a hexagonal BN sheet: identical influences of zigzag/armchair edges. Phys Chem Chem Phys 15(3):803–806
150. Wang J, Zhao R, Liu Z et al (2013) Widely tunable carrier mobility of boron nitride-embedded graphene. Small 9(8):1373–1378

Index

A
Agri-film, 17
8-Amino-1,3,6-naphthalenetrisulfonates (ANTS), 51
1-Amino-8-hydroxy-3,6-disulfonaphthalene (H-acid), 51
Aminotrimethylenephosphonic acid (ATMP), 21
Ammonium 1-anilinonaphthalene-8-sulfonate (ANS), 37
4-(4-Anilinophenylazo)benzenesulfonate (AO5), 55
Anion exchange, 69
Antacid/anti-pepsin agent, 139
Anticancer efficiency, 151
Antiferromagnetism, 199
Antifogging, 19
Antireflection (AR), 6, 18
Asphalt, aging, 13, 15
Assembly, 1
Atomic layer deposition (ALD), 110
Azobenzenes, 180, 185
Azomethine-H, 58

B
Bandgap engineering, 229
Band structure engineering, 213
Benzophenones, 11
Benzoquinone (BQ), 125
Benzotriazole, 12
5-Benzotriazolyl-4-hydroxy-3-sec-butylbenzenesulfonic acid (BZO), 12
Bio-LDH, 137, 141
Biomedical applications, 137
Bis(R)-4,4′-bipyridinium, 180, 200

Bis[2-di(b-hydroxyethyl)amino-4-(4-sulfophenylamino)-s-triazin-6-ylamino] stilbine-2,2′-disulfonate (BBU), 51
Bis(8-hydroxyquinolate-5-sulfonate)zinc, 41
Bis(8-hydroxyquinolate)zinc, 33
Bis(N-methylacridinium) (BNMA), 30
Bis(2-methyl-3-thiophenyl) perfluorocyclopentene, 186
Bis(phosphonomethyl)glycine (GLYP), 21
Bis(2-sulfonatostyryl)biphenyl (BSB), 53
Bond dissociation energy (BDE), 219

C
Calcein, 51
Carbyne, 214
Chemiluminescence, 61
China clay, 18
4-Chloro-2-nitrophenol (CNP), 123
Cisplatin, 166
Co-intercalation, 37
CO_2 reduction, photocatalytic, 131
Congo red, 122
Coumarin-3-carboxylate (C3C), 43
Crystal structure, 69

D
Defect-induced luminescence, 6
Diacetylene (DA), 56
Diagnosis, 137
Diarylethene, 180
Dihexadecylviologen, 204
Dihydroxyanthraquinone (DAQ), 192

Dimethylviologen, 200
Dioleoylphosphaidic acid (DOPA), 153
Disuccinatocisplatin (DSCP), 166
Di(*R*)viologen, 200
DNA–LDH nanohybrids, 143
Dodecylbenzene sulfonate (DBS), 37
Doxorubicin, 167
Drug delivery, 137
Dynamic tuning, 50

E
Edge effect, 225
Electrochemiluminescence, 61
Electroluminescence, 61
Electron transfer, 177
Energy conversion, 105
Environmental remediation, 105
2,2′-(1,2-Ethenediyl) bis[5-[[4-(diethylamino)-6-[(2,5-disulfophenyl) amino]-1,3,5-triazin-2-yl] amino]benzene-5 sulfonic acid] hexasodium salt (BTBS), 48, 57
2-[4-(4-Ethylphenylazo)phenoxy]ethyl (trimethyl)ammonium, 197
Excited-state intramolecular proton transfer (ESIPT), 58

F
Ferrimagnetism, 199
Ferromagnetism, 199
Few-layer graphene (FLG), 227
Fluorescein (FLU), 37, 52
Fluorescein isothiocyanate (FITC), 156
Fluorographene, 219, 224
Förster resonance energy transfer (FRET), 7
Free-radical addition, 213, 218
French chalk, 18

G
Gene delivery, 137
Graphene, 213
  photohalogenation, 219
  superlattice, 213, 230
Graphene oxide (GO), 119
Graphyne, 214

H
Heat-preservation, 19
Hectorite, 181, 204
1-Heptanesulfonic acid sodium (HES), 37, 52
Host–guest, 10, 177, 181
Host layers, 107
Hydrotalcite, 139, 191
p-[2-(2-Hydroxyethyldimethylammonio) ethoxy]azobenzene bromide, 194, 197
2-Hydroxy-4-methoxybenzophenone-5-sulfonic acid (HMBA), 11

I
Imaging
Intercalation, 19, 35, 69, 138, 177
Interlayer coupling effect, 226
Interlayer guest, 112
Ion exchange, 36
IR absorption materials, 17
Isomerization, 177

J
Janus graphene, 213, 224

L
Lanthanides, 7, 160
  contraction, 88
Layer-by-layer (LbL) assembly 25
Layered double hydroxides (LDHs), 1, 70, 105
Layered rare earth hydroxides (LREHs), 69
Li-fluor-taeniolite, 194
Light-induced sensors, 57
Low-density polyethylene (LDPE), 17
Luminescence, 1

M
Magadiite, 195
Magnetic nanoparticles (MSPs), 167
Magnetism, 199
Malachite green (MG), 123
Mechanofluorochromism, 57
Merocyanine, 186
Metal–organic framework (MOF), 181
Metal phthalocyanine sulfonates (MPcS), 113
Methotrexate (MTX)-LDH, 150
Methylene blue (MB), 122
Methylenedisalicylic acid (MDSA), 11
Mica, 181
Mixed metal oxides (MMO), 6, 109
Molecular dynamics (MD), 32
Montmorillonite, 186

Index

N
Nanographenes, 225
Nanohybrids, 137
Nanomedicine, 137, 140
Nanoribbons, 216, 225, 226
Nanosheet crystallites, exfoliated, 100
Naphthalene acetate, 36
Niobate, 195
NLO applications, 193

O
One-dimensional photonic crystals (1DPC), 6
Optical pH sensors, 51
Oxide phosphors, 97

P
Pd(II)-tetrakis(4-carboxyphenyl) porphyrin (PdTPPC), 60
PEGylated phospholipid-coated LDH (PEG-PLDH), 153
1-Pentanesulfonate (PS), 58
Perfluorophenylazides (PFPA), 222, 229
Peroxyl radical, 125
Perylene tetracarboxylate (PTCB), 35
8-((p-(Phenylazo)phenyl)oxy-octanoate, 199
pH sensor, 51
N-Phosphonomethyliminodiacetic acid (PMIDA), 21
Phosphor, 69
Photoalkylation, 220
Photoarylation, 220
Photocatalysis, 105, 114
Photocatalytic oxidation, 220
Photochemical hole burning (PHB), 192
Photochemistry, 213
Photochromism, 57, 177, 179
Photohalogenation, 219
Photoisomerization, 57
Photoluminescence, 6, 62, 69, 92
Photoregulation, 177
Photostability, 19
Photostabilizers, 13
Piezochromic luminescent (PCL), 56
Pillared layered framework, 69
Platinum (Pt), 117
Pollutants, photocatalytic degradation, 122
Polycyclic aromatic hydrocarbons (PAHs), 225
Polydiacetylene (PDA), 56
Polyoxometalates (POMs), 36, 113
Polypropylene, photostability, 13
Poly(tert-butyl acrylate-co-ethyl acrylate-co-methacrylic acid) (PTBEM), 33
Polyvinylsulfinate (PVS), 31
Pressure sensors, 56
Prussian blue, 199

Q
Quantum dots (QDs), 47

R
Rare-earth-doped luminescence, 7
Rare earth oxides, 71
Reactive oxidative species (ROS), 124
Reduced folate carrier (RFC), 157
Relative humidity (RH), 6
Rhodamine B isothiocyanate (RITC), 156
Rhodamine 6G (RhG), 123

S
Saponite, 181, 204
Scattering, 8
Second harmonic generation (SHG), 193
Sensors, 1
Separate nucleation and aging steps (SNAS), 4
Silver/silver halide, 118
Silylation, 154
Smectite, 181, 200
Sodium salicylate (SS), 11
Solvothermal reaction, 74
Spiropyran, 180, 187
Static adjustment, 38
Stilbazolium, 180, 191
Sulforhodamine B (SRB), 37, 122
Sulfosalicylic acid (SSA), 11
Surfaces, property modulation, 229
Surfactants, 181

T
Targeting, 137, 150
Temperature sensors, 53
Terephthalic acid (TA), 109
[Tetrakis(4-carboxyphenyl)-porphyrinato] zinc(II) (ZnTPPC), 60
Tetrakis(4-sulfonatophenyl) porphyrin (TPPS), 60
Tetraphenylporphine, 204
Theranostics, 137, 162
Therapy, 137
Thermochromic luminescence (TCL), 53
Thiodisalicylic acid (TDSA), 11
Titanoniobate, 195

Trichlorophenol, 122
Trimethylspiro[2H-1-benzopyran-2,2′-indoline] (H-SP), 185
Trinitrotoluene (TNT), 51
Tris[2-(4,6-difluorophenyl)pyridinato-C^2,N] iridium(III), 34
Tris(8-hydroxyquinolate-5-sulfonate) aluminum, 33
Tunable multicolor emissive film, 46

U
Ultrathin films (UTFs), 25
UV shielding, 1, 8

V
Viologens, 180, 200
Volatile organic compounds (VOCs), 6, 51

W
Water splitting, photocatalytic, 126

Z
Zeolites, 181
Zinc phthalocyanines (ZnPc), 37, 41
Zirconium phosphonate viologen salt, 201
ZnAlIn–MMO nanocomposite, 111

Printed in the United States
By Bookmasters